DISCARDED

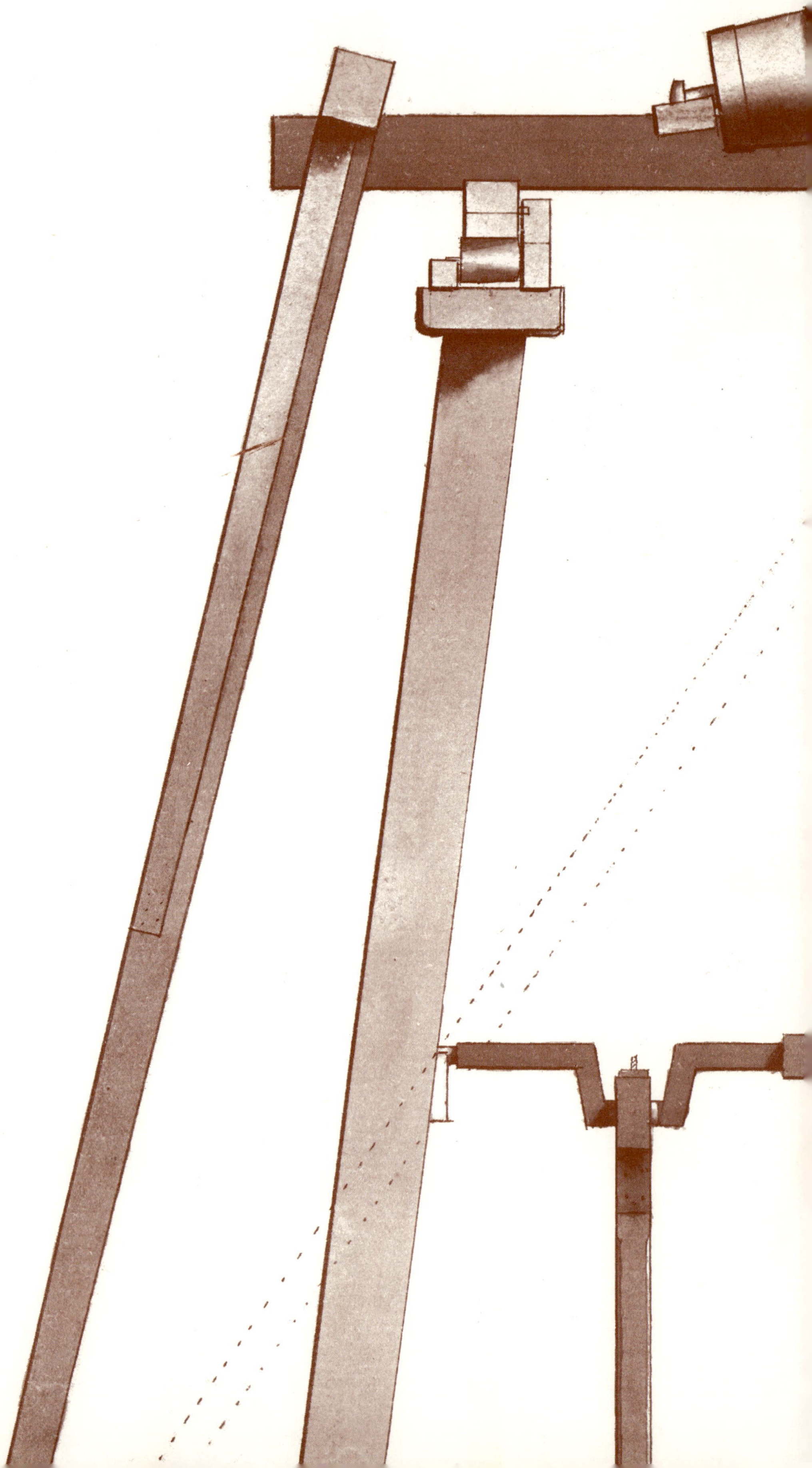

The Development of the Factory

Jennifer Tann

Steam Eng
Steam Eng

A. Water pump
B. grinding

Figured above
3 Regular Engine

A. Water pump
B. grinding

May 23 W will have 2
observe that B & Co do not have
immediate charge of — the Engine
3 Regular Engine what by their
Engine but is formerly by them

The Development of the Factory

Jennifer Tann

Cornmarket

Designed by Nicholas Jenkins
© Cornmarket Press Limited 1970
42/43 Conduit Street
London W1R ONL

The illustrations in this book are reproduced from originals in the collection of Boulton and Watt papers in Birmingham Reference Library

Printed in England by Kingprint Limited, Richmond
ISBN 0 7191 2134 5

Contents

Preface	1
Introduction	2
1. The Evolution of the Factory in the 18th Century	3
2. Factory capacity & Fixed capital	27
3. Animal Power in the Factory	47
4. Water Power in the Factory	59
5. Steam Power in the Factory	71
6. The Millwrights	95
7. Factory Heating	109
8. Factory Lighting	123
9. The Evolution of the Fire-Proof Factory	135
10. Factory Architecture	149
Bibliography	170
Index	171

Illustrations

Cotton
Louson, Arbroath, Angus, 1807	6
Wilkes & Jewsbury, Measham, Leicestershire, 1802	8
John McCracken, Belfast, 1806	10
Walker, n.d.	12
G. & J. Robinson, Papplewick, Nottinghamshire c.1785	13
John Bartholomew and Co, Glasgow c.1796	16
Mr. Bannatine, Glasgow, 1790	16
P. Drinkwater, Manchester, 1789	17
A. & G. Murray, Manchester, 1805	17
Markland, Cooksen and Fawcett, Leeds, 1792	18
Owen, Scarth & Co, Chorlton, Manchester, c.1795	20
McConnel & Kennedy, Manchester, 1809	19
Birley & Hornby, Chorlton, Manchester, 1809, 1813	22
Philips and Lee, Salford Twist Mill, 1805	21
Thomas Harrison, Stalybridge, Cheshire, 1823	24
George Cheetham & Sons, Stalybridge, Cheshire, 1833	23
Great Western Cotton Factory, Bristol, 1839	26

Wool and Worsted
Nussey, Birstall, Leeds, 1796	28
H. Hicks & Sons, Churching Mill, Eastington, Gloucestershire, 1822	32
H. Hicks & Sons, New Mill, Eastington, Gloucestershire, 1816	30
Wormald, Gott & Wormald, Park Mill, Leeds, 1806	34
Davison & Hawkesley, Arnold Mill, Nottinghamshire, c.1792	36
J. Cartwright, Revolution Mill, Retford, Nottinghamshire, 1789	36
Bedworth Mill, Collyhurst, Bedworth, Warwickshire, c.1800	38
John Wood, Bradford, 1833	40, 41

Flax
George Wilkie, Dundee, 1799	42
Marshall, Hives & Co, Shrewsbury, 1811	44
Benyon, Benyon & Bage, Shrewsbury, 1811	43

Silk
James Hoddinot, Nr. Shepton Mallet, Somerset, 1806	45

Rope
Staniforth, Liverpool, 1811	46

Bleaching, Dyeing, Printing
Ground plan of a bleaching mill, n.d.	48
Joseph Baker & Co, Raikes, Bolton, Lancashire, 1790	50
Duffy, Byrne and Hamill, Donnybrook and Ballsbridge, Dublin, 1809	51
Goodwin, Platt, Goodwin & Co, Southwark, London, c.1790	52
Daintry, Ryle & Co, Manchester, 1805	53
Daintry, Ryle & Co, Manchester, 1805	54

Starch
Stonard & Curtis, London, 1784	55

Oil Seed Crushing
Cotes & Co, Hull, c.1784	56
Brooke & Pease, Hull, 1795	57
Cotes & Co, Hull, 1802	58

Corn
Llanyre Mills, n.d.	60
Thompson & Baxter, Hull, 1788	61
Union Company, 1789	64
Unknown mill, n.d.	64
Albion Mills, London, c.1794, 1802	69, 70
Deptford Corn Mill, 1825	66

Iron and Engineering
Knight's slitting mill, Whittington, n.d.	74
Tredegar Ironworks, 1802; Wolford rolling mill, n.d.;	76
Mr. Ryelands slitting mill, n.d.	76
Soho Manufactory, Birmingham, 1788	77
Soho Foundry, Birmingham, c.1796	79
Gardner and Manser, King and Queen Foundry, Rotherhithe, 1790	82
Woolwich Smithery, Commissioners of the Navy, 1814, 1833	84
Williams and Jones, Britannia Nail Works, Birmingham, 1814	86

Mints
Soho Mint, Birmingham, c.1800	90
British Mint, 1806	88
Danish Mint, 1805	93
Russian Mint, 1800	92
Brazil Mint, 1811	94

Snuff and Tobacco
Fish & Yates, 1786	95, 96

Brewing and Distilling
Goodwyn & Co, London, 1784	98
S. Whitbread, London, 1784-5	100
Gyfford, London, 1787	102
Madder, Dublin, 1809	104
Constitution Brewery, London, 1807	
Barclay and Perkins, Anchor Brewery, Southwark, 1786	106
Gosport Brewery, Hampshire, n.d.	106
Aitcheson & Brown, 1787	107
John Bushby jun., 1837	108

Factory Heating
Gardom Pares & Co, Calver Mill, Derbyshire, 1807	110
Robert Arkwright, Bakewell Mill, Derbyshire, 1813	112
H. Hicks & Sons, Churching Mill, Eastington, Gloucestershire, 1816	114
Samuel Oldknow, Mellor Mill, Derbyshire, 1809	116
Wedgwood and Byerley, drying house, Etruria, Staffordshire, 1807	119
David Dale, New Lanark Mills, Scotland, 1796	117

Sugar Refining
Boyes and Carlill, Hull, 1805	120
Peter Whitfield Brancker & Co, Liverpool, 1815	122

Gas Lighting
Philips & Lee, Salford Twist Mill, 1806	124
Strutt, Derby Calico Mill, 1806	126
Strutt, Milford Mills, Derbyshire, 1806	128
Watson Ainsworth and Co, Preston, 1806	130
Greg and Ewart, Manchester, 1809	132
John Maberley, Aberdeen, 1814	133
Wormald, Gott and Wormald, Leeds, 1808.	134

Fire Extinguishing Apparatus
Belper North Mill	136
Sketches from Goodrich's journals showing the structure of Belper North Mill	136, 138
Creighton's drawing of the Salford Twist Mill, 1801	138
Thomas Orrell, cotton factory, Stockport, 1819	140
Birley and Marsland, cotton spinners, Manchester, 1804	142
William Creighton's list of fire extinguishing apparatus supplied by Boulton and Watt, n.d.	144

Glass
British Plate Glass Co, Ravenshead, Lancashire, 1788	146
British Plate Glass Co, Ravenshead, Lancashire, 1846	146
Chance Bros, Birmingham, 1846	148

Saw Mills
A wind driven saw mill designed by Rennie, n.d.	152
Don Fernando de Torres, Spain, 1790	150

Pottery
Hamilton's flint mill, Fenton Low, 1807	163
Wedgwood and Byerley, Etruria, 1800	164
Josiah Spode, Stoke-on-Trent, 1810	165

Paper
Plan of a paper mill, ground floor and upper floor, n.d.	166
W. & R. Balstone, Maidstone, 1806	167
Koops, Tait & Co, Chelsea Paper Mill, c.1800	168

Yarranton's proposed factory village at Milcote Warwickshire	3
Derby Silk Mill	4
Soho Manufactory	154
Cromford Mill, Derbyshire	158
Masson Mill, Matlock Bath, Derbyshire	156
Mellor Mill, Derbyshire	154
Miller's Dale Mill, Derbyshire	160
Tower Mint, London	162
East Front of Enoch Wood's factory at Burslem	154

Preface

I owe a great debt of gratitude to the City Librarian of Birmingham Mr. W.A. Taylor for allowing me access to the Boulton and Watt Papers. Miss D.M. Norris, Miss. D.H. McCulla, Mr. A. Andrews, Miss N. Jenkins, Miss M. Kurkiewicz, Miss M. Hinton and the staff of the Local Studies Section of the Birmingham Reference Library have been unstinting in their help. Mr. A. H. Westwood, Assay Master, kindly permitted me to use the papers in the Assay Office.

I thank the staff of the Brotherton Library, Leeds; Leeds City Library; Shrewsbury Public Library; Derby Public Library; Derbyshire Record Office and the Science Museum for their help and I would like particularly to thank the library staff at the University of Aston for locating material for me.

I gratefully acknowledge permission to reproduce drawings: Mr. F.M. Fitzroy Newdigate of Arbury for the drawing of Bedworth Mill; The Director, The Science Museum for the drawings from Goodrich's Journal; Derbyshire Archaeological Society for the watercolour of Arkwright's Cromford Mill now in Derbyshire Record Office.

Many colleagues and friends have helped in innumerable ways and I can only name a few: I am grateful to Mr. B.L. Latham for relieving me of some of my teaching load for part of the summer term; Dr. S.D. Chapman kindly allowed me to see two of his papers before publication and has helped in many other ways; Mr. R.A. Chaplin has generously shared his knowledge of country house architecture; Mr. John Goodchild of Cusworth Hall Museum has drawn MSS in his possession to my attention; Mr. Malcolm Nixon drew the elevation of Masson Mill; Mr. James Thompson has suggested new lines of enquiry and the social sciences and architecture students of the University of Aston and the extramural students of the University of Birmingham have helped more than they know. Professor J.R. Harris, Mr. James Thompson, Mr. R.A. Chaplin and Dr. S.D. Chapman have kindly read and commented on sections of this book, I thank them all; naturally the remaining errors are mine.

I gratefully acknowledge two grants from the Pasold Research Fund towards my research on the Boulton and Watt papers. Finally I thank my family for help in many ways.

University of Aston 1970

Introduction

The change to the factory system of production which took place in Britain between the middle of the 18th and the middle of the 19th centuries posed a series of new and difficult problems for the entrepreneur. The most immediate ones were associated with building the factory and were of two kinds, technological and managerial. Technological problems included the choice of a site for the factory, the layout of the factory, the type of machinery to be used, the form of construction to be adopted, the methods of lighting and heating to be employed, the choice of a power system, an estimation of the amount of power required and the application of that power to the best advantage.

The managerial problems were wide ranging. Capital was required, overlookers and perhaps a manager had to be found, a labour force had to be recruited and kept and discipline had to be maintained. In the water power phase of the Industrial Revolution factories were often built in remote places for an adequate labour force and the entrepreneur was forced to build housing, an inn, a shop and possibly a chapel and school in order to attract labour. In other words he not only had to build a factory but a village as well.[1] The school and chapel did not bring a direct financial return but many entrepreneurs made a fair profit from the rents of housing. However the provision of school and chapel and 'superior' housing was regarded as a means of improving the moral character and hence the discipline of the working population so that they were often regarded as a useful investment in an indirect sense. As Ure said 'It is ... excessively the interest of every mill-owner to organise his moral machinery on equally sound principles with his mechanical, for otherwise he will never command the steady hands, watchful eyes, and prompt co-operation essential to excellence of product.'[2] The way in which this was done depended upon the individual entrepreneur: 'Thus Quakers showed some fine feeling for their workers but made high demands of moral conformity on them;' philanthropic masters like Owen or the Fieldens attempted to humanize not only their works, but also those of others; 'and the hypocrites employed clergymen, sometimes paid with the workers' pence, to teach their hands how to suffer starvation wages without protest.'[3]

Discussion of the recruitment of labour, of factory villages and factory discipline and of the far reaching social implications of industrialisation is beyond the scope of this book. In the following pages discussion centres on the development of the factory as a physical structure in the late 18th and early 19th centuries.

This study is based to a large extent on the Boulton and Watt Papers. Although this collection has been used for some excellent studies of the firm of Boulton and Watt it has not been used anything like so extensively for a study of other firms. It is hoped that by using letters and drawings of firms both large and small, ones that survived and ones that collapsed, a balanced view of factory development can be achieved. One of the themes which emerges from the following pages is that it was the same few manufacturers who adopted the costly innovations such as fire-proof buildings, who installed gas lighting, steam or warm air heating and fire extinguishing apparatus; they were the giants, the ones who are most likely to have left some record of their activities behind, yet in many respects they were uncharacteristic. They appear to have found little difficulty in recruiting capital yet there were many smaller manufacturers who found difficulty in obtaining long-term loans, to whom a fire-proof factory or gas lighting would have seemed an unobtainable luxury. In this respect some of the most valuable letters are from those manufacturers who decided against buying a Boulton and Watt steam engine which was a good deal more expensive than an atmospheric engine or a simple water wheel.

About 20,000 drawings exist amongst the Boulton and Watt papers. Most of these are drawings of steam engines and engine parts but scattered amongst the drawings are some uncatalogued plans, sections and elevations of factories. Their survival is fortuitous for most of them were prepared by the entrepreneur's millwright or engineer and were loaned to Boulton and Watt to obtain their advice on the placing of the engine house. The standard of draughtsmanship varies greatly from the fine detailed drawings of John Sutcliffe the Halifax millwright (page 18) or the workmanlike drawings of Boulton and Watt's agent Creighton (page 19) to the rough sketch plans prepared by unknown millwrights (page 12). The drawings in this book were selected for their importance as historical documents but they also have aesthetic appeal for a functional drawing, like a functional building, can have high visual quality. They have been grouped by subject although in some cases the grouping is arbitrary. There are few elevations of factories amongst the Boulton and Watt drawings so these have been supplemented where appropriate by contemporary engravings and a measured elevation.

A date after the title of a drawing is the date of the drawing and not necessarily the date of the erection of the factory.

On the majority of drawings the dimensions of the buildings are written in; some also have a linear scale. On others the scale is indicated in words and since most of the drawings have been reduced, the written scale is meaningless. In these cases an approximate scale has been noted in the description of the drawing.

One of Boulton and Watt's methods of copying a drawing was to take a press copy of it. This was called a reverse drawing and was a mirror image of the original. Where one of these drawings has been reproduced the word 'reverse' appears on the drawing or by the portfolio number in the description.

---

1. W. Ashworth, 'British Industrial Villages in the Nineteenth Century,' *Econ. Hist. Rev.* 3 1950-1; Sidney Pollard, 'The Factory Village in the Industrial Revolution,' *English Historical Review*, 89, 1964; 'Industrial Monuments at Milford and Belper,' *Archaeological Journal* 117, 1961 (1963).
2. Andrew Ure, *The Philosophy of Manufacturers*, p.417.
3. Sidney Pollard, 'Factory Discipline in the Industrial Revolution,' *Econ. Hist. Rev.* 16, 1963.

1. The Evolution of the Factory in the 18th Century

1. A. Ure, *The Philosophy of Manufactures*, 1835, p.13.
2. Andrew Yarranton, *England's Improvement by sea and land*, 1677, pp 51, 128.
3. Peter Mathias, *The Brewing Industry in England 1700-1830*, pp.7-11.
4. Sidney Pollard, *The Genesis of Modern Management*, pp.55-60.

The concentration of the means of production

'The term factory, in technology, designates the combined operation of many orders of work-people, adult and young in tending with assiduous skill a system of productive machines continuously impelled by a central power'. Ure's[1] classic definition of a factory may now seem narrow for in his view the essential elements of a factory system were machinery and power. He excluded all establishments 'in which the mechanisms do not form a connected series, nor are dependent on one prime mover' such as iron works, dye works, soap works, breweries, distilleries, engineering works — the list could be longer. In excluding these works he excluded some of the most remarkable factories of the 18th century. A fundamental characteristic of the factory system was the concentration of the means of production at the factory site. Even if the industry was labour intensive the step was no less important because of this for the basic organisational change had occurred. Three of the greatest 18th century entrepreneurs Wedgwood, Boulton and Gott established factories at a time when power driven machinery could perform only limited tasks in their industries.

There are well-known isolated examples of 16th and 17th century factories; for example William Stumpe's weaving factory at Malmsbury or Jack of Newbury's factory where he is alleged to have employed 1,000 artisans. Not so well-known is Andrew Yarranton's proposal[2] in the mid 17th century to build a factory estate at Milcote near Stratford-on-Avon with flax spinning schools, tape engines, 'furnis houses to boyl the yearne' and houses for the workers (page 3). Peter Mathias has shown[3] that by the late 1690s considerable capital had been invested in London breweries. Edmund Halsey owned a large brewery with mills and pumps worked by animals and 'its management [was] already accustomed to dealing in large sums'. The Phoenix Brewhouse was valued at £11,700 in 1698. The total fixed capital investment in numerous ironworks belonging to the Crowleys and Foleys was in the region of £50-60,000 in the 17th century.[4] But these were exceptional establishments. It is not until the 18th century that a more general move towards factory production can be seen.

The embryo factory

The beginnings of the concentration of the means of production are discernible in the textile industries from the early 18th century when some manufacturers began to gather hand operated machines into workshops. These embryo factories did not at first constitute a threat to the domestic manufacturer although there was sometimes opposition to them. In the Midlands merchant hosiers began to assemble knitting frames in warehouses and workshops. Alderman W. Wilson of Nottingham, for example, had 'a dozen or so handframes' in an attic over his workshop and counting house. Samuel Fellows, a framework knitter of London who migrated to Nottingham in *c* 1706,

Yarranton's proposed factory village at Milcote, Warwickshire

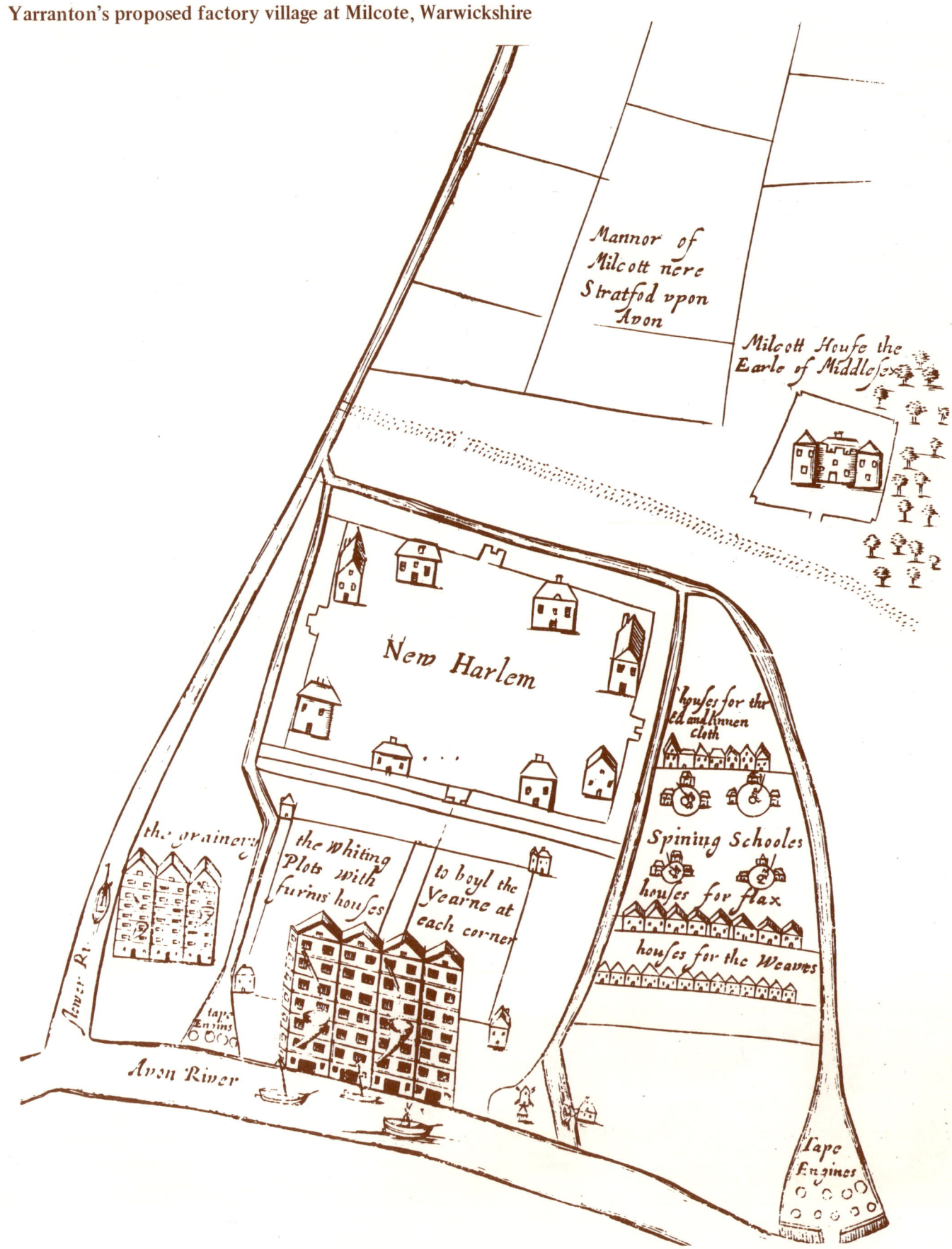

Derby Silk Mill

4

5. S.D. Chapman, *The Early Factory Masters*, pp.34-40.
6. C. Aspin & S.D. Chapman, *James Hargreaves and the Spinning Jenny*, p.36. The spinning jenny was invented between 1764 and 1769 and patented in 1770.
7. Bolton. Civic Centre Museum, Irving Bequest.
8. W. Radcliffe, *Origins of Power Loom Weaving*, 1828, quoted in S.D. Chapman op.cit. p.56.
9. W. Radcliffe, *Origins of Power Loom Weaving*, p.65.
10. J. Rollins, 'The Forge Mill, Redditch,' *Industrial Archaeology*, 3. 1966.
11. Jennifer Tann, *Gloucestershire Woollen Mills*, p.141.
12. John Somervell, *Water Power Mills of Westmorland*.

had established a workshop employing more than forty apprentices by the early 1720s. Fellows bought an invention for imitating Spanish eyelet-hole mitts and established 'a large factory' in Nottingham. A Chesterfield hosier, John White, also learned a method of making similar work and built a factory near his home.[5] In these 'factories' the machines were hand operated and there was no power system.

By the 1760s there was a shortage of yarn in the cotton industry. The demand for cotton cloth had increased and while the weavers had been assisted by Kay's flying shuttle, spinning was still performed on the one thread wheel. The obstruction was partly overcome by Hargreaves' spinning jenny which at first had eight spindles. The jenny did not directly influence the factory movement in the way that Arkwright's water frame did for, unlike the water frame, it was hand operated; the early jenny was primarily a domestic machine which remained in the operative's home. But by the 1780s machines with eighty to one hundred spindles had been produced. These were too large for domestic use and were installed in workshops. Hargreaves and his partner Thomas James built Hockley Mill in Nottingham for their first jennies and the factory contained fifty machines.[6] But the jenny was of less value to the Midlands hosiery industry than to the Lancashire cotton industry for jenny-spun yarn was coarse and generally used as weft whereas warp yarn was needed for fine hosiery. Jenny spinning was widely distributed in north west Derbyshire and Lancashire and when Crompton made his census of spinning machines in 1811 he found many jennies, especially around Stockport, where a number of manufacturers had no other kind of spinning machinery.[7]

Crompton's spinning mule was hand operated at first. Mules were installed in town and country workshops where they often replaced earlier jennies, new mule workshops were built and William Radcliffe described the period 1788-1803 as the golden age of the small rural spinner.[8]

The germ of the factory was contained in three other kinds of textile workshop; the hand loom weaver's shop, the merchant's finishing shop and the calico printer's shop. The Yorkshire clothier who began to employ labour beyond his immediate family to weave in the workshop adjoining his house; or the Spitalfields silk weaver who employed a number of journeymen weavers in workshops built at the back of his house was moving towards factory production. The Yorkshire merchant who established finishing shops in Leeds, Halifax or Huddersfield to dress the unfinished cloths bought from the country clothiers was developing a factory organisation. Small calico printshops containing only a few printing tables could be found in the Wandle valley in Surrey and scattered over Lancashire and Cheshire and the adjacent parts of Yorkshire in the late 18th century. The embryo factory emerged in the building of workshops behind the master potter's house in North Staffordshire between 1740 and 1760. Hand chain shops containing up to thirty hearths were built near the merchants' warehouses in the Black Country, while Birmingham toy manufacturers began to bring their outworkers under one roof.

Many workshops were purpose-built but a distinction must be made between the dwelling house containing a workshop for one family and the larger workshop, the embryo factory, which contained several chain makers' hearths, nail forges, knitting frames, jennies or looms. The great expansion in the cotton industry led to much workshop building but also to the conversion of existing premises. Houses belonging to the aristocracy and gentry were converted to cotton spinning and print works and farms, barns, houses, chapels were also used. William Radcliffe[9] recalled that after 1788 'every lumber room, even old barns, outhouses or outbuildings of any description were repaired, windows broke through the old blank walls, and all fitted up for loomshops.' A feature common to all these examples is the absence of power. But one form of power was in most cases easily installed, the animal wheel. The installation of a horse wheel enabled the manufacturer to graduate to a more capital intensive organisation. The horse wheel could drive carding and spinning machinery, boring machinery, clay mixing machinery at a pottery or dash wheels at a bleachworks. A horse wheel was a useful and successful bridge between the unmechanised workshop and the full-scale factory driven by water or steam power. The manufacturer who began in this way was able, by living frugally, to plough back his capital until he was in a position to move to a water or steam powered site.

Another focus for the development of the factory was the water mill. In the early stages of industrial expansion an existing building was often adapted to a new purpose. Corn mills were converted to fulling and scribbling mills, to carding and spinning mills, or to forges or saw mills. An iron forge became a needle polishing mill at Redditch,[10] Worcestershire, and a fulling mill became an iron slitting mill in Gloucestershire.[11] Flexibility in the use of early water mills is well illustrated in John Somervell's[12] study of the changing uses of south Westmorland mills.

The purpose-built factory
The climax of the factory movement, the emergence of the purpose-built factory, covered a period of well over a century and a half, for the chronology of factory building generally followed the innovation of powered machinery within each industry, although almost all industries had their early exceptions. On the one hand silk throwing was power driven before 1720 whilst on the other hosiery frames were only beginning to be steam driven in the late 1850s. The iron forge, foundry and furnace were organised on a factory basis by the 1780s whereas nail and chain manufacture were only slowly mechanised and

Cotton
Louson, Arbroath, Angus 1807
These drawings show a small country spinning mill of the kind that was erected in the 1770s and 1780s to spin on Arkwright's principle. There were six frames of twenty four spindles.
Portf: 764

6

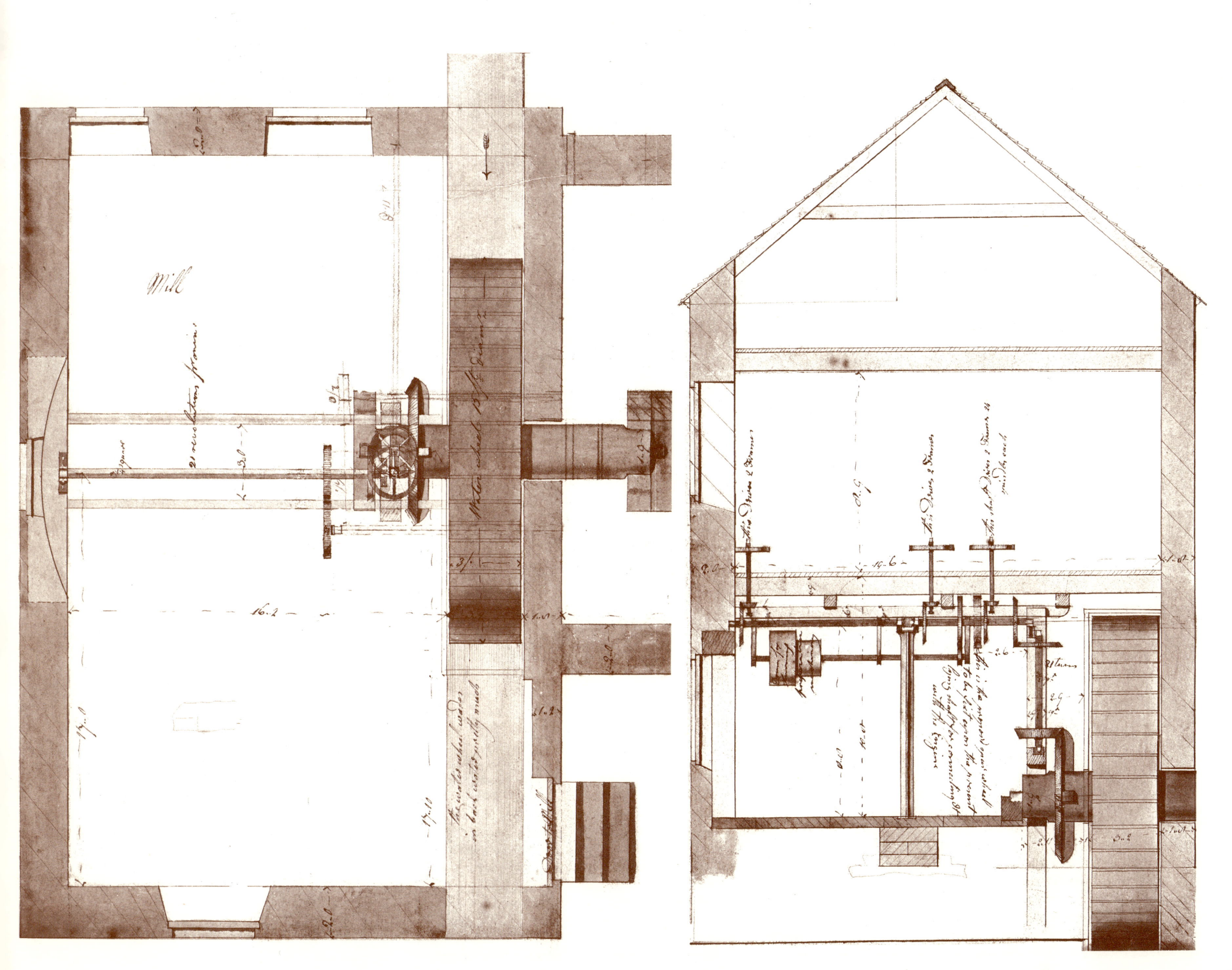

13. Samuel Timmins, *Birmingham and the Midland Hardware District*, 1866, p.391.
14. F. Williamson, 'George Sorocold of Derby' *Derbyshire Arch. Journ.* 10, 1937, pp 56-64.
15. Sir Frank Warner, *The Silk Industry of the United Kingdom*; S.D. Chapman, op.cit. pp.40-42; Cyril T. Boucher, *James Brindley Engineer 1716-1772*, pp.31-38.
16. Pp. 1816, 111, 217.
17. Birmingham Reference Library, Wyatt MSS; A.P. Wadsworth & J. de Lacy Mann, *The Cotton Trade and Industrial Lancashire*, pp.431-448.
18. E. Butterworth, *Historical Sketches of Oldham*, 1856, pp.112-3.
19. A.P. Wadsworth & J. de Lacy Mann op.cit. p.488.
20. E. Butterworth op.cit. p.116.
21. Ibid p.116.
22. Ibid p.118.
23. C. Aspin & S.D. Chapman op.cit. pp.29-39.
24. S.D. Chapman op.cit. pp.67-8.

the gradual transfer to factory production has taken until the 20th century to be completed. In some industries outworkers have survived to the present day. The Birmingham Small Arms Factory was established in 1862 but in 1866 'only in the important establishments are all the branches carried on on the premises of the gunmaker'[13] and some high quality sporting guns are still made in domestic workshops in Birmingham.

Silk

The first factories in the modern sense of the word were erected for silk throwing. In 1702 Thomas Cotchett built a silk mill on the River Derwent in Derby. It was 62 ft long, 28 ft 5 in wide and three storeys high. There were four machines on each of the lower two storeys and accommodation was provided for forty-eight doublers. In 1718 Thomas Lombe obtained a patent for a silk throwing machine and began to build a larger mill alongside Cotchett's which was completed in 1721. Lombe's mill was 110 ft by 39 ft and 5 storeys high. Adjacent to this mill was a doublers' shop 139 ft long capable of containing 306 machines.[14] With the expiry of Lombe's patent in 1732 a number of silk mills were built in different parts of the country; there was one at Sherborne, Dorset by 1740; one at Macclesfield by 1743; Stockport by 1752; Sutton-in-Ashfield, Derbyshire in 1753; Chesterfield in 1757; Sheffield in 1758 and by 1768 there were six in Stockport. By the end of the 18th century there were at least three silk mills near Bishop's Stortford, Herts.[15] Derby remained a centre of the silk industry and Pilkington implied that there were eleven mills in the town employing 900 people by the end of the 18th century. When Strutt gave evidence to the 1816 Commission he said that the hours of work were twelve hours per day 'This has been the invariable practice at the original silk mill in Derby [and] in this neighbourhood for more than 100 years' as if not only the precedent of scale but of organisation had been set by Lombe and Cotchett.[16]

Carding Mills

The experiments of Hargreaves and Arkwright were preceded by those of Lewis Paul and John Wyatt who in the 1730s developed a method of spinning by rollers which was patented in 1738. This patent was followed ten years later by one for a carding machine. Paul and Wyatt set up a spinning machine in a Birmingham warehouse and received enquiries from Tamworth, Colchester, Worcester and Romsey, while Paul spent some time in Nottingham trying to establish mechanised spinning there. Edmund Cave was granted a licence and after working the machine by hand in a London Warehouse moved to a water powered mill at Northampton. Five frames of fifty spindles each, together with carding machinery were installed but the mill, under the management of the millwright Thomas Yeoman, was a commercial failure. One other mill was set up by Daniel Bourne at Leominster, probably in 1748. This was burnt down in 1754 and the Northampton mill was advertised for sale in 1756.[17] Butterworth[18] records that when the Northampton mill was abandoned a carding cylinder was taken to Brock Mill near Wigan in about 1764 by a Mr. Morris.

The carding machine played 'an important if not a decisive part in the development of the cotton factory system'.[19] Wadsworth and Mann suggest that there were a number of carding machines in Lancashire based on Paul's or Bourne's designs in the 1770s. In 1772 John Lees of Oldham invented a feeder for the carding machine and established a small factory operated by a horse.[20] In 1775 Arkwright patented a carding engine which included the crank and comb, essential for cleaning the 'teeth' of the carder. Although factories based on Arkwright's system of roller spinning overshadowed all others there was also an increase in the number of small rudimentary factories based on the carding machine. Butterworth records[21] that 'The greater part of the earliest cotton mills were moved by horse power' and the majority of these were carding mills. The carding mill was another bridge to the larger factory; it required little capital and a small labour force and as long as the mule was hand operated there was a market for rovings. The carding mill was a means by which the artisan or farmer or tradesman could become an entrepreneur. This was uncommon in the Midlands where the cotton trade was in the hands of a merchant élite but in Lancashire and north Derbyshire there seems to have been a greater opportunity for the artisan to become a small manufacturer. Butterworth commented 'Several of the smallest of the original mills in Oldham and the neighbourhood commenced business with not more than eight or ten hands each, upon an average'.[22]

Power Spinning

A critical part in the development of the factory system was played by Arkwright who in 1769 patented a method of spinning by rollers — the water frame. He intended his machine to be used for warp and weft but in practice the frame was used for warp only and the jenny and water frame remained complementary until the jenny was superceded by the mule. Arkwright's first factory was built in Nottingham in 1769. It was a four storey building, 117 ft long and about 27 ft wide and was powered by horses. During the early 1770s Hargreaves' and Arkwright's Nottingham factories were rivals but Arkwright employed two-thirds more workers than Hargreaves and Thomas James, Hargreaves' partner, was forced to become a licensee of Arkwright.[23] In 1771 Arkwright built his first water powered factory at Cromford and a second was built in 1777. Although Cromford became the centre of Arkwright's empire, he was concerned with different partners in cotton factories in Derbyshire, Yorkshire, Lancashire, Staffordshire, Worcestershire and Scotland, in some cases offering his patent rights for a share in the concern.[24] Robinsons at Papplewick, James at Nottingham

Wilkes & Jewsbury, Measham, Leicestershire, 1802
Joseph Wilkes, agricultural improver, turnpike builder, coal mine owner, cotton and corn mill builder, built this mill in the 1780s. He was one of the first cotton spinners to enquire after a Boulton and Watt rotary steam engine but did not buy one for the cotton mill until 1802.
Portf: 234

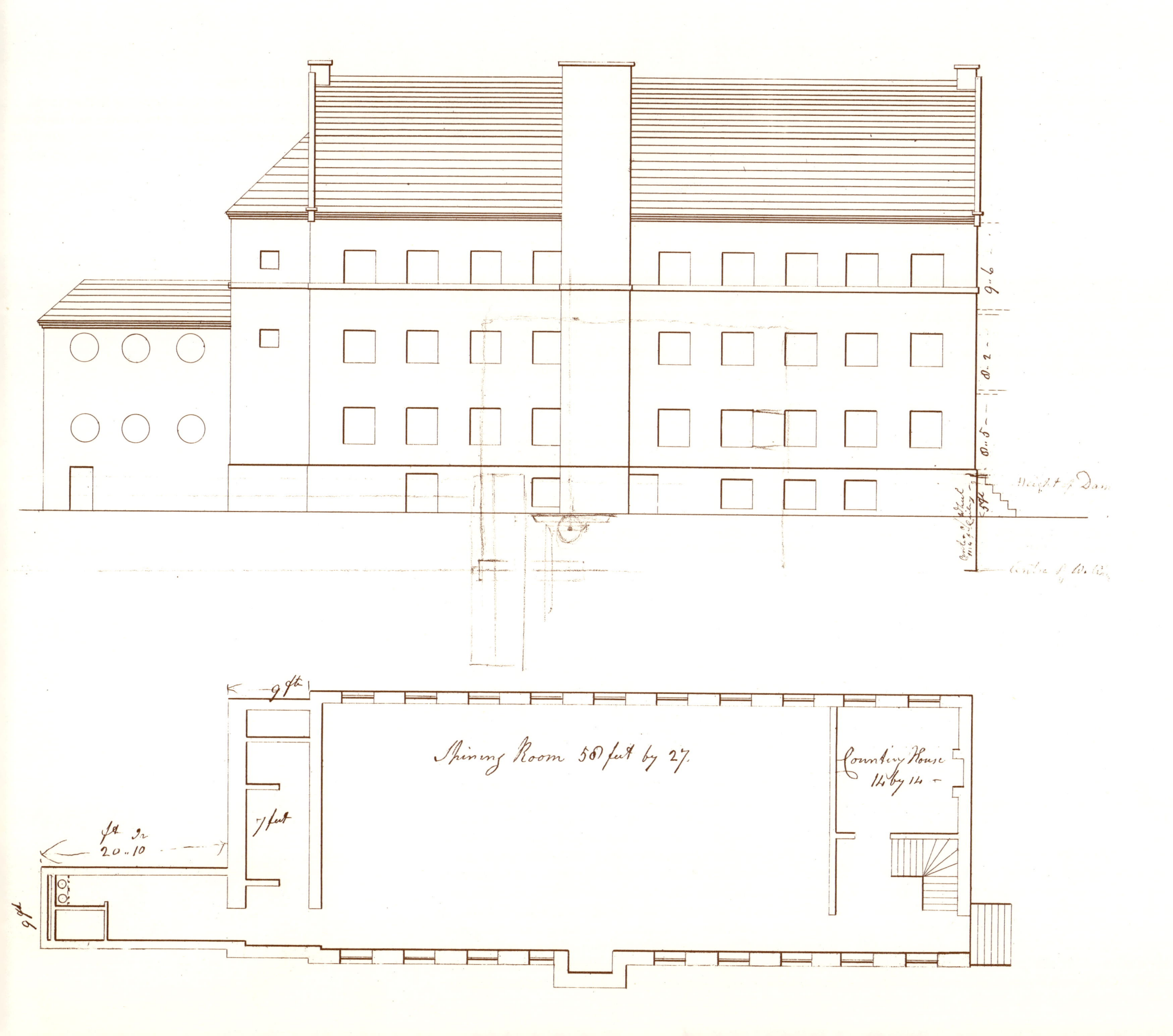

Spining Room 58 feet by 27.   Counting House 14 by 14

25. Ibid pp.73-4.
26. M. Boulton to J. Watt 15th March 1786, Letter Book, A.O.
27. S.D. Chapman op.cit. p.77.
28. W.H. Chaloner, 'Robert Owen, Peter Drinkwater and the Early Factory System in Manchester 1788-1800,' *Bulletin John Rylands Library*, 37, 1954-5, p.94.
29. Richard L. Hills, *Power in the Industrial Revolution*, pp.126-128.
30. S.D. Chapman, 'The Pioneers of Worsted Spinning by Power,' *Business History*, 7, 1965.
31. S.D. Chapman, *The Early Factory Masters*, pp.101-124.
32. J. Cooksen to Boulton and Watt 7th February 1792. Guildhall Library, Sun Insurance c.s. 8/636966 13/653223; 17/666225.
33. Jennifer Tann, *Gloucestershire Woollen Mills*, pp.127, 138-199.
34. W. Smith, *The History and Antiquities of Morley*, 1876, p.215.
35. John Goodchild, 'The Ossett Mill Company' *Textile History* 1, 1968.

and Gardom and Pares of Calver Mill, Derbyshire were licensed to use water frames. The latter firm paid £2,000 premium for the use of the water frame and £5,000 for the use of the carding engine plus an annual royalty of £1,000.[25] Such an expensive royalty hardened the opposition to Arkwright. As Boulton said later 'if he had been a more civilized being and had understood Mankind better he would now have enjoy'd his patent. Hence let us learn wisdom by other Men's ills.'[26] A large number of cotton mills were erected in Lancashire and the Midlands before 1781, the date of the first court decision against Arkwright, and this defeat 'precipitated a rush to make fortunes in warp spinning'.[27] When the patents were annulled in 1785 the way was clear for future expansion.

Crompton's mule was at first a hand operated machine and as such could be used in the spinner's home. But it was also taken into factories almost from the outset. Peter Drinkwater built a large factory in Picadilly Manchester in 1789 (page 17) in which carding machinery and roving machinery was driven by steam and spinning was performed by hand on mules of 144 spindles.[28] Robert Owen began manufacturing on his own account using hand spinning mules. William Kelly of New Lanark first applied water power to the mule in 1790 but it remained partly hand-operated until the inventions of Roberts made it self-acting between 1825 and 1830. Double mules were being constructed by McConnel and Kennedy by 1796[29] and by 1800 the manually controlled power mule had been improved to the point where its versatility in producing threads of both fine and coarse counts meant that it was being widely adopted. Some manufacturers installed mules in their water frame factories but from the 1790s onwards larger steam driven mule factories were being built.

The spinning machines of Hargreaves, Arkwright and Crompton were designed for spinning cotton but early attempts were made to adapt these machined to the spinning of worsted. Worsted spinning mills were built in Lancashire and Yorkshire and in the Midlands[30] but technical problems dogged the early experimenters and no Arkwright-equivalent emerged from amongst the entrepreneurs interested in worsted spinning.[31] Markland, Cooksen and Fawcett of Leeds having originally intended to build a worsted spinning factory turned their attention to cotton spinning since its success was proven.[32]

Woollen Factories

Although the embryo factory emerged early in the 18th century in the woollen industry the large factory containing machinery driven by a central power source developed later than in the worsted or cotton industry. Machinery for cotton manufacture was adapted for wool with varying degrees of success and not always quickly. Roller spinning on Arkwright's principle was not successful for wool and power spinning was not adopted until the 1820s to 30s when the mule was adapted to woollen manufacture. The transfer to the factory system in the West of England was assisted by the fact that the larger merchant clothier generally organised his operations from a fulling mill and was therefore in a good position to develop his water power site as mechanisation progressed and he began to appreciate the advantages of supervision. Some West of England factories date from the 1790s and were built to contain preparing machinery, fulling stocks and gig mills (for raising the nap of the cloth). Some of these factories also contained jennies and, occasionally, hand looms. Townsend Factory, Dursley contained two pairs of stocks, one gig mill with burling shops above and the 'new workshops' contained a shear shop, four carding machines, four billies and six jennies in 1795. But the majority of West of England factories date from the period 1813-30 when hand looms were being brought into the factory and power spinning was being adopted.[33]

In Yorkshire the transfer to factory production took longer and the factory was arrived at by several routes. There were a few late 18th century and early 19th century factories built by merchants in which some machinery was power operated but in which the spinning and weaving was done by hand. Benjamin Gott's Bean Ing Factory at Leeds, built in 1792 (page 34) is an example, built on an unprecedented scale and larger than anything in the West of England. On the other hand there were the small water driven fulling and scribbling mills; these were generally owned not by the cloth manufacturers but by fullers and scribbling millers who worked on commission. Although they were well placed to develop their water power facilities they often lacked capital and the experience of manufacturing on their own account. One way around this problem was multiple tenure as an historian of Morley[34] noted:

'Many of the mills are occupied under a plurality of tenure, the various occupiers having just risen to the dignity of employers of labour and making up by their personal work and supervision for smallness of capital, and other disadvantages, and in course of time, if things go well with them they will build for themselves and leave their present temporary holdings to the next aspirants for mastership.'

Another solution was the company mill which was generally built and run by a group of clothiers on a co-operative basis.[35] But the quasi-independent country clothier survived well into the 19th century. He generally conducted his manufacture from a workshop or warehouse and was only rarely the owner of a fulling or scribbling mill. When the power loom was introduced the larger clothier who had steadily increased his means of production was in a position to consider the adoption of steam power and the building of a factory near his upland warehouse or workshop. 'The manufacturer's [clothier's] mill, when weaving came to be done by power, was built in the upland hamlets and villages where his weavers lived and where his own home-

John McCracken, Belfast, 1806
Cotton spinning by power made rapid strides in Belfast in the late 18th century. This mill was built in two stages but the new part was designed to match the old.
Portf: 380

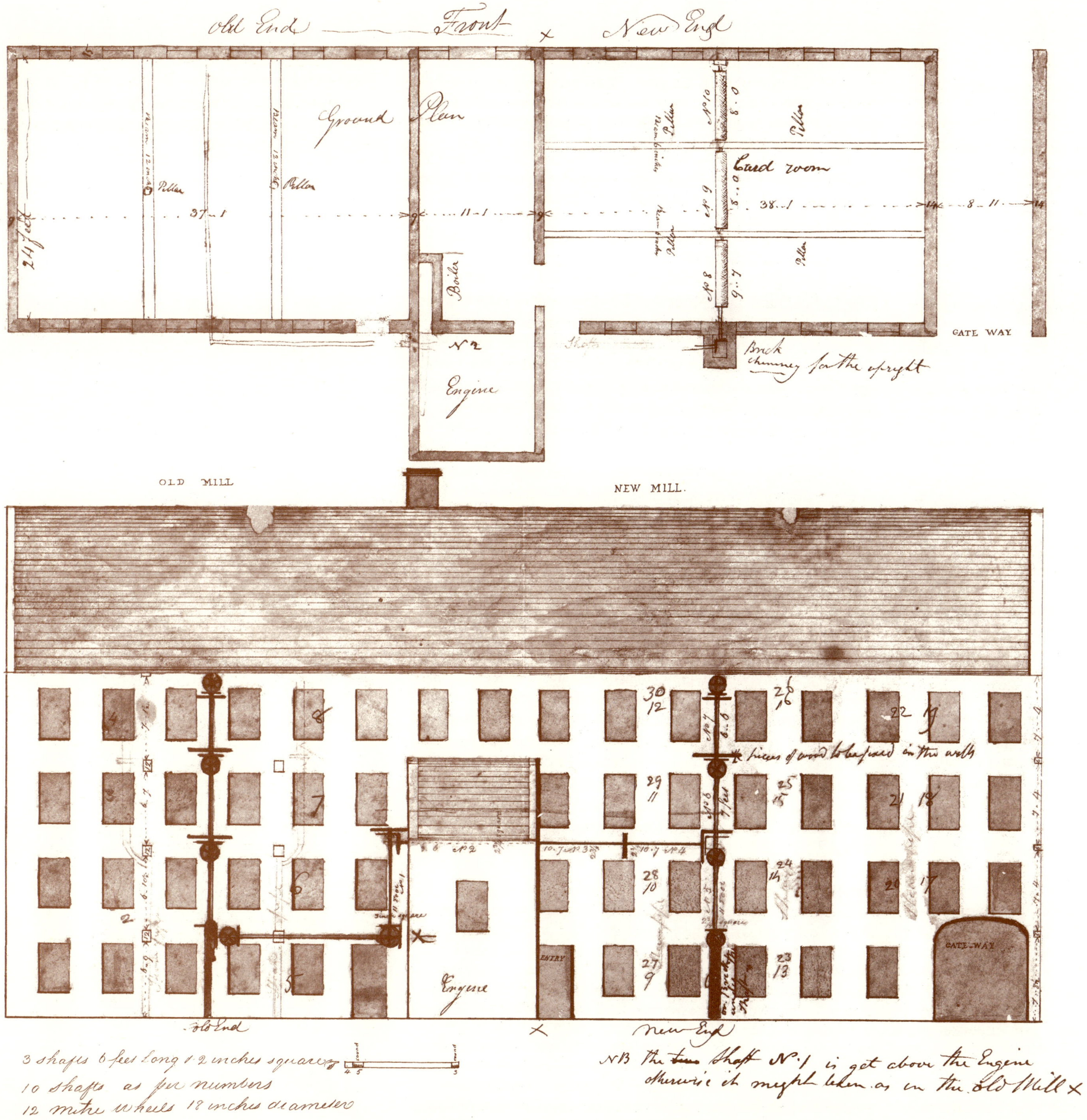
Old End — Front × New End
Ground Plan
Card room
24 feet
37..1
11..1
38..1
8..11
Boiler
No 2
Engine
Brick chimney for the upright
GATE WAY
OLD MILL
NEW MILL.
pieces of wood to be placed in the walls
Engine
ENTRY
GATE-WAY
Old End
New End
3 shafts 6 feet long 12 inches square
10 shafts as per numbers
12 mitre wheels 18 inches diameter
NB The Shaft No 1 is got above the Engine
otherwise it might taken as in the Old Mill ×

36. W.B. Crump & Gertrude Ghorbal, *History of the Huddersfield Woollen Industry*, p.82.
37. Felkin's *History of the Machine-Wrought Hosiery and Lace Manufacture*, Centenary Edn. 1967, p.464.
38. T.S. Ashton, *Iron and Steel in the Industrial Revolution*, p.42.
39. *A Description of Coalbrookdale in 1801 A.D.* (Ironbridge Gorge Museum Trust Limited) p.8.
40. T.S. Ashton op.cit. pp.45-53.
41. Sidney Pollard op.cit. pp.75-79.
42. M.J. Wise, 'On the Evolution of the Jewellery and Gun Quarters in Birmingham,' *Trans. Inst. British Geographers*, 15, 1950 p.68.
43. J. Ward, *The History of the Borough of Stoke-on-Trent*, 1843.

stead was, whereas the spinning mill grew out of the old fulling mill and the later scribbling mill and was consequently down by the streams.'[36]

Hosiery

In the hosiery industry few attempts were made to power drive the stocking frame before the 1840s and Felkin[37] noted that by 1844 'the hosiery machinery had scarcely begun to be gathered into large factories. The number of frames under one roof averaged then rather more than three only and even now [1857] the absorption of narrow hand machines into large masses can scarcely be said to have more than commenced.' In the early to mid 19th century the hosiery market was dull and while the work performed on the frame remained basic there was little advantage to be gained by gathering workers under one roof for the purposes of supervision. But the static hosiery market was stimulated by new designs introduced by Leicester hosiers in *c* 1845 and several manufacturers were then prompted to build factories to house power driven stocking frames.

Iron

The adoption of the coke smelting process in the 18th century gradually led to the transformation of the ironworks from a relatively small scale enterprise to a large factory complex. In the mid 18th century the larger works seem to have consisted of separate groups of blast furnaces, forges, and rolling mills situated along a river valley, but combined works developed in the late 18th and early 19th centuries. Tern Works, Shropshire, described on page 59 is a little known example of early 18th century factory development in the iron industry. The growth of Coalbrookdale resulted in new furnaces being put into blast at Willey (1732), Horsehay (1755) and Ketley (1756). In 1794 the Coalbrookdale Co was valued at £62,575 and Reynolds' concerns were estimated to be worth £138,067.[38] An anonymous writer[39] described the Lower Furnace at Coalbrookdale in 1801 as 'forming a considerable pile of buildings' consisting of seven moulding rooms, three mills for boring and grinding, four shops for carpenters and smiths and a warehouse. The Wilkinson concerns at Bersham and Bradley received great impetus from John Wilkinson's improvements in the boring of cast cannon. Wilkinson cast the important parts of Boulton and Watt's steam engines until Soho Foundry was built following a split between the two establishments. The Walkers of Masborough near Rotherham, starting from small beginnings at a small foundry in a nailer's smithy near Sheffield, built up their concern until it was worth £87,500 at the outbreak of the American war.

Carron Ironworks set up in 1759 by Roebuck and Garbett of Birmingham in conjunction with William Cadell, a local ironmaster, was the centre of a number of important experiments. It was modelled on Darby's works at Coalbrookdale. Smeaton installed a boring machine and augmented the power systems, introducing his cast iron blowing cylinders in *c* 1768. Although this was a combined forge and foundry it is clear that founding was the more important department and like Wilkinson's works,

gun founding was practised on a large scale. As T.S. Ashton[40] has said 'Coalbrookdale, Broseley, Masborough, Carron all became schools of instruction in the arts of smelting and refining and each of them gave birth to a lusty progeny bearing a close likeness to the parent firm'. The Parkers, relatives of the Darbys moved to Tipton, Staffordshire, where they set up Coneygre Furnace; Gilbert Gilpin, former clerk to John Wilkinson, set up his own ironworks at Coalport; J. Guest moved from Broseley to Dowlais.

From the 1790s the spread of puddling encouraged coal-iron works producing a wide range of iron goods, steam engines and other engineering products; Butterley Co, the Smiths of Chesterfield, and Low Moor and Bowling near Bradford are examples. Most spectacular of all was the rise of the south Wales iron industry. Cyfartha employed 1,000 men in 1804, there were three furnaces in 1796, nine in 1830 together with eight steam engines, three forges, a foundry, eight rolling mills and a boring mill. Dowlais had twelve furnaces, twenty-six fineries, six mills and employed, including the coal mines, *c* 5,000.[41]

The Birmingham small metal industry was mainly conducted in workshops but there were also large scale businesses run by such manufacturers as John Taylor, John Gimblet, Edward Thomason, and Boulton and Fothergill. Soho Manufactory was undoubtedly the most important factory in the small metalware trade. The initial buildings cost approximately £10,000 and additions were made towards the end of the 18th century; few of the processes of toy manufacture were mechanised although Boulton used water power and, in dry seasons, horse power to work rolling machinery. Boulton's motives in building a factory on this scale were primarily the desirability of supervising his labour force as well as the increase in production made possible within a factory organisation. Boulton's own declared aim was 'to obtain a school of designers who should give to the products of the Soho Factory an artistic style and finish not obtainable elsewhere', this could only be achieved through close supervision in a factory. Thomason, who had been trained at Soho, built a toy factory in Church Street, Birmingham in 1796.[42]

Pottery

The concentration of the means of production began early in the 18th century in the pottery industry with the buildings of workshops behind the master potter's house. But the first 'brick-built factory roofed with tiles',[43] as a local historian quaintly put it, was the Ivy House Works in Burslem, built by Thomas and John Wedgwood in 1750. Josiah Wedgwood began at Etruria in 1769. This was the model on which the other major pottery firms based their factory layout. It was planned on a rational basis with each process being conducted in a separate room or rooms in sequence. Wedgwood, like Boulton, desired to regulate the quality of his products by supervision and the training of workpeople but he also availed himself of

Walker, Manchester, *c.* 1790

This drawing is undated but it is one of the most informative of all the drawings of cotton spinning mills for it shows the layout of machinery on Arkwright's principle with the central shafting and frames to either side.
Portf: Misc Mills

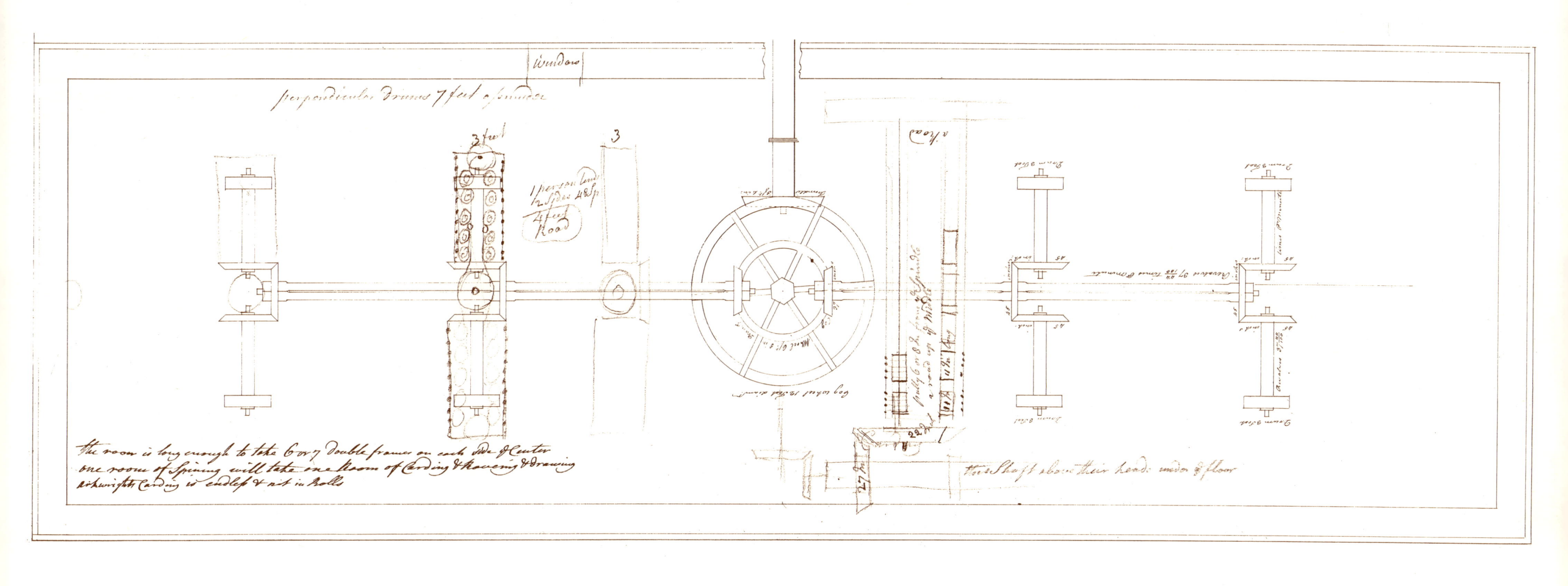

12

44. John Thomas, 'Josiah Wedgwood as a Pioneer of Steam Power in the Pottery Industry,' *Trans. Newcomen Soc*, 27 1936-7, p.17.
45. Peter Mathias, op. cit. pp.xxiii-xxiv.

steam power in the grinding of flint and colour stuffs, in clay tempering and in operating the potters' wheels.[44] With this degree of mechanisation a far higher output could be achieved in a well planned factory than in a part-domestic system. Spode, Enoch Wood, Samuel Alcock and others built large factories on the Etruria model and rapid growth was maintained until the mid 19th century.

Brewing

Expansion had been under way in the brewing industry before the mid 18th century but the scale of production was determined by crucial marketing conditions.[45] Not only had there to be a demand but that demand had to be confined to an area covering not more than four to six miles. The high overland transport costs meant that 'the most dynamic entrepreneur was locked inside his local market, save in special circumstances.' Large breweries could only develop in the intensive market of a city and London was such a market. Here 'a brewery became potentially more suited for large-scale production than most other contemporary manufacturing concerns.' This was an exceptionally capital intensive industry. Beyond the confines of the large towns the small common brewer and the innkeeper who brewed his own beer continued to flourish well into the railway age.

G & J Robinson, Papplewick, Nottinghamshire *c.* **1785**
The Robinsons owned six mills along the Leen valley. These drawings are probably of one of the factories at Grange Mills. Robinsons were the first cotton spinners to use a Boulton and Watt rotative engine.
Portf; 9

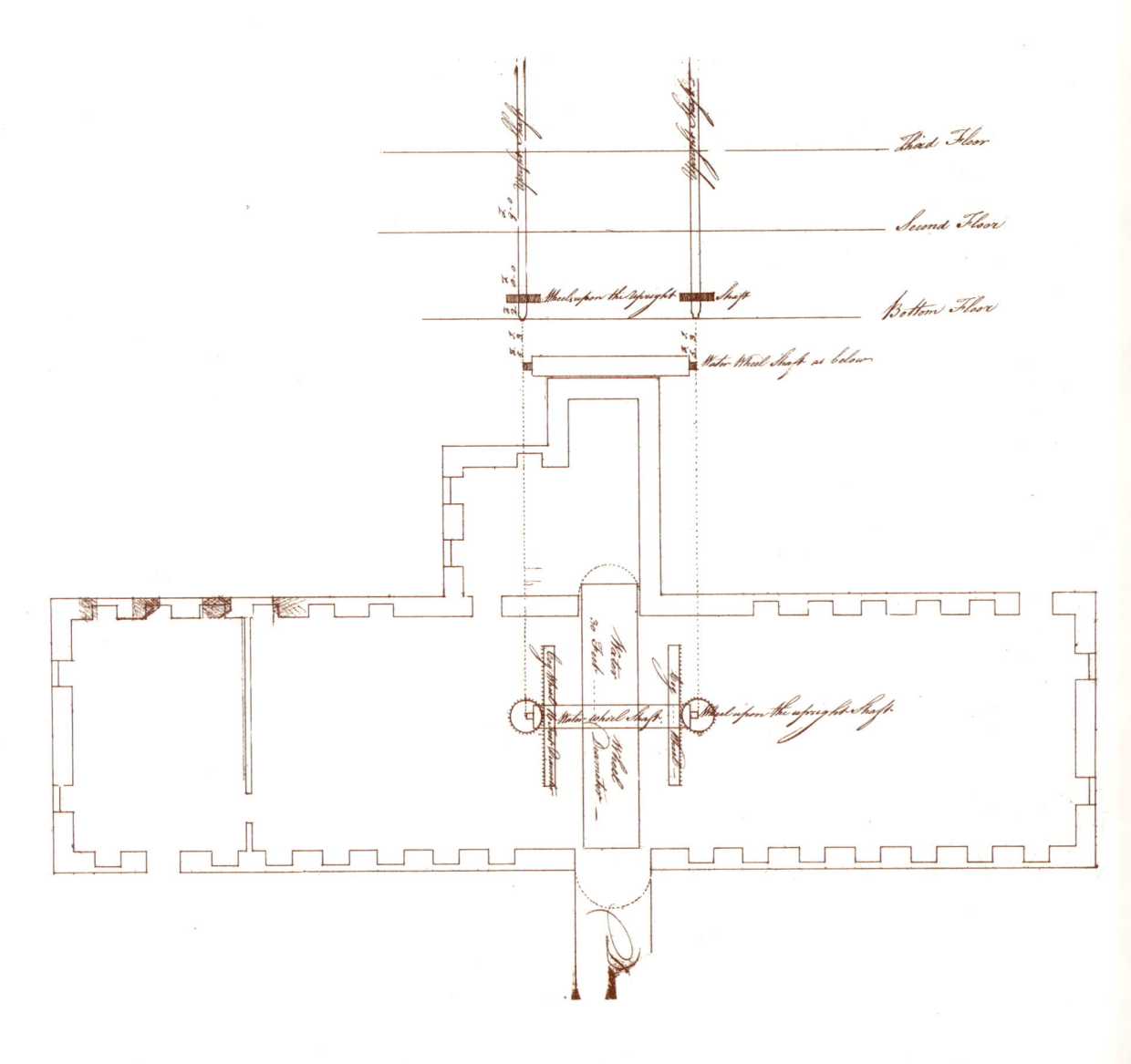

Length 122 Feet & a half

from Mr Bannatine with letter in Decr 1799

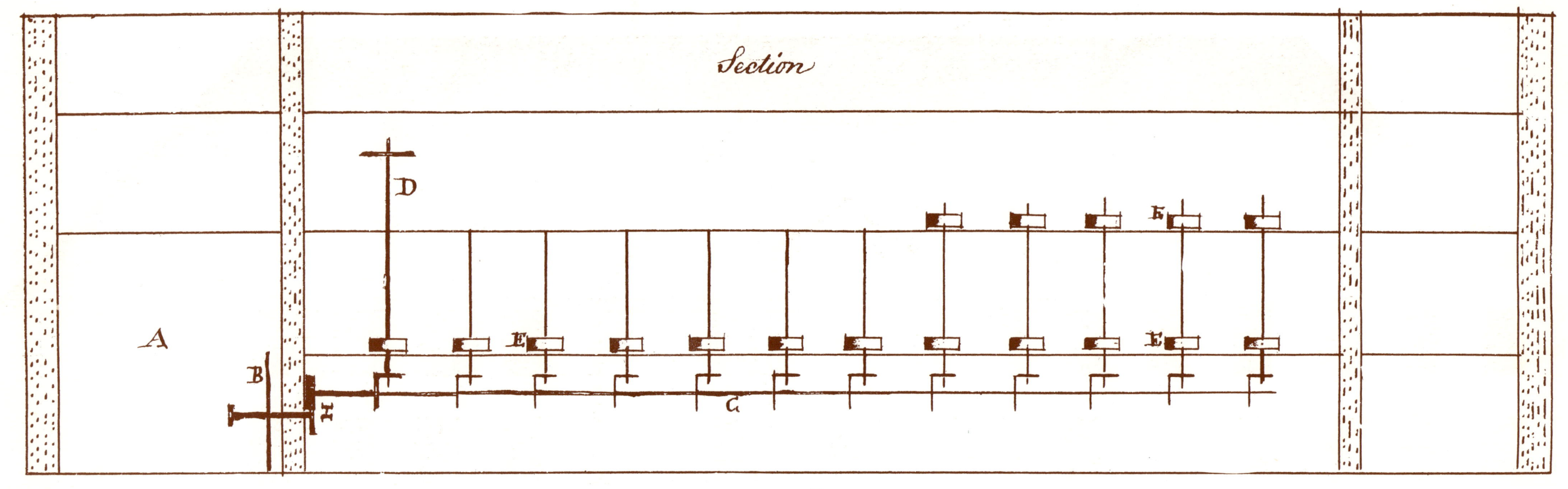

Section

Scale of Feet

Plan

Mr Bannatine, Glasgow, 1790
A cotton spinning factory on Arkwright's principle.
Portf: Misc Mills

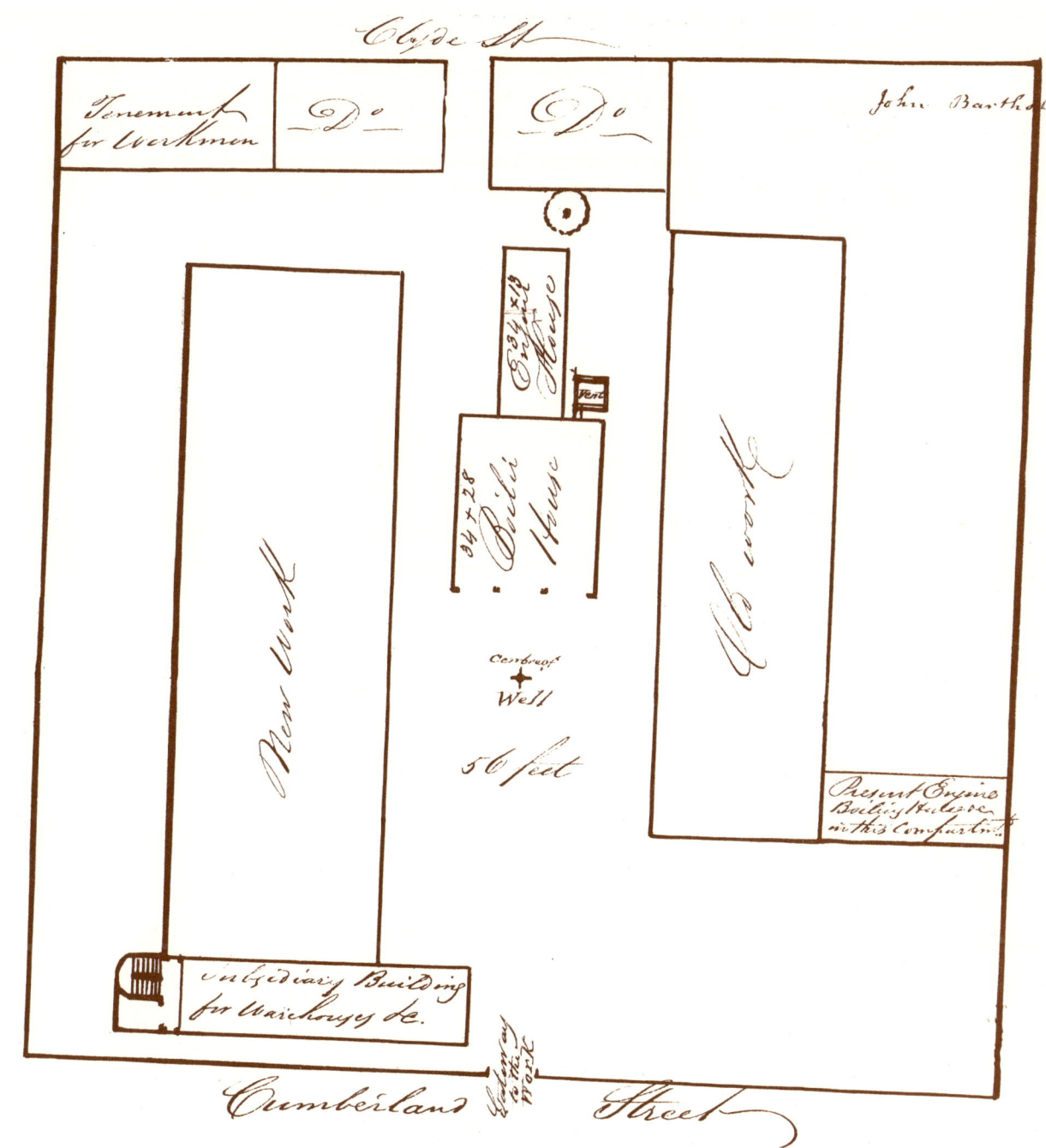

John Bartholomew and Co, Glasgow c. 1796
The drawing shows a large Scottish cotton works with industrial housing as part of the complex.
Portf: 464

16

P. Drinkwater, Manchester, 1789
Drinkwater was the first Manchester cotton manufacturer to use a Boulton and Watt rotary engine in his spinning factory. He corresponded with Boulton and Watt on the layout of his factory and the problem of providing 'necessaries' for the mill hands. The steam engine was used to drive preparing machinery whilst the spinning was done on hand mules.
Portf: 44-45

A. & G. Murray, Manchester, 1805
The Murrays, like McConnell and Kennedy, moved from Scotland and began as machinery manufacturers before going into cotton spinning. This drawing is of interest for it gives the dates of factory building.
Portf: 364

Markland, Cooksen and Fawcett, Leeds, 1792
These drawings by John Sutcliffe the Halifax millwright are a fine example of 18th century draughtsmanship. Markland Cooksen and Fawcett intended using the mill primarily for worsted spinning but eventually used it mainly for cotton spinning.
Portf: 91

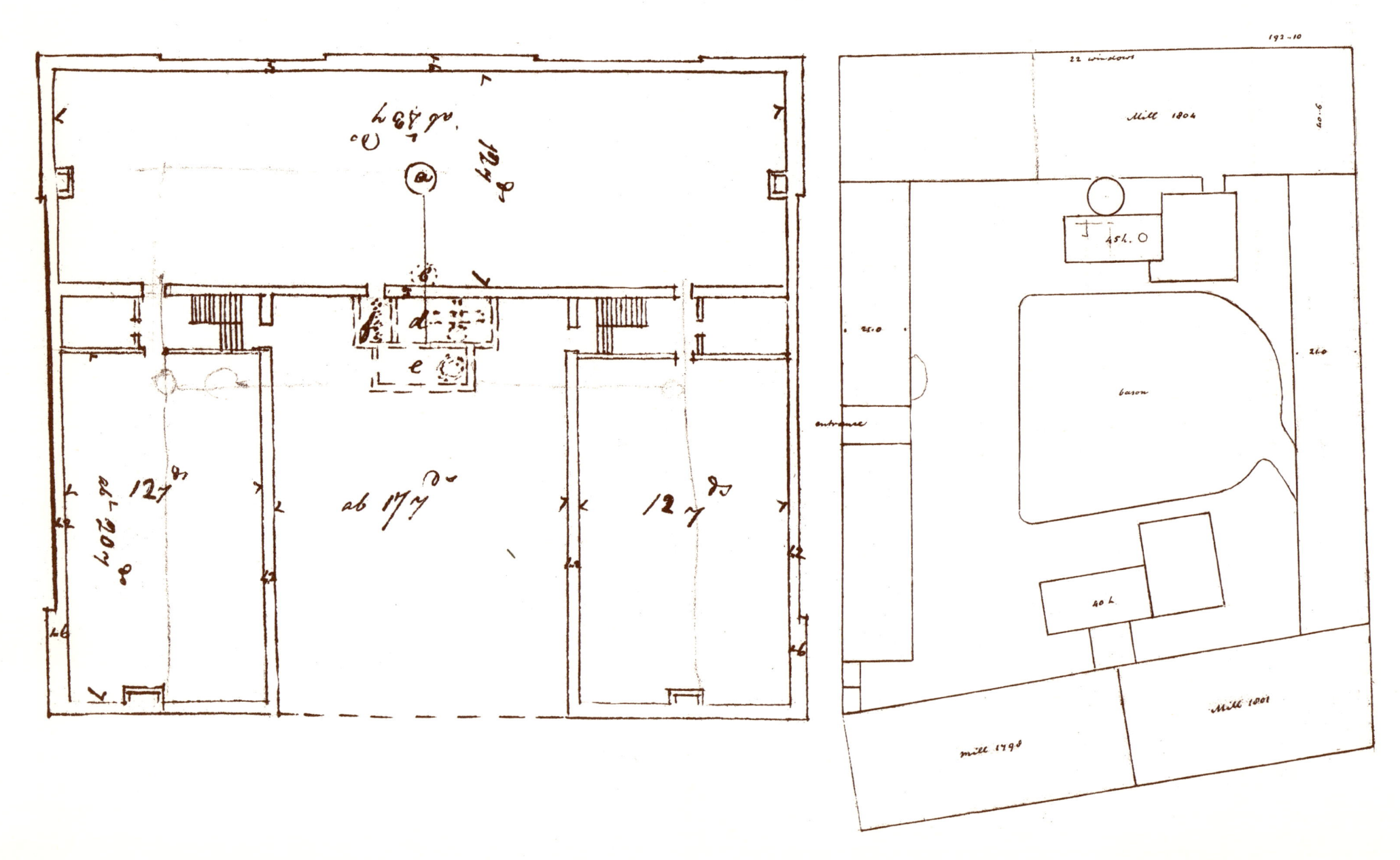

The Ground Plan

Halifax 21 Feb 1796

McConnel & Kennedy, Manchester, 1809
McConnel and Kennedy began in business as machine makers and mule spinners with a capital of £6-700. They moved to Union Street in 1797 where they built the 'Old Factory' and in 1801 they began to build the first half of the 'new mill'. The second half was built in 1806. This was a large seven storey building of fireproof construction.
Portf: 303

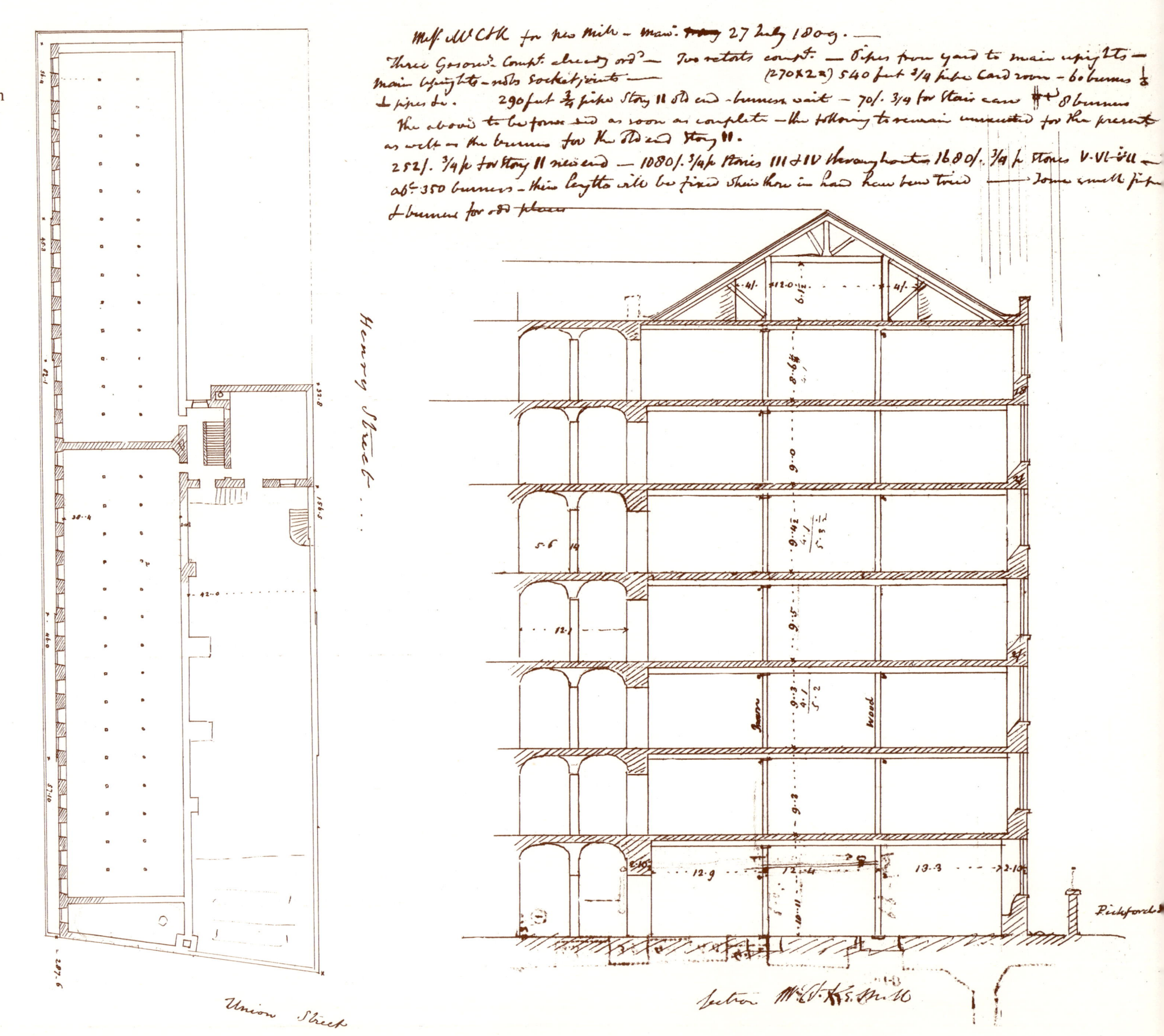

Owen, Scarth & Co, Chorlton, Manchester *c.* 1795
The implication to be drawn from this plan is that Owen Scarth & Co's and Marsland's intended factory were planned as one large industrial unit. Marsland probably built Owen's factory for the building was not insured by Owen and he was so short of capital that Marsland had to pay half the cost of Owen's steam engine. It was from this factory that Owen moved to New Lanark.
Portf: 124

Longitudinal section of Messrs Philips & Lees Mill &c.—

Ground Floor or Cellar Floor

Only these two Floors to be lighted at present.

The horizontal pipes to run as close under the spring of the arches as possible.

½ Inch to the Foot.— 14 Sepr. 1805.—

Plan of New Mill.—

= 1500 Candles in this Mill
The Floor C & D to equal 400 Candles, that is, 200 each—

6, 6, 6, 6, 6, 6 situation of the lights in the Floors C & D —

Stair case not to be lighted at present

21

Philips and Lee, Salford Twist Mill, 1805
This mill, built in 1802 and designed principally by Lee was one of the first cast iron framed buildings. It was built for mule spinning.
Portf: 242

Birley & Hornby, Chorlton, Manchester, 1809, 1813
The plan was drawn for the purpose of installing gas lighting and shows the large factory complex around a court. The section dates from 1813 and is in Creighton's hand – he wrote to Boulton and Watt saying 'Mr Birley will be obliged by any remarks you may have to make.'
Portf: 418, 449

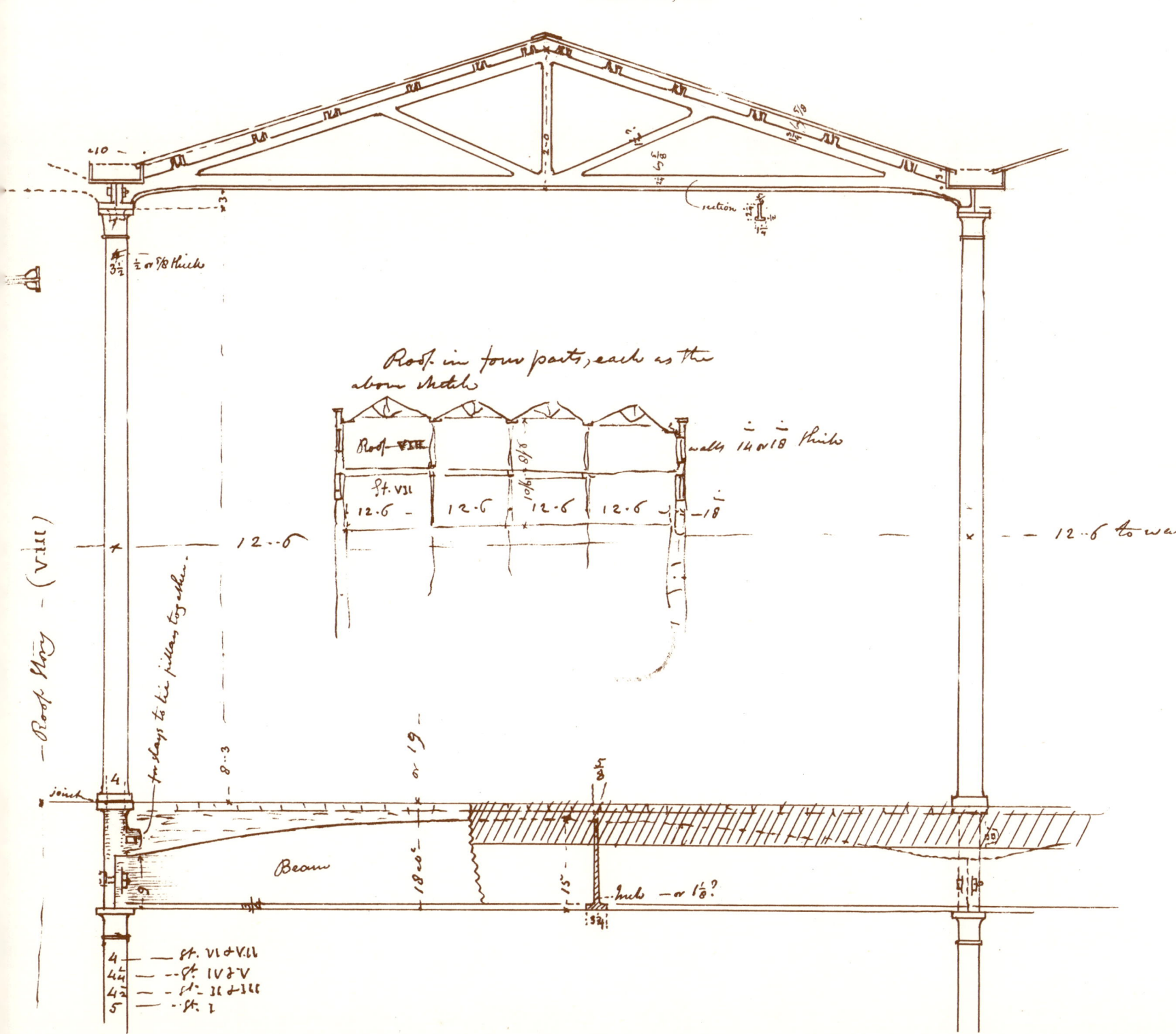

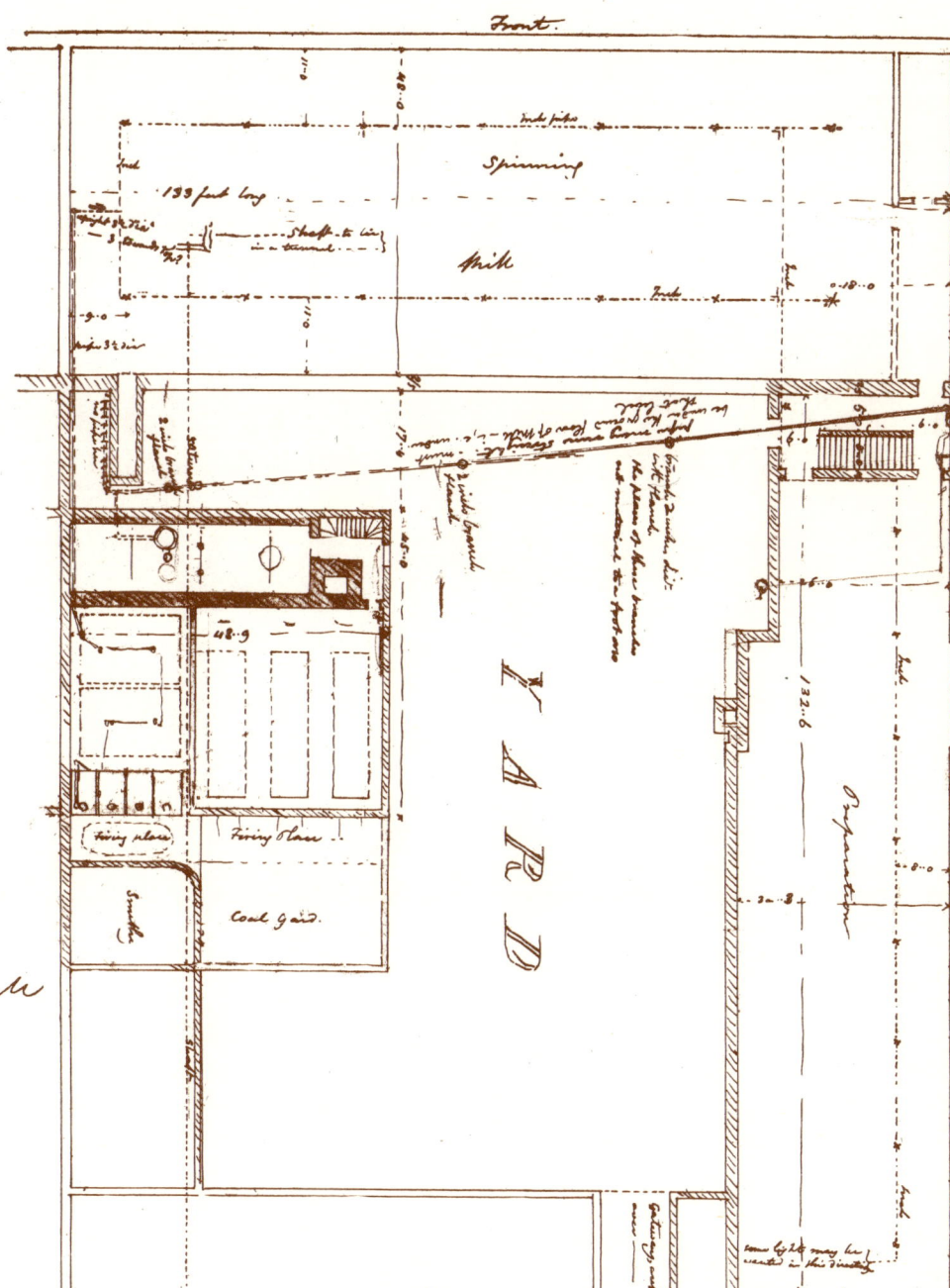

George Cheetham & Sons, Stalybridge, Cheshire, 1833
This was a vertically integrated fireproof factory. The note at the bottom of the drawing is valuable for it gives the approximate horse power required by the different manufacturing processes.
Portf: 1049

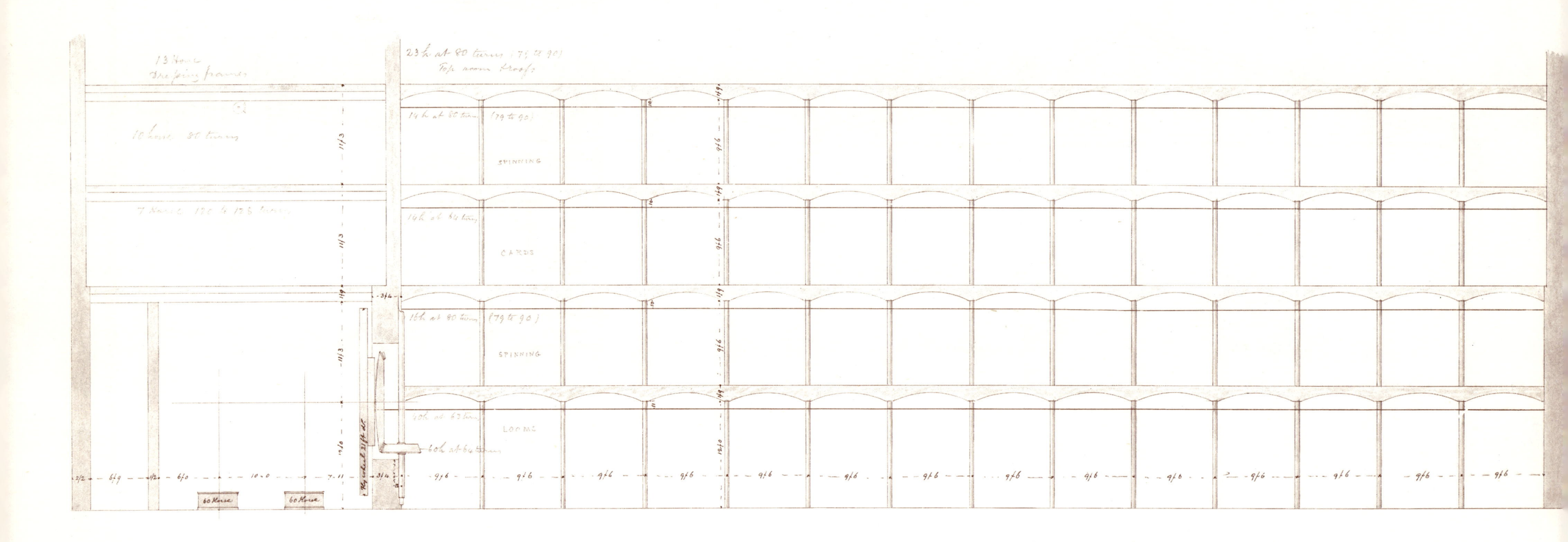

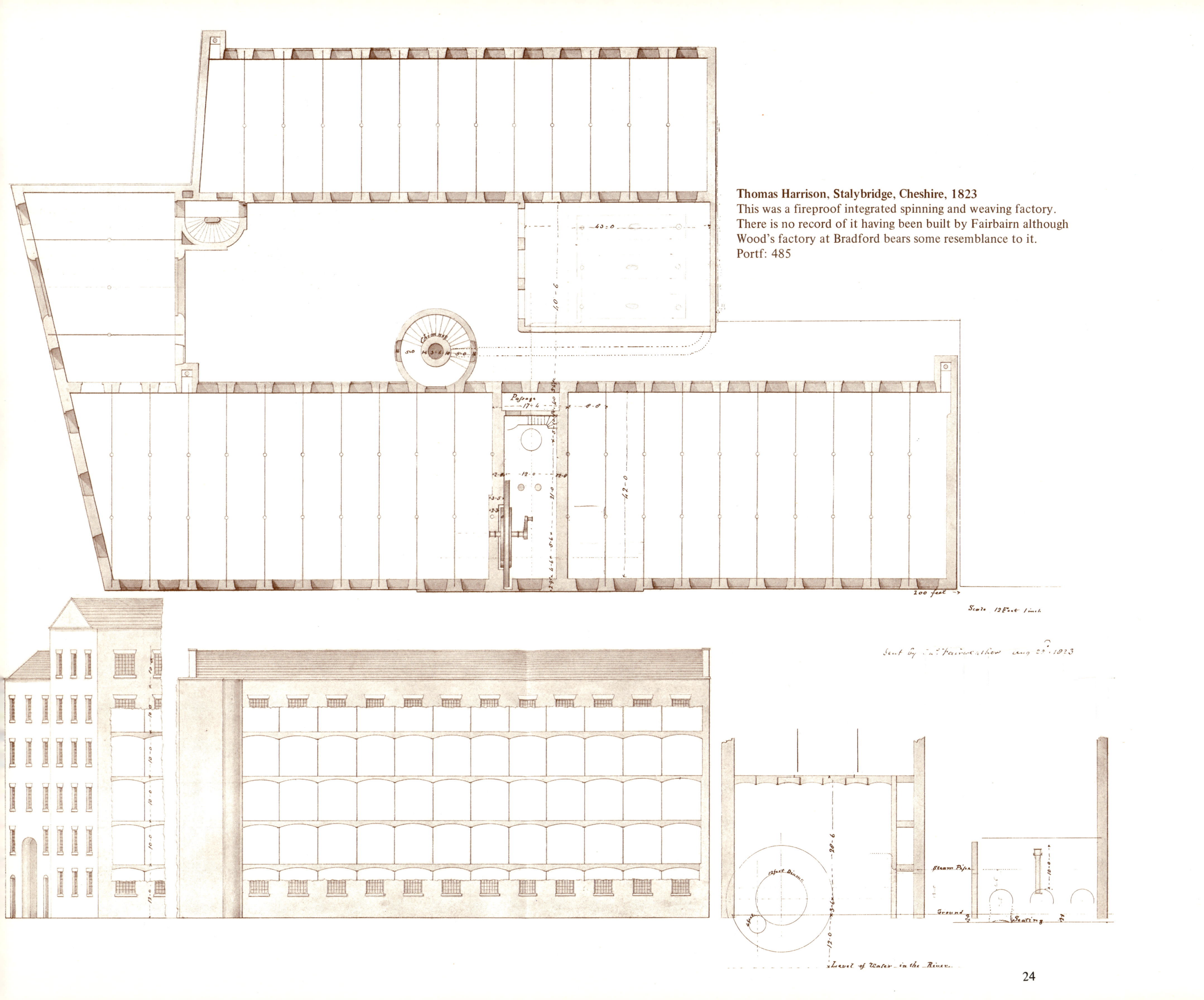

Thomas Harrison, Stalybridge, Cheshire, 1823
This was a fireproof integrated spinning and weaving factory.
There is no record of it having been built by Fairbairn although
Wood's factory at Bradford bears some resemblance to it.
Portf: 485

Great Western Cotton Factory, Bristol, 1839
There were several early attempts to establish cotton spinning in Bristol, this was the last and they all failed. The Great Western Cotton Company built this estate including housing and a school in 1839-40. A feeder canal was built to supply raw cotton from the port of Bristol but it was found cheaper to obtain cotton from Liverpool.
Portf: 516

26

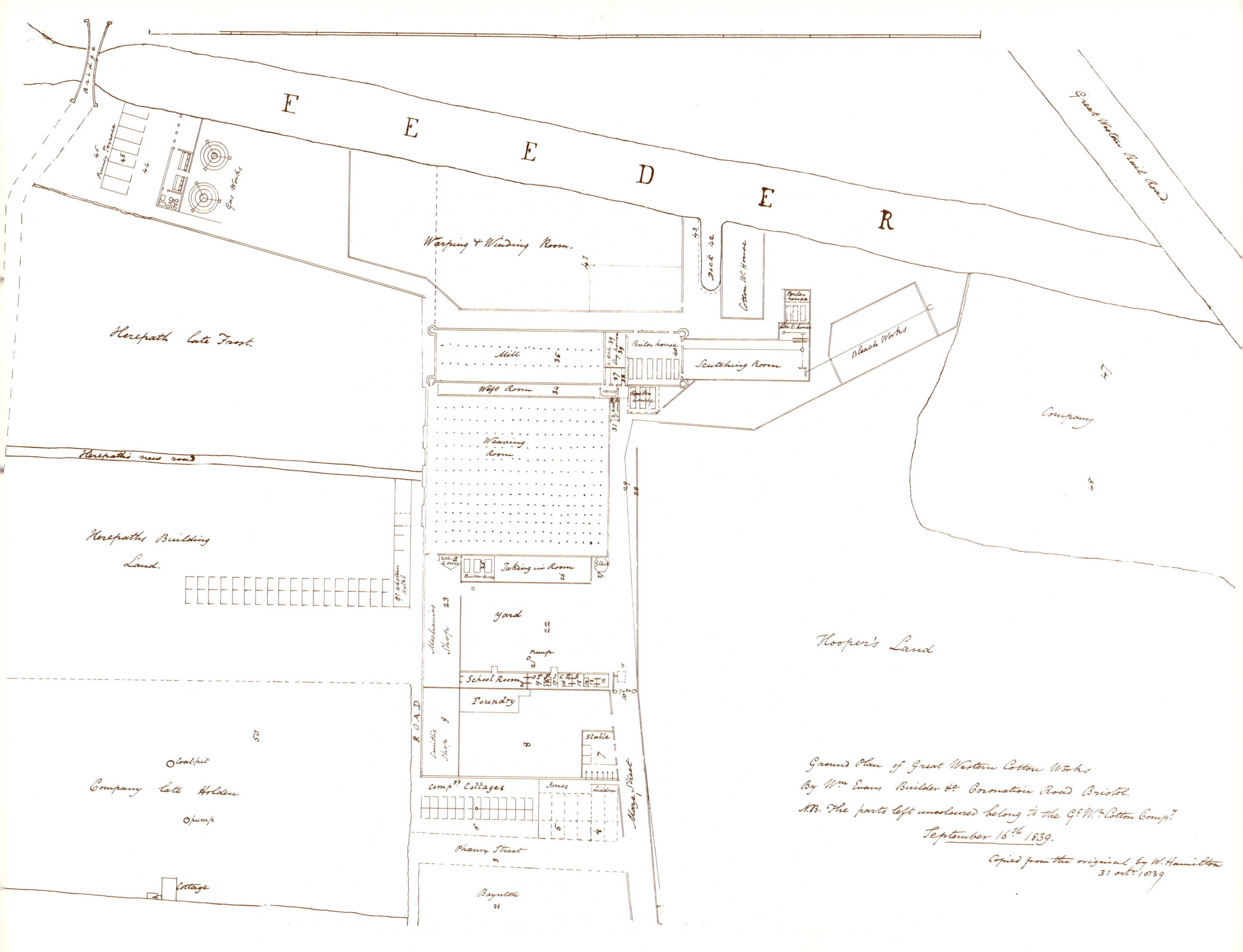

2. Factory capacity and Fixed capital

1. P.P. House of Lords, 1828, 515, p.289.
2. Richard L. Hills, *Power in the Industrial Revolution*, pp.191-2.
3. See chapter 4.
4. R. Denison to Boulton & Watt 2nd June 1791.
5. P. Ewart to J. Watt 3rd July 1792.
6. P. Ewart to Boulton & Watt 6th January 1792.
7. P. Drinkwater to Boulton & Watt 3rd June 1789; Richard L. Hills op.cit. p.192.
8. John Kennedy, *Brief Notice of my Early Recollections, Miscellaneous Papers*, Manchester 1849, p.17.
9. C. Lees to Boulton & Watt 27th November 1795.
10. R. Paley to Boulton & Watt 20th June 1797.
11. Guildhall Library, Sun Insurance Registers, CS I/695749.
12. Ibid 148/1015260.
13. Ibid 148/1018850.
14. Ibid 150/1030005.
15. J. Watt jun. to M.R. Boulton 23rd June 1802.

A fundamental point to be considered by the entrepreneur thinking of building a factory was the output that he and his partners hoped to achieve when the factory was complete and running to full capacity. In practice there were a great many difficulties and a number of factories probably never reached capacity. Some manufacturers underestimated the cost of erecting a factory and were forced to moderate their plans when the building programme was well under way. In these cases the completed building was something of a compromise. An investigation of the letters sent to Boulton and Watt reveals clearly that entrepreneurs with exceedingly little technical knowledge were prepared to risk large sums of money in manufacturing ventures. Technical knowledge was not everything and many of these entrepreneurs brought other skills or assets to the firm such as management ability, a good knowledge of markets acquired through a career as a merchant, or capital. Nevertheless, the entrepreneur either had to be prepared to learn the technical processes or to go into partnership with someone who was, or to appoint a manager who was skilled in the technical aspects of manufacturing. Benjamin Gott belonged to the first group: 'I was brought up as a merchant and became a manufacturer rather from possessing capital than from understanding the manufacture I paid for the talents of others in the different branches of manufacture.'[1] The active partner in the firm of Philips and Lee was George Lee who, after discussion with his partners, was responsible for building and equipping Salford Twist Mill. Peter Drinkwater had no knowledge of manufacturing and employed Robert Owen to manage his mule spinning factory at Picadilly, Manchester.

It has been suggested[2] that the size of a factory was determined mainly by the amount of power that was available at the chosen site. It is clear that during the water power era, however, either the entrepreneur or his millwright frequently misjudged the potential power of a stream[3], or the quantity of power available was a secondary consideration. Transport costs are another suggested factor limiting factory size. It seems likely however, that when a new factory was being built the size was determined not by the potential power nor by the transport system but by the capital available for investment on a long-term basis and the maximum output that the entrepreneur thought he could achieve within that limit. The possibility of a power shortage was rarely an immediate problem, for few factories reached capacity in the first few years of production. When a shortage of power did become a problem the entrepreneur investigated supplementary sources.

In the first thirty years or so of power spinning the majority of large manufacturers had workshops in which machinery was made on the premises and it was several years before the factory was filled. Robert Denison of Nottingham said that it would be impossible for him to utilise the full powers of his steam engine for several years.[4] Lodge of Oakenrod near Rochdale had a cotton mill containing 804 spindles and intended to install 3,000;[5] Lawrence and Yates' mill contained about 2,500 spindles but the factory was large enough to contain 6,000.[6] Drinkwater illustrated the point forcefully: 'Circumstanced as a Cotton Manufacturer is, it is impossible to exercise its [the steam engine's] full power at once — the Factory at Shudehill [Arkwright and Co] . . . has been more than 7 years filling with machinery and I am doubtful whether with convenience it could have been filled much sooner and . . . That I am now building will and must labour under the same inconvenience so far that I fear I shall not be able to use more than the power of 3 or 4 horses within the compass of two years or more after the engine is set agoing.' It was nine years before the clockmakers who made most of the machinery at Styal Mill disappeared from the wages books.[7]

The letting of 'room and power' was common. When a large factory was built by a manufacturer he often let part of it, realising that he would be unable to fill the entire building with machinery for some years. By so doing the owner of the factory was able to cover the interest due on the capital cost of the building by the rents of his tenants. John Kennedy[8] in his first partnership with Sandfords and McConnel moved into 'a building in Canal Street (Manchester) called Salvin's Factory from the name of the owner who occupied a portion of it himself, letting off the remainder to us.' Charles Lees of Stockport had tenants in part of his mill.[9] Richard Paley of Leeds let part of his building together with the frames in it to a quarterly tenant who was unable to pay the rent:[10] 'I understand they have not produced prime cost of their twist setting aside the rent as they are people of very little or no property. I am afraid I shall have the whole thrown on my hands again.' John Kenyon insured a wool and cotton carding mill near Bury 'in several tenures' in 1799.[11] In 1824 James Rowbottom insured his machinery in Wardles silk factory in Macclesfield.[12] In the same year Richard Fowler insured his machinery on the fifth floor of Soho silk mill on Macclesfield Common[13] and David Downes insured his fixed machinery in two rooms of a silk mill in Dog Lane, Macclesfield.[14] Some factories were erected by speculative builders who had little knowledge of or little direct interest in industry. It was obviously a lucrative business. Watt jun. described[15] a Mr. Taylor of Rochdale, a man of considerable property. 'He is retired from business and employs his spare capital in building mills, which he lets out so as to produce good interest.' Robert Owen found a builder to put up a factory for him in Manchester; he occupied a part and let the rest.

It is clear that either a partner or a manager had to have a working knowledge of the processes of manufacture but many entrepreneurs launched into factory building with little knowledge of the size of building required to achieve the desired

Wool and Worsted
Although wool was the raw material in both the woollen and worsted industries the processes of manufacture were different and the development of the factory took a different course in each case. In worsted manufacture spinning was the first process to be mechanised and spinning factories on Arkwright's principle were set up in the 1780s and 1790s. The late 18th century woollen factory on the other hand contained only carding and scribbling machinery, fulling stocks and gigs for dressing the finished cloth, and the factories were smaller.

Nussey, Birstall, Leeds, 1796
The drawing shows a fine Palladian style three storey woollen factory. The steam engine worked stocks and carders or gigs. Portf: 138. Scale approx 1in : 12ft.

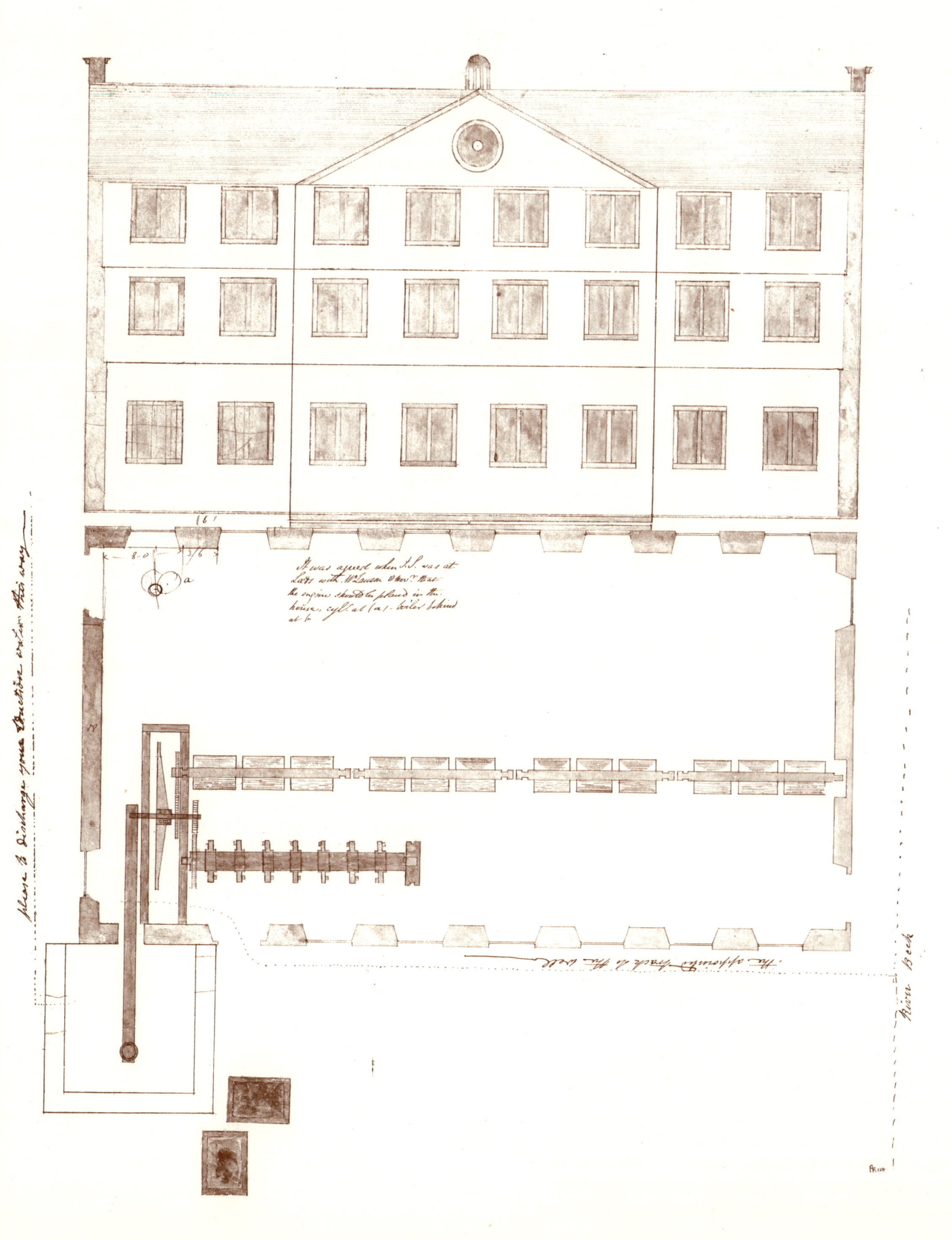

16. J. Southern to Clennell 22nd September 1796, Foundry Letter Book.
17. J. Southern to Walker 17th September 1799 Foundry Letter Book.
18. J. Sutcliffe, *Treatise on Canals and Reservoirs*. . . . .
19. S. Wyatt to M. Boulton 6th January 1788, Albion Mill Box, A.O.
20. A.O. Westworth, 'The Albion Steam Flour Mill', *Economic History*, 2 1930-3, p.383.
21. Sun Fire Insurance Registers CS 1/620008.
22. Ibid 1/622055.
23. Guildhall Library, Royal Exchange Registers 32a/158829.
24. Mr. Low millwright of Nottingham estimate of a corn mill December 1794, Albion Mill Box, A.O.
25. Memo for Mr. Weinman February 1796, Box 3, W.
26. P. Atherton to Boulton & Watt 20th January 1781.
27. J. Southern to C. & G. Stanton 4th January 1792 Foundry Letter Book.
28. P.P. 1816, III, pp.134, 141.
29. Moon to Boulton & Watt 5th March 1783, 23rd June 1783.
30. W. Carr to Boulton & Watt 6th March 1790.
31. P. Ewart to Boulton & Watt 6th January 1792.
32. J. & T. Wilkes to M. Boulton 19th October 1783; Southern to Markland, Cooksen & Fawcett, 16th July 1792, Foundry Letter Book.
33. W. Douglas to Boulton & Watt 1st January 1792.
34. J. Southern to P. Ewart 22nd October 1791, Foundry Letter Book.
35. J. Watt jun. to M.R. Boulton 19th August 1803.

output or of the amount of power required, sometimes with little knowledge of whether it was even possible to harness power for a particular process. Boulton and Watt received many enquiries about the amount of power required to drive different machines. Their own knowledge of the iron trade and their contacts within the trade supplied most of the answers in the field of metals; their experience at Albion Mills enabled them to assist other corn millers and the knowledge gained by installing steam engines for the Robinsons and Arkwright enabled them to draw up a rough rule of the horsepower per spindle requirements in cotton spinning. For other industries they had to admit their ignorance. 'We are not acquainted with machinery applied to the Hat Manufactory, but will do our best to view some and then give our opinion on how our engines could be applied to drive the machines.'[16]

Corn Milling

In corn milling Boulton and Watt generally allowed 8 hp per pair of millstones grinding at the rate of 8 Winchester bushels per hour. There were, however, variations in the performance of customers' mills.[17] John Sutcliffe the Halifax millwright estimated that 25 hp was required to turn 5 pairs of stones at between 90 and 100 rpm.[18] Wyatt believed that it would be possible to grind 2,000 quarters of wheat per week at Albion Mills in 1788, but he thought it unlikely that this amount could be sold and said that 1800 'will be the extent of my wishes.'[19] Two sets of 8 pairs of stones were driven by two 50 hp steam engines. With this scale of enterprise the wagons loaded with wheat reached from the mill door to nearly half-way across Blackfriars Bridge. The nominal capital of the firm was £60,000, a third of which was represented by the factory building.[20] Albion Mills were exceptional and fixed capital investment in corn mills was generally well below £2,000. Coltham mill at Hiblington, Middlesex was insured for £1,200 in 1794;[21] a water corn and flint mill at Sandon, Staffordshire was insured for £400 in 1793.[22] A combined corn and carding mill at Holbeck, Leeds was insured as follows in 1797:[23]

Mill: ground floor corn milling, others for carding, slubbing and spinning wool	£ 600
Steam engine	200
Water wheel etc.	1,200
House	400
	£2,400

The building estimates for corn mills survive, one by Thomas Lowe the Nottingham millwright and one by Benjamin Gott the Leeds woollen manufacturer. Lowe's estimate, for the millwork only, was for a corn mill containing three pairs of French Burr stones and two wire machines for dressing the flour 'to be made compleat and ready to send to Copenhagen.' Lowe gave his cost as £348 10s and Boulton compared this with what Gardener of Liverpool asked — £530; and Rennie — £600.[24] Gott's estimate was for a corn and fulling mill.[25] He noted that 16 hp would be required for the corn mill to work two pairs of stones, dressing machinery and sack hoist. 5 hp would be needed for 3 pairs of falling stocks and 1 pushing stock, 9 carders and scribblers and a willy.

Stones etc.	£ 100
9 scribblers and carders	600
20 hp engine	1,095
Millwright's work for corn mill	250
Millwright's work for 4 stocks	220
Millwright's work for scribblers and carders	200
Mill 20 yds by 11½ yds	700
Pipes	100
2nd boiler	100
	£3,365

Cotton Spinning

In 1781 Peter Atherton wrote to Boulton and Watt from London[26] 'If you are acquainted with what number of cotton spindles any given size of your engines now turns or will turn when full employed, or if you can procure such information, or if you can let me know what size of your engines will be equal power of a proper sized bucket water wheel . . . it may assist in governing what sizes of engines may be necessary for the intended purposes.' Arkwright's experience using horses at his Nottingham mill set the formula for the ratio of spindles to horsepower in the early water frame cotton factories. When Charles and George Stanton asked Boulton and Watt what difference there was between the power required for spinning cotton and for carding Boulton and Watt admitted that they did not know but added 'Sir Richard Arkwright used to reckon 9 horses to 100 spindles'[27] and thereafter 1 hp per 100 spindles was adopted as a working formula for coarse spinning. This included power for the preparation and allowed a little in reserve. Arkwright's licencees appear to have been permitted to install 1,000 spindles and the early cotton spinning mills on the Arkwright pattern were roughly similar in size and value. Peel acknowledged the indebtedness of the early cotton spinners to the inventor of the water frame 'Arkwright originated the buildings, we all looked up to him and imitated his mode of building.'[28] 1,000 spindles could be contained in a building approximately 70 to 80 ft long by 30 ft wide. When Arkwright wished to increase his scale of production he extended the length of his factory but Peel preferred to increase his by the multiplication of the standard units.

By the time that Arkwright's patent had been annulled a number of factories contained more than 1,000 spindles, and by the 1790s factories containing 3,000 spindles were becoming fairly common. Col Mordaunt's factory at Halsall contained nearly 500 spindles in 1783 and was 'in a flourishing state', and it was intended to increase the number to 1,500.[29] Moon, the manager, estimated that 1 hp turned three frames of 48 spindles each. W. Carr[30] of Liverpool intended having 1,500 spindles and 24 or 26 carding machines in 1790. Lawrence and Yates had c 2,500 in 1792;[31] Wilkes of Measham, and Markland, Cooksen and Fawcett of Leeds intended installing 3,000 spindles.[32] William Douglas had 3–4,000 spindles at Pendleton;[33] Daintry, Ryle and Co, of Manchester built a factory to contain 5,000 and Simpson of Manchester intended increasing his spindleage to 5,000.[34] By 1803 Cark Mill had 4,250–4,500.[35] Although by the 1790s cotton spinners were considering installing up to 5,000 water frame spindles many small firms were still in business. In Scotland the small firm survived well into the 19th century and Loudon's mill illustrated on p. 6 is a good example. But many of the figures quoted above are for the projected number of spindles and there is no means of knowing whether these early spinning factories worked to capacity.

H. Hicks & Sons, New Mill, Eastington, Gloucestershire, 1816
Although 'new' this factory was built of the traditional
materials: wood floors and beams and wooden pillars
Portf: 1334

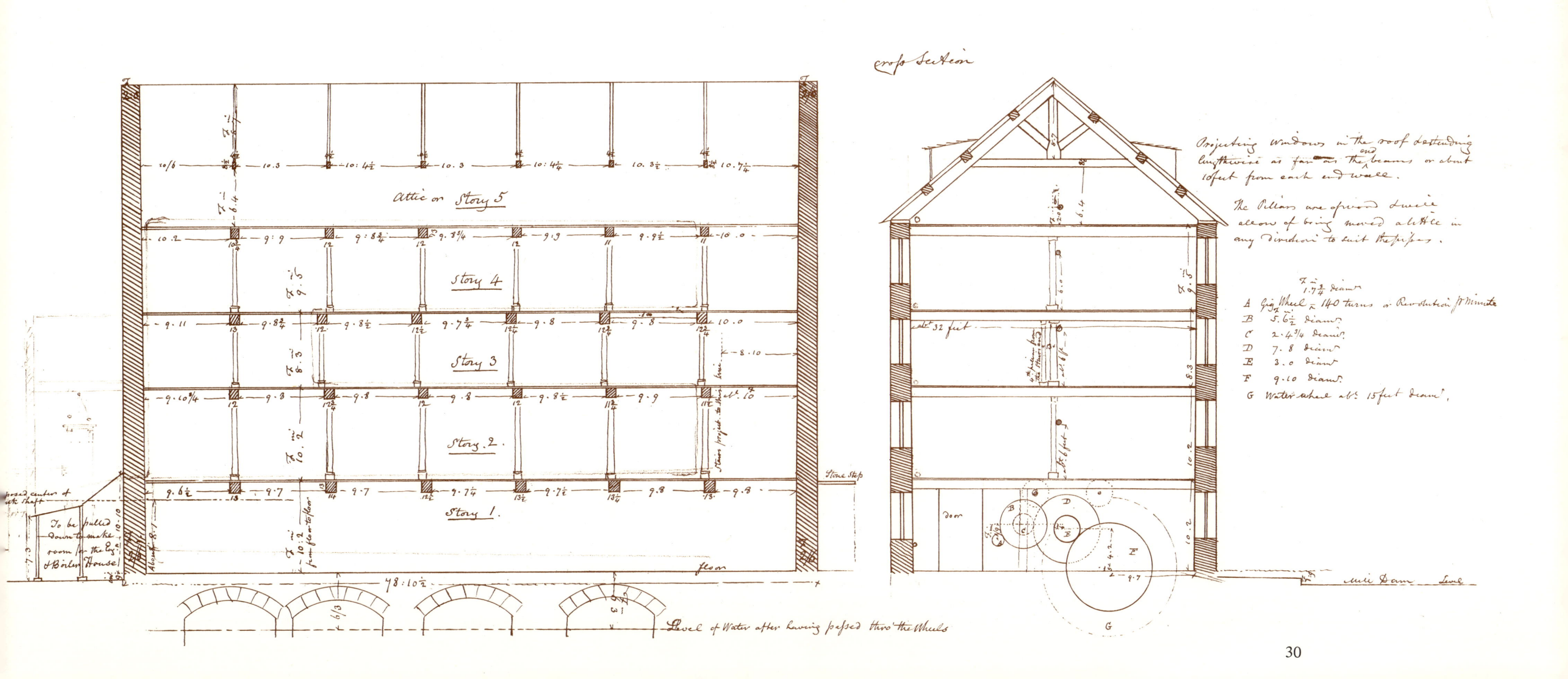

The small cotton spinning mill containing only a few hundred spindles represented a small investment. Two mills at Milnthorpe, Westmorland were insured as follows in 1787:[36]

Mill	£ 50
Utensils and Stock	410
Mill	250
Utensils and Stock	590
	£1,300

A mill in Edale, Derbyshire was valued at £1,000 as follows:[37]

Mill	£ 333
Water wheel, gears etc.	67
Clockmaker's work	433
Stock	167
	£1,000

The Arkwright type cotton mill containing roughly, 1,000 spindles represented a fixed capital investment of approximately £1,000 for the building and £1,000 for the machinery. The total valuation of the mill, including stock was generally about £3,000; Arkwright's own mills at Cromford, Bakewell and Belper (1779) were all insured for this sum. Robinson's Linby mill was valued at £3,600.[38] Robinson's Papplewick Mill was producing between 300 and 500 lb spun cotton per day ranging in counts from 22s to 40s or about 2,000 lb per week. Profits were high and a calculation made in 1784 showed that they were running at 100% and '1s over for Carriage and Extra Expenses.'[39] Peter Ewart, a millwright and former employee of Boulton and Watt's, was approached to go into partnership with some Boulton bleachers who were intending embarking in cotton spinning in 1791. They claimed that with 1,000 spindles their gross annual receipts would be £11,000 and with profits averaging 57% the annual profit of the concern would be about £6,300.[40]

At some of his mills, Rocester for example, Arkwright increased the scale of production by doubling the length of the mill. The value of these factories was generally about £5,000 and the annual output in the region of 100,000 lbs with an average count of 20s, i.e. twenty hanks to the pound.[41] The larger steam driven water spinning factories containing 3,000 spindles or more represented an investment in the region of £10,000. Peter Atherton's four mills were valued at £43,500 in c 1795. Philips and Lee's factory at Salford was valued at £10,000 and so was Daintry Ryle and Co's.[42] In general the larger steam powered water frame factories produced less in proportion to the investment than did the moderate sized mule factories of a comparable size and date. Markland, Cooksen and Fawcett's steam powered factory at Leeds was insured for £12,000 and produced about 150,000 lbs a year. A mule factory of approximately the same cubic capacity cost a little over £6,000 to build and had an annual production of 300,000 lbs of coarse yarn.[43]

One of the great advantages of the mule was that apart from being capable of producing finer counts of yarn, one horse power could drive many more spindles. It was this factor which stimulated the building of 45 ft wide steam mule factories in the 1790s and made the mule factory built of traditional materials more economical than a large water frame factory. In contrast, until the 1830s, the large fireproof mule factory costing up to £40,000 was less economic to build and run having a higher capital cost per spindle than either the more modest mule factory built of traditional materials or the large water frame factory. In the 1790s one horse power could drive ten times more mule spindles than water frame spindles. As Southern told John Lum of Bolton 'It has been the custom about Manchester to reckon a horse equal to the turning 1,000 mule spindles with preparations.'[44] Lum who had 9,600 spindles therefore only required a modest 10 hp engine. The Underwood Spinning Co had 13,000 mule spindles in 1798[45] and Dixon Greenhalgh and Welchman of Bolton requested a 45 hp engine to turn 45,000 mule spindles in 1802.[46]

But constant refinements were being made in spinning machinery and the formulas could not remain unaltered for long. Moon said in 1782 that 'great acquisition made here on Mr. Arkwright's principle of spinning of the frictions therein being so very much alleviated.' He added optimistically in another letter that he thought he might only require one tenth of the power he had originally expected to need.[47] However, this was not an advanced establishment for the owner of the mill was still convinced that a wind mill would be more efficient than a steam engine. Peel appears to have devised a spinning frame of circular form which required about half the power per spindle of Arkwright's frame. Beverley Cross and Co of Leeds installed frames on this pattern — their factory when filled contained 38 to 40 84 spindle frames and they allowed 1 hp per 150 spindles when installing their steam engine.[48]

Mules too were improved. McConnel and Kennedy wrote in 1795 that '216 is made to run as light as 144 used to then'[49] (two years previously). As early as 1790 Boulton and Watt wrote 'We have erected several engines for Cotton Mills, but the machinery differs so much in different places that we can not direct you as to power.... It seems to depend on the perfection of the machinery and the velocity with which the spindles turn. Fine twist also takes less carding per spindle than coarse.'[50] By 1814 McConnel and Kennedy were allowing 1 hp for only 250 mule spindles.[51]

There is little information available on the cost or capacity of weaving factories but in 1824 George Lee prepared an estimate of the cost of building a factory to contain 450 cotton tape looms at Tean Hall, Staffordshire. The building was estimated to cost £6,400; the engine and engine house £3,600; millwork £2,000 and gas lighting £1,500. Depreciation, interest on capital, and the cost of fuel was estimated at £2,250 per year.[52]

Worsted Spinning

The Arkwright frames were used as a rough guide by early worsted spinners. John Cartwright having originally intended to fill his mill with looms decided to concentrate on spinning after heeding Watt's advice 'of attending to only one of these objects at a time.' He wrote 'As in the spinning a far greater quantity of machinery stands within a given space than in the weaving and of a sort much more loaded with friction, we must now proceed on very different calculations of power. Each of our floors in the present building will contain 1728 spindles so that three of them will contain above 5,000 while the fourth will be occupied with preparing machinery.' He estimated that the factory would probably contain 6,000 spindles when working to capacity. Cartwright presumed that a 60 hp engine would be required but Boulton and Watt wrote 'The powers necessary for 1,000 spindles we gave you were for cotton spinning, the wool spinning we know nothing about, but should suppose it will be much the same.'[53] Cartwright's Revolution Mill represented a capital investment of £13,000[54] although the building itself was valued at only £2,700.

Woollen Factories

The approximate power required to work fulling stocks for woollen cloth was known by most millwrights and clothiers since fulling had been mechanised long before the industrial revolution. It was possible to measure the work done accurately since the weight of the hammers multiplied by the height to which they were raised and the number of strokes per minute gave their performance in foot pounds raised per foot per minute. Ewart described Lodge's mill in which 'each stock has two hammers — each hammer weighs about 378 lbs and the head is raised about 2 ft perpendicular every stroke and makes 30 strokes per minute.'[55] Ebenezer Aldred of Wakefield found that a fulling mill containing two pairs of stocks required a water wheel 16 ft diameter and 4 ft wide with a fall of 5 ft.[56] Benjamin Gott allowed 5 hp for four pairs of stocks and nine carders.

In April 1786 William Cross of Halifax asked what power was required to work a scribbling engine. Southern replied 'I know nothing of the power required to work scribbling engines, you say two are equal to 1 pair of corn mill stones but the power requisite to turn these varies exceedingly.'[57] However, by 1796

---

36. Guildhall Library, Sun Fire Insurance Registers 374/538189
37. Stanley D. Chapman, *The Early Factory Masters*, p.130.
38. Guildhall Library, Royal Exchange Registers 4/75060-1, 7/86104-5 8/87827.
39. Notts. Co. Rec. Off DD41 79/63. I am indebted to Dr. S.D. Chapman for this reference.
40. P. Ewart to Boulton & Watt 16th February 1791.
41. S.D. Chapman, 'Calculation of Fixed Capital Formation in the British Cotton Industry 1770-c1815', *Economic History Review*, 23, 1970.
42. Ibid.
43. Ibid.
44. J. Southern to J. Lum 28th January 1800, Foundry Letter Book.
45. J. Southern to Gillies, 5th November 1798, Foundry Letter Book.
46. Portf. 299.
47. Moon to Boulton & Watt 19th September 1782, 23rd October 1782.
48. Beverley, Cross & Co to Boulton & Watt 23rd June 1792.
49. Richard L. Hills op.cit. p.204.
50. J. Southern to Beverley, Cross & Co, 25th June 1792 Foundry Letter Book.
51. M.M. Edwards, *The Growth of the British Cotton Trade*, p.188.
52. Staffs Rec.Off. D644/8/1.
53. J. Cartwright to Boulton & Watt 28th December 1788.
54. S.D. Chapman, *The Early Factory Masters*, p.131.
55. P. Ewart to Boulton & Watt 3rd July 1792.
56. E. Aldred to Boulton & Watt 9th July 1783.
57. J. Southern to W. Cross 25th April 1786 Office Letter Book.

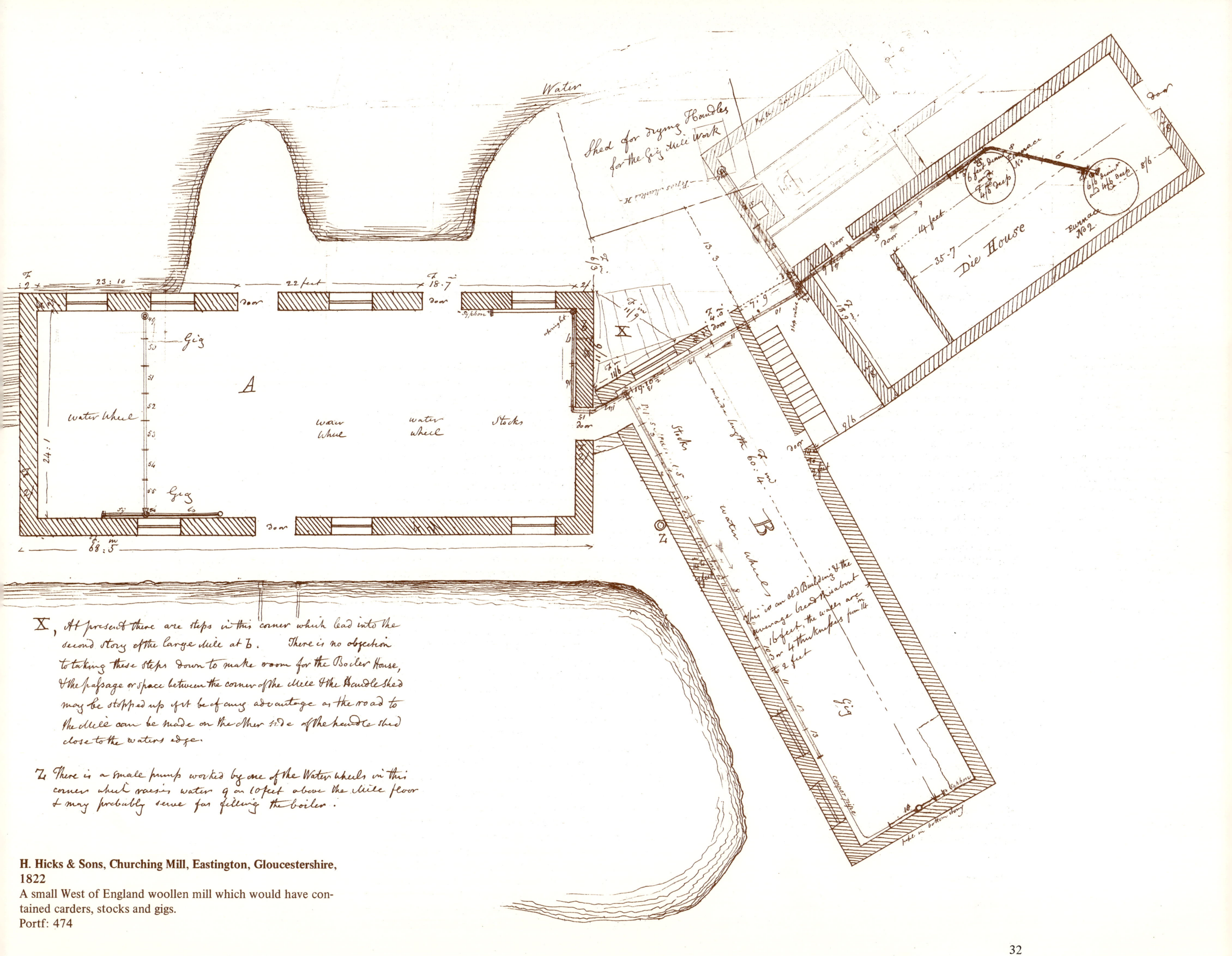

H. Hicks & Sons, Churching Mill, Eastington, Gloucestershire, 1822
A small West of England woollen mill which would have contained carders, stocks and gigs.
Portf: 474

he replied confidently to David Roberts of Painswick, Gloucestershire that Yorkshire carding machines took 2/3 hp each. He added that perhaps it would be sensible to observe how many horses could do the work.[58]

Estimates of the output of woollen mills vary greatly and there seems little correlation between output and factory size. There is no means of discovering how far short of maximum capacity the mills were operating when the output was recorded, but it seems likely that the majority of West of England factories never reached full capacity. Halmer Mill in Gloucestershire contained three water wheels, four pairs of stocks, three gig mills and was capable of milling and roughing twelve ends (half pieces) of cloth weekly.[59] Upper Cam Mill, Gloucestershire measured 137 ft by 31 ft and had 'three floors for driving machinery.' There were three water wheels and a 20 hp Boulton and Watt engine driving eight pairs of stocks, five gig mills 'with corresponding machinery' and the factory was capable of making '60 ends of fine Saxony broadcloth per week.'[60] Walk Mill, Kingswood was 76 ft by 24 ft and 5 storeys high, another building was 75ft by 24 ft and 4 storeys high. This factory was 'capable of manufacturing 10 ends of fine cloth weekly,'[61] and is an obvious case of underemployment of a factory building. Dunkirk Factory, Somerset[62] was 72 ft 6 in by 32 ft and 6 storeys high and was advertised as having a capacity of thirty cloths a week. Christian Malford Mill, Wiltshire was a six storey building 136 ft by 24 ft and contained two water wheels which worked eleven scribbling and carding machines, two wool tuckers, sixteen shearing frames, two brushes, two gigs and had a capacity of twenty pieces of broadcloth per week. Bridge Mills, Trowbridge had a capacity of eighty to a hundred ends of broadcloth per week with a 20 hp engine working nineteen scribbling and carding engines, thirty shearing frames, seven pairs of stocks and four gigs.

The West of England woollen factory represented a capital investment of between £1,000 and £3,000. Huntingford Mill was insured for £1,000 in 1824.[63] In 1818 Halmer Mill was insured for £2,700.[64] Newcombe's Mill was mortgaged for £1,350 in 1804[65] and St. Mary's Mills were insured for £2,500 in 1812.[66] The factory complex of an exceptional Gloucestershire entrepreneur, Edward Sheppard, was reputed to represent an investment of £50,000 by 1825.[67] The fixed capital at Gott's factory at Leeds was valued at £28,000 in 1801[68] but the small Yorkshire scribbling and carding mills were valued at between £400 and £500. A fulling mill at Haslington, Lancashire was insured as follows in 1794:[69]

Mill	£400
Machinery	100
Stock	100
	£600

There was little difference between the valuations of these small mills and the workshops of the carder or jenny spinner. Baynes and Co of Mill Holme near Skipton insured a stone building housing hand mules for £150, their machinery for £430 and stock for £120;[70] a Devon domestic serge manufacturer, William Upcott of Cullompton, insured his house, garden, shop, sorting chamber and offices for £50 and his utensils and stock for £450.[71] Thomas Cuming of Totnes insured his house,] office, tenement and workshop for £570 and his utensils and stock for £50.[72]

Flax Factories

William Brown in his tour of Leeds made in 1821 noted that in flax spinning 'each horse power .... is reckoned to drive 2 frames of 84 spindles, with the requisite quantity of machinery for preparing.'[73] There had been no reduction in the power required per frame although the spindleage had possibly increased since Kendrew set up his patent flax spinning factory at Darlington. A Scottish flax spinner noted in 1796 that the 12 hp steam engine at Darlington drove 24 frames.[74] Regional variations in output were noted by Brown 'Each horse power .... in Dundee produces 25 spindles per day, whereas in Leeds it produces only 19 — this difference is caused chiefly by the greater twist being given to the yarn in Leeds.' Factory layout differed too. In Leeds 'few of them are without some defect or other in the height, length, width or shape of the rooms and where irregularity exists in the building complication and confusion must be the consequence in the machinery — shafts and belts must be running in all directions and cards and frames standing in all positions. In Leeds this is certainly much more the case than in Scotland ... the rooms in general are much larger than those in Scotland some of them containing 60 to 80 frames or from 20 to 30 cards and employing several overseers each.' He continued 'The average quantity of 16 lea or 3lb yarn spun per spindle per day of 12 hours in Leeds may be stated at 14 leas or cutts — and about the same quantity of 8 lea or 6 lb tow yarn — and other kinds in proportion. In Scotland these quantities are about doubled but in Leeds one spinner manages two sides or frames, as universally, and I think fully as easily as one manages a single side in Scotland.' The price of hackling in Leeds was nearly one and a half times more than in Scotland and this was one of the reasons why Brown reckoned the profits of spinning in Leeds to be 'considerably lower than in Scotland.'

Marshall's 200 ft long Shrewsbury factory covering 3,460 sq yds probably contained 700 spindles and 200 twisting spindles in 1798-9. Additions were made to the factory but it always ran below capacity and this is what seems to have accounted for the higher production costs in Shrewsbury compared with Marshall's Leeds factory; between 1804 and 1815, the Shrewsbury mill profits represented a return of £3.5 per spindle compared with £6.7 per spindle at Leeds. Water Lane Mills in Leeds running to full capacity made 100,000 bundles of yarn per year by 1800. By 1803 the Leeds and Shrewsbury mills together contained nearly 7,000 spindles and at the dissolution of the partnership in 1804 the Shrewsbury mill was valued at £64,000 and the Leeds mills at £74,000.[75]

In comparison, other flax mills in England and Scotland were much smaller. Coates' mill at Ripon contained 9 frames of 30 spindles and only three were at work in 1800 'spinning line 13 lees per lb very badly .... Mr. Coates is not at all calculated for such a concern and it must be a very losing one.' Benson and Braithwaite's mill near Ambleside had 3 spinning frames 'in all only 72 spindles and 1 carding engine' and produced between ½ and 1 ton of tow per week. Marshall added 'One would imagine that if it answered they would increase their machinery as they have plenty of power.' The largest flax mill seen by Marshall in the early 19th century was one at Egremont near Whitehaven which belonged to Hornby, Bell and Birley. They had 1,500 spindles and spun 8 ton of flax per week. They 'are building another mill that will do as much more ... their machinery is rough and clumsy and not better than the Darlington Machinery.' Marshall noted that Scottish spinners did not do fine spinning and added 'I do not find any of them are increasing much or making much profit by it.' The thread manufacturers of Scotland produced between 10 dozen and 400 dozen per week.[76]

The small rural flax mills represented only a fraction of Marshall's fixed investments at Leeds and Shrewsbury. Cragg and Jenkinson, flax spinners of Milnthorpe, Westmorland, insured a mill and warehouse for £500 and stock and utensils for £1,200 in 1793.[77] Robert Bissett and Co of Haugh Mill near Levern, Fife, insured their mill in 1799 as follows:[78]

Mill	£ 600
Millwright's work	350
Clockmaker's work	950
Stock	500
Heckling house	50
Stock etc. in warehouse	500
	£2,950

---

58. J. Southern To D. Roberts, 29th June 1796, Foundry Letter Book.
59. Glos. City Library RF 65.30.
60. Ibid RV 65.3(4), R 65.7
61. Ibid RX 180.3(1).
62. I am indebted to Mr. K. Rogers of Wiltshire Rec.Off. for the following details from his forthcoming book *Wiltshire Woollen Mills*.
63. Glos. Rec.Off. D 654/TII/4TI; Glos. City Library RZ 41.1(1-3).
64. Glos. City Library Boxes 14-16.
65. Glos. Rec.Off. D 38A/12.
66. Jennifer Tann, *Gloucestershire Woollen Mills*, p.188.
67. Ibid p. 135.
68. Sidney Pollard, 'Fixed Capital in the Industrial Revolution,' *Journ. Econ. Hist.* 24, 1964, p.303.
69. Guildhall Library, Sun Fire Insurance Registers I/619120.
70. Guildhall Library, Royal Exchange Registers 32a/154867.
71. Sun Fire Insurance Registers 361/558826.
72. Ibid 361/558866
73. W. Brown, 'Information Regarding Flax Spinning at Leeds, 1821', typescript Leeds City Library.
74. G. Roy to Boulton & Watt 27th October 1796.
75. W.G. Rimmer, 'Castle Foregate Flax Mill, Shrewsbury 1797-1886', *Salop Arch. Soc. Trans.* 56, 1957-8, pp.52-3; W.G. Rimmer, *Marshalls of Leeds Flax Spinners 1788-1886*, p.47.
76. Brotherton Library, Marshall MSS 62, pp.4,5,8,36.
77. Guildhall Library, Sun Fire Insurance Registers CS 1/622671.
78. Ibid CS 1/695967.

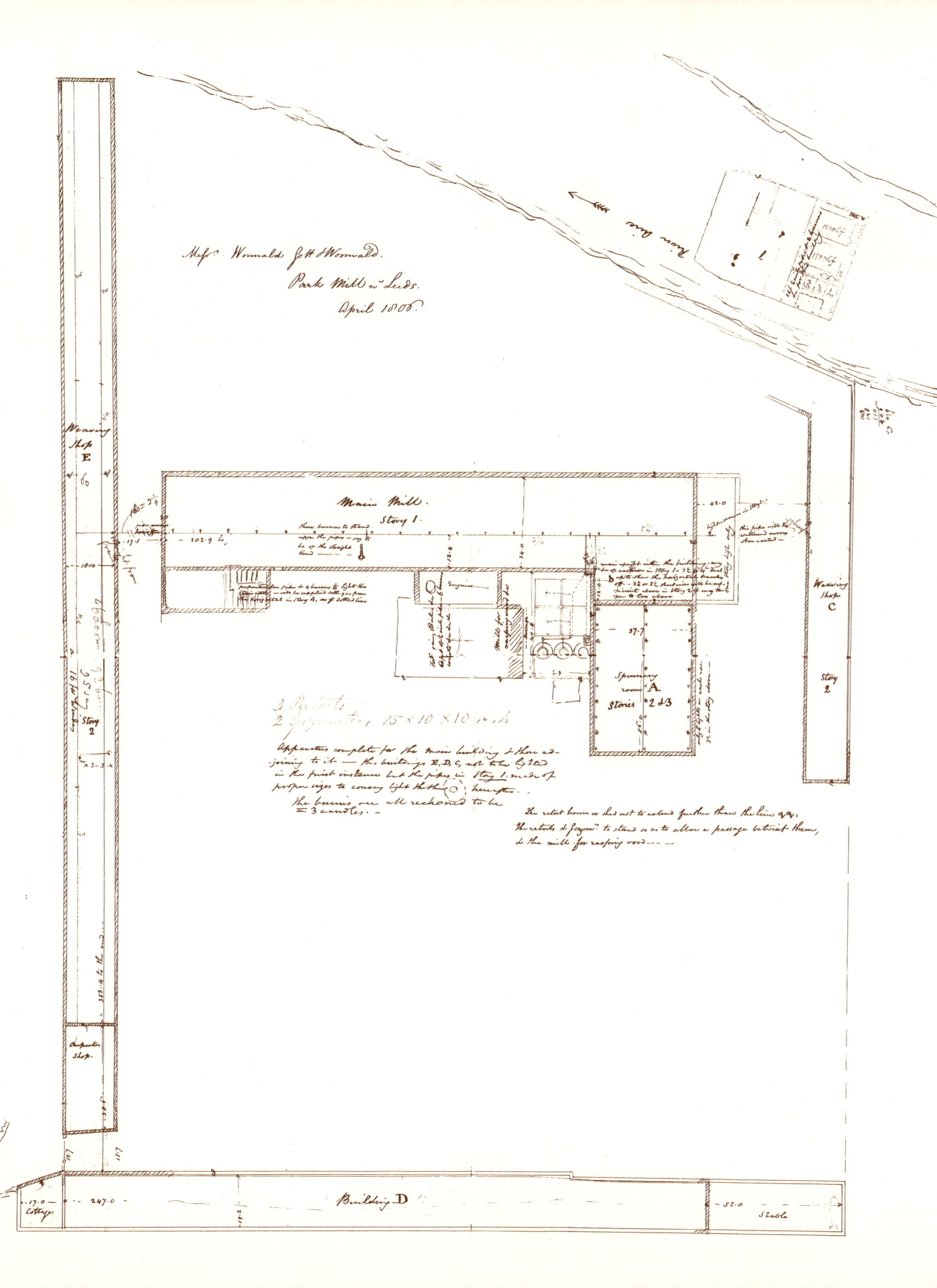

Wormald, Gott & Wormald, Park Mill, Leeds, 1806
This factory, built in 1792 was the largest woollen factory in the country and exceptional for the time. The size of Nussey's or Hicks' factories were typical for the period 1780-1810. Gott, described by his rivals in the blanket trade as 'that lion Gott,' did not apply power to any more machinery than his contemporaries but he considered it advantageous to have his workers under supervision.
Portf: 812

79. Marshall MSS 62 p.33.
80. T. Houghton to Boulton & Watt 13th August 1786, 4th October 1787.
81. J. Southern to Chapman 20th April 1795, Foundry Letter Book.
82. 5th October 1805, Misc.Mills.Box.
83. I owe this information to Mr. G.R. Morton.
84. Alan Birch, 'The Haigh Ironworks 1789-1856: A Nobleman's enterprise during the Industrial Revolution,' *Bulletin John Rylands Lib.* 35, 1953, p.327.
85. Engine Book.
86. Scottish Record Office, Seaforth Muniments GD 46/17/8. I am indebted for this reference to an article by J. R. Hume & J. Butt, 'Muirkirk 1786-1802, 'The Creation of a Scottish Industrial Community,' *Scottish Hist. Rev.* 45, 1967.

Marshall carefully noted the building costs and anticipated profits of a small Scottish lint mill in which the flax preparation processes were carried out.[79]

'say Building 7 yds by 15 2 stories   £ 200
land 2 acres   200
sheds etc.   100
2 breaking and 2 scutching machines   200
10 horse steam engine   400
                                     ———
                                     £1,100

Rent 10 pr ct £110 per week   £2. 2. 6.
coals 3 tons 10/-   1. 10. 0.
overlooker   1. 1. 0.
repairs oil etc.   1. 1. 0.
6 men 12/- 14 boys 6/-   7. 16. 0.
Engine Man   1. 0. 0.
                                   ————
                                   £14. 10. 6.

1 Ton line pr wk at 1/6 pr st   12. 0. 0.
1 Ton   "   "   " 2/- " "   16. 0. 0.
Suppose profit of mill 1/- pr st
£400 per an'.

Paper and Sawmills

One of the first paper manufacturers to establish a formula for the ratio of power to machinery was T. Houghton of Hull. He considered that a 10 hp engine would work two paper engines but when he reported on the success of the scheme a year after installing his engine he remarked that the rag machine worked well when the knives were sharp but when the edges wore off the engine could only drive one machine. He added that a water corn mill at Barrow, Lincs, which he had converted to a paper mill was more successful, for a 4 hp wheel drove a rag engine containing eight stone of rags.[80]

Boulton and Watt had little knowledge of the saw milling trade and in 1795 Southern asked a Mr. Chapman on Watt's behalf for details of the power required to work a given number of saws.[81] John Grieve reported on two Scottish saw mills at Leith and Fisherrow where atmospheric engines of about 6 - 7 hp worked saws which cut 100 ft of 1 ft thick timber per hour, replacing eight men with common saws. Grieve added 'The waste of power in applying these engines to sawing timber seems to lie chiefly in the friction of the saw frames.'[82]

Metallurgical Industries

By the end of the 18th century the average output of a blast furnace was about 2½ to 3 tons per day.[83] The output of the two furnaces at Haigh Ironworks, built in 1789-90, was sometimes 17 tons per week but 'the average was nearer 14 tons per week'.[84] Blowing engines for furnaces ranged from the 20 hp engine for Joshua Walker's Milton Furnace near Rotherham and Wilkinson's 30 hp engine for New Willey, Salop, to engines of nearly 70 hp, such as those at Dowlais and Pen-y-Darren Ironworks.[85] There is insufficient data on the number of furnaces blown by these engines but William Wilkinson estimated that a blowing engine with a steam cylinder of 42 in diameter with a 7 ft stroke, and engine of approximately 30 hp, would provide enough power to blow two furnaces or one furnace and four finery hearths. In 1787 John and William Wilkinson were approached to consider taking a financial interest in an ironworks to be built at Muirkirk, Ayrshire on land which Lord Dundonald was interested in developing. William Wilkinson, assured that the Landowner, Commodore Keith Stewart would invest also, provided a detailed costing of an integrated bar-iron plant on the Shropshire pattern:[86]

'In Shropshire we compute an Engine capable to blow Two Furnaces, or one Furnace with Four Finerys, having Two Boylers and Regulating Bellows suitable will cost at least Three Thousand Pounds —
A Furnace with Bridge house, Casting houses, and useful conveniences for coaking the coals, and roasting Kilns for the Mine cost about fifteen hundred pounds. —
This is nearly the case at Snedshill where every thing was made with Care, and all superflous expense avoided — — The coalwork Mines being opened some years ago, all that Expense as well as Houses for the Workmen were avoided, and which must have been considerable — and if it had been to do when the Work was erected would have cost at least Two Thousand five hundred Pounds.
An Engine calculated to carry Two Hammers with the Machinery compleat, should estimate a like sum of Three thousand Pounds. A Forge capable to contain four Finerys, a Chaffery with Two air Furnaces, calculated to stamp pot Thirty Tons of pigs weekly and to admit of shingling the Potted Iron and draw the same in to Bars; with convenient Buildings for making Pots, grinding Clay, Making Bricks for the dayly consumption in the Works — Workmens houses, a Warehouse, smiths Hearth and the bare useful Conveniences I should esteem in Shropshire or Staffordshire where Resources are at hand — would cost at least Two Thousand five hundred pounds.
The great prices paid for Labour of all kinds in these Two Counties are counterbalanced by the Resources to be found to be there in procuring suitable materials with Expedition; and although they may appear high at the first instance yet when the Time gaind is considered few Places are better situated for erecting these Kinds of Works — and altho several ironworks have been begun in South Wales within these few years, and those by Gentlemen in the Trade, yet the delays and Difficulties they have met with have been such that there is not one Work in that country which has not cost double the sum it would have done in Shropshire.
I should estimate that a Blowing Engine with a Steam Cylinder of 42 Inches Diameter and Blowing Cylinder of 6 feet making a 7 feet stroke into a large Water regulating Bellow; will have sufficient Power to blow Two Furnaces, or one Furnace and 4 Finerys with great Ease — even altho the coals be of a more difficult Nature to consume than those in Shropshire.
And that an Engine whose cylinder is 48 Inches diameter mak-

Davison & Hawkesley, Arnold Mill, Nottinghamshire, *c.* **1792**
Davison and Hawkesley were pioneers of worsted spinning on Arkwright's principle and their factory of *c.* 1792 was much longer than the typical woollen factory of the period. Water power being found inadequate they installed an atmospheric engine to drive their machinery.
Portf: 1379

J. Cartwright, Revolution Mill, Retford, Nottinghamshire, 1789
This was built by the elder brother of Edmund Cartwright, inventor of the power loom. But John Cartwright had no knowledge of industrial techniques and having built his factory had little idea of what it should do. He considered power loom weaving as this drawing shows and shortly afterwards he discussed cotton spinning, deciding finally to concentrate on worsted spinning.
Portf: 41

Ground Plan of the Mill &c.

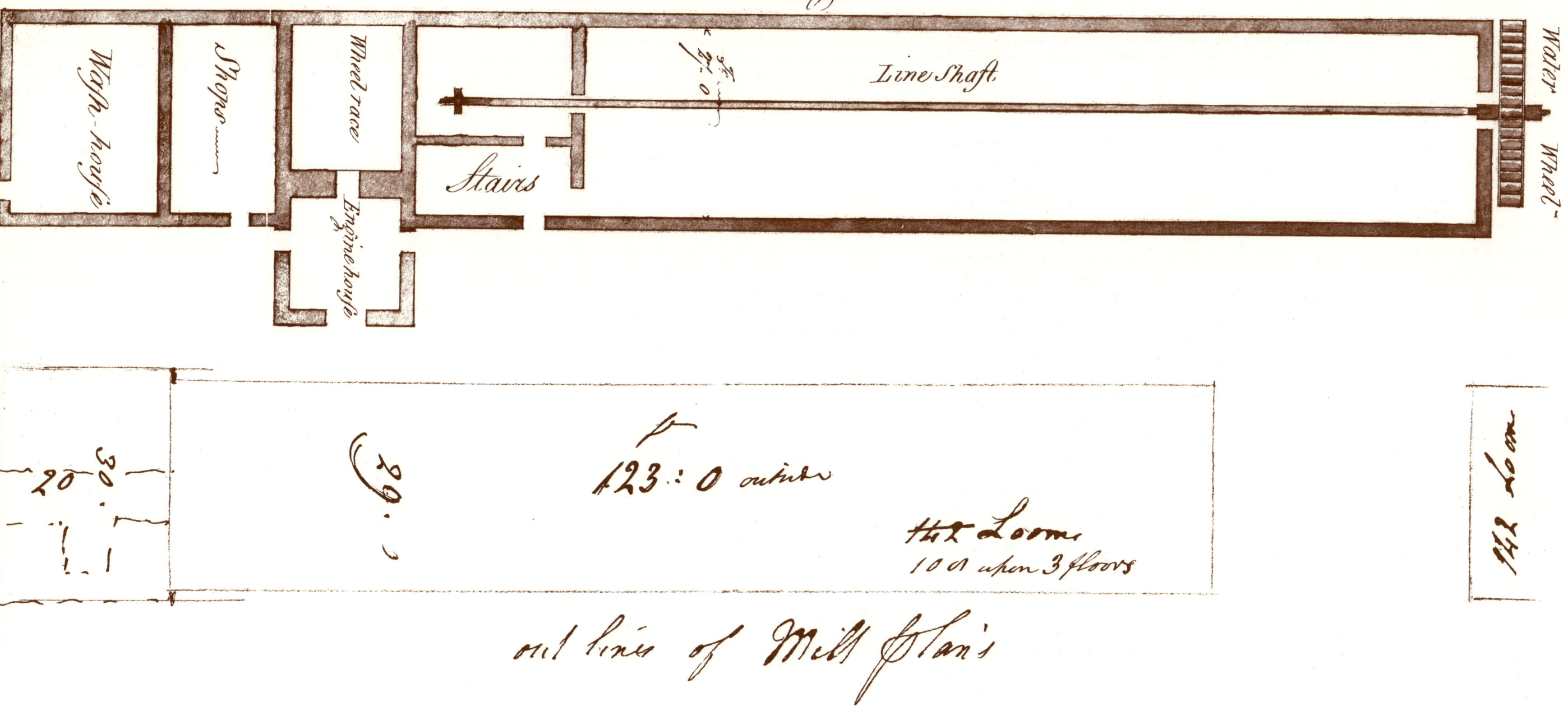

Wash-house | Shops | Wheel race | Engine house | Stairs | Line Shaft | Water Wheel

123 : 0 outside

Loom
100 upon 3 floors

out lines of Mill Plans

87. J. Southern to Micklethwaite 30th December 1795, Foundry Letter Book.
88. J. Southern to J. Foster 22nd August 1799, Foundry Letter Book.
89. J. Southern to Todd, Fletcher & Co 13th May 1792, Foundry Letter Book.
90. M. Boulton to T. Gibbons 17th March 1798, A.O.
91. Jennifer Tann, *Gloucestershire Woollen Mills*, pp.141-2.
92. M. Boulton to T. Gibbons 17th March 1798, A.O.
93. M. Boulton Notebook, 16, pp.45-7 A.O.
94. J. Southern to Sam Lucas 17th July 1792, Foundry Letter Book.
95. M. Boulton Notebook 6, pp.106-116.
96. J. Rose to M. Boulton 21st December 1800, A.O.
97. M. Boulton Notebook 16 pp.45-7, A.O.
98. Sidney Pollard, 'Fixed Capital in the Industrial Revolution in Britain,' *Journ.Econ.Hist.* 24, 1964, p.303.
99. S.D. Chapman, 'Calculation of Fixed Capital in the British Cotton Industry 1770-c.1815,' *Econ. Hist. Rev.* 23, 1970.
100. L.S. Pressnell, 'The Rate of Interest in the 18th Century' in L.S. Pressnell ed. *Studies in the Industrial Revolution*, p.199.
101. S. Pollard op.cit. p.309.
102. R.M. Hartwell, 'The Yorkshire Woollen and Worsted Industries 1800-1850,' D. Phil thesis, Oxford, 1955, pp.371-3.
103. S. Pollard op.cit. p.311.
104. Ibid pp. 299-300.

ing a 6 feet Stroke with suitable flys (say 18 feet diameter) will be sufficiently powerful to work Two Hammers, the one for stamping and the other for shingling and drawing the Produce of one Blast Furnace.

Harrowgate June 6th 1787   WW '

An engine of 30 hp or more was recommended for iron rolling and slitting; Southern noted that although a smaller engine would turn the rolls the rate of work would be so slow that frequent re-heating would be necessary.[87] Boulton and Watt recommended a 30-32 hp engine to work two pairs of rolls and a pair of slitters for John Foster of Selby.[88] Southern told Todd Fletcher and Co that if they had a 40 hp engine for rolling and slitting three furnaces would probably be required to keep the engine working to full capacity if each furnace made five heats in twelve to thirteen hours and produced about 10 cwt at each heat.[89] Hunt and Cliffe of the Brades near Tipton, Staffs had a 50 hp engine for rolling sheet iron and steel.[90] Fromebridge Mill, Gloucestershire, a water driven 'rolling mill, tilting mill and Block mill all under one roof with 'a wire mill and offices adjacent' and 'a brass nealing house' was insured for £1,000 in 1790.[91]

A forge required less power than a rolling mill although Boulton seems to have underestimated the power required when he told Thomas Gibbons that a 10 to 12 hp engine would be sufficient for a moderate sized forge.[92] A water wheel 16 ft diameter and 5 ft wide with a 14 ft head was sufficient to lift a 6 cwt hammer 2 ft high to make 100 strokes per minute.[93] But as Southern told Samuel Lucas of Sheffield different hammers required different amounts of power, although as a general rule Boulton and Watt allowed 13 hp per hammer.[94]

Boulton visited some Black Country forges in 1775.[95] At the stamping hearth at Cradley Forge 'they weekly bring 8 tons of pigs into nature and stamp it when they have water but they often want water'; at the hammers 'when they have water they make 10 to 12 ton a week at one hammer . . . the iron is tolerable mill iron about as good as Mr. Knights not better the nailors dont complain of it . . . . He [the owner] says they make 250 ton a year but I suppose one.' At Bromwich Forge 'in the last 14 weeks there was made by 3 Workmen Finers at one hearth 90 ton 10 cwt of pigs into nature iron . . . they have generally 3 to 3½ cwt at a heat in the Hearth and which is brought into nature in about 2¼ hours . . . . At one hammer they draw 10 ton a week into mill bars.'

John Rose of Nechells Mill, Birmingham, told Boulton that with a 16 hp engine to assist his waterwheel he could roll 8 tons of copper per week, excluding other business.[96] A waterwheel of 19 ft diameter and 11 ft head was said to be necessary to work a copper rolling mill with 15 in diameter rolls making 20 to 25 revolutions per minute.[97]

Fixed capital requirements in the first phases of the factory movement varied from industry to industry and from firm to firm. On the one hand a water driven cotton spinning mill on the Arkwright pattern represented an investment of *c* £3,000 and on the other a country corn mill cost *c* £800. The giants within any industry were exceptional; Gott with his fixed capital of £28,000 or the Albion Mill Co with an estimated fixed capital of £20,000. The large amount of capital invested in factory buildings and machinery has been stressed by several historians who have stated that the proportion of fixed to circulating capital rose significantly during the late 18th and early 19th centuries. Yet Sidney Pollard has shown that although in absolute terms both fixed and circulating capital increased, in relative terms the proporation of fixed capital, although rising, was generally well below 50% of the total.[98] Soho Foundry buildings and machinery represented only 8.8% of the total inventory valuation in 1822; Gott's fixed capital investment in 1801 was £28,000 but his circulating capital was £65,000; a total of £60,000 was estimated to be required by the Albion Mill Co, of which a third was to be invested in buildings and machinery.

Circulating capital, frequently obtained by a network of credit, tended to be largest where partners in the firm had merchant connections. S.D. Chapman has shown that 76% of the principal partners in 43 leading north of England cotton spinning firms in the 1790s were merchants.[99] 'In general businessmen formalised in a considerable degree the credit they granted among themselves',[100] although it depended on the state of the market and the stability of the firm. The existence of this credit system also enabled the smaller firm to go into production with 'only a fraction of the capital it ultimately used and in this way to accumulate enough to enlarge the base of its operations.'[101] R. M. Hartwell noted that up to 1830 in the Yorkshire woollen industry 'the variable costs were spread in a network of credit which eased the burden of the manufacturer: he bought his raw materials on credit, and although he sold his cloths on credit he could get advances on consignments.'[102] Not only were raw materials and fuel paid for on credit but even wages were based on credit. Always in arrears, they were often 'paid' in rent, coal or purchases from the truck shop. 'The firm paying wages in truck, in book entries in arrears had no financial resources of any kind; it lived by borrowing in the course of trade and it borrowed in effect also from its own wage earners.'[103]

The relative ease with which circulating capital was obtained has tended to obscure the problems of recruiting the smaller long-term loans for buildings and machinery. It has been said that 'actual historical investigations have usually found that the provision of fixed capital in industry appears to have presented no great problem.'[104] 'Of serious shortage of capital we hear strangely little; presumably the traditions of thrift and mutual

Bedworth Mill, Collyhurst, Bedworth, Warwickshire, *c.* **1800**
Built by Sir Roger Newdigate and run as a worsted mill by Henry Lane.

38

confidence deep rooted in a commercial and often dissenting society worked to the advantage of the entrepreneur.'[105] This is almost certainly true of many partnerships based on merchant capital. The net partnership capital of a number of firms was often less than the trade credit of the former mercantile concern. In 1796 Benjamin Gott's debtors stood at £146,000 and his creditors at £120,000 compared with a partnership capital of £69,000. Thomas Allingham who became the active partner in the King and Queen Foundry at Rotherhithe told Boulton and Watt 'I have met with a monied gentleman who is desirous of engaging with me in the plan I acquainted you with.'[106] John Foster of Selby had money to invest and wished to put it to good purpose. He built a flint mill at Ferrybridge Pottery and then told Boulton and Watt 'As my intention is to employ some money to the best advantage I could wish you to inform me at same time the same particulars respecting an engine for spinning of cotton as also for slitting and rolling iron to enable me to determine which of them may be pursued to most advantage.'[107] George Lee of the Salford Twist Mill wrote 'I am convinced that eventually spinning can only be a lucrative trade upon an extensive scale in the Hands of capitalists and in the present state of affairs there is less limitation in this than in any other kind.'[108]

On the other hand there is also much evidence in the Boulton and Watt MSS of the shortages of capital for fixed assets. Certain reservations must be made about the source however; the customer for a steam engine no doubt considered it an advantage to appear to be short of capital in the hope that Boulton and Watt would reduce the engine premium and in almost all cases the evidence is qualitative rather than quantitative. When Boulton and Watt considered setting up a rival foundry to Matthew Murray's at Leeds in 1802 Watt jun. reported: 'it does not appear that any of the present ones would be eligible connections; they are men without character and without means.'[109] John Cartwright's partners were either unwilling or unable to advance as much capital as had at first been proposed for Revolution Mill at Retford: 'Some time past, when our design was first formed it was expected the work would have been undertaken with a capital equal to the complete erection at once but it being now determined to proceed upon a less scale and only to extend the works into a second wing when the returns will enable us.'[110] Oldknow's estate developments took all the capital he could obtain so that he was unable to pay his engine premium: 'I assure you what with a world of fixed property and the drains of the Peakforest Canal (none of which can be turned into money just now). . . . I crave your kind indulgence a little longer.'[111] As Lee commented 'Oldknow has extended his concerns beyond his capital in the late commercial storm has reason to feel his error.'[112] Daniel Lees wished to alter his atmospheric engine by adding Watt's patent air pump and condenser 'were it not on account of the badness of trade and my having spent so much of my capital in Machinery etc. etc. I would desire Thomas Livesey to put the pump the very next week . . . . I have a will to apply the air pump and pay the whole premium now but I have not the power, and that is the truth.'[113] Benjamin Lees asked if a steam engine could be paid for in instalments 'as the expense of building the factory and the Engine and machinery are all together very great.'[114] Samuel Marsland had to lend Robert Owen half the cost of his steam engine because Owen was unable to pay for it and Boulton and Watt had to ask Barnes to suppress any inclination amongst their customers to prolong credit because the Soho Foundry was requiring money 'faster than we can easily get it.'[115] The Coalbrookdale Co found their concerns 'more extensive than is desirous to us' in 1794 and decided to contract the firm by disposing of Madeley Wood Furnaces and collieries.[116] This decision prompted John Petty Dearman to write to Watt highlighting another characteristic of the late 18th century — the transfer of capital from one industry to another — although in this case the proposal was not altogether serious. 'I remember thou once ask'd me whether I had any inclination to turn cotton spinner for which I was obliged to the[e] and now give me leave to ask whether thyself or my friend James have any inclination for iron works.'[117] Watt jun. and Lee's discussion of a flax spinning venture is another example (page 141).

Raising long term loans and the ability to pay interest on the capital was one of a complex set of managerial problems that the entrepreneur had to face. Costing was another. Sidney Pollard, in a study of accounting in the Industrial Revolution[118] came to the conclusion that there was little evidence of accounts being used as a direct aid to management. This view has been modified by N. McKendrick[119] who cited the case of Josiah Wedgwood who, during a period of slack trade, calculated the cost of production of each article made at Etruria. The greater dependence upon hand processes in pottery manufacture compared with some other industries may have made this possible but two leading entrepreneurs in the cotton and engineering industries were finding difficulties in the 1790s, twenty years after Wedgwood analysed his costs. One of the problems in the more highly mechanised industries was how to account for machinery made on the premises. James Watt jun., busy equipping Soho Foundry, asked George Lee's advice. Lee's reply is significant on two counts; it substantiates Professor Pollard's view (Wedgwood excepted) that accounts were not being used as an aid to management in the Industrial Revolution but shows that, contrary to his view, at least in the more advanced concerns there was a feeling of need for improved accounting methods:[120]

'I wished I could communicate to you any information respecting the mode of keeping manufacturing Books — in the construction of machinery we never yet could reduce it to regular piece work or divide the labour of Making and Repairing it in such a manner as to determine the distinct cost of each. In the Manufactory I have attain'd rather more accuracy but yet far short of my hopes and wishes.'

The inability of, perhaps, the majority of entrepreneurs to plan the building and equipping of their factories methodically and their failure to develop accurate accounting techniques were amongst the factors which led to the belief that the success or failure of a firm depended on the quality of the partners and on supervision. As Lee said, after acknowledging his inability to discover a system of accounting to suit his factory: 'The first object appears to me to be close personal inspection for which there is no substitute.'[121] Boulton, having discussed the shaky financial position of the Albion Mill Co told Wyatt 'It is of the greatest importance to the concern that a <u>Rational Sensible Active Spirited</u> man of Business who knows wheat meal, millers and markets well and that is an honest punctual <u>firm</u> man whom you may converse and advise with on all occasions'[122] be appointed to manage the works (Boulton's underlining). To Watt he wrote 'The best engine and mill in the world is not sufficient without exact and shrewd management. without money and management the mill will be a bad thing, but if there is no deficiency in these two points, I am persuaded it will be a very good one',[123] for as he had previously said to Wyatt 'tis the trade that is to produce the profit and not the buildings.'[124]

---

105. Charles Wilson, 'The Entrepreneur in the Industrial Revolution in Britain,' *History*, 42, 1957 p.115.
106. S. Pollard op.cit. p.305; T. Allingham to Boulton & Watt 22nd November 1784.
107. J. Foster to Boulton & Watt 19th February 1792.
108. C.A. Lee to J. Watt jun. 8th January 1800.
109. J. Watt jun. to M.R. Boulton 19th June 1802.
110. J. Cartwright to Boulton and Watt 15th September 1788.
111. S. Oldknow to Boulton & Watt 22nd December 1798.
112. G.A. Lee to J. Watt jun. September 1793.
113. D. Lees to Boulton & Watt 6th April 1797.
114. B. Lees to Boulton & Watt 17th January 1791.
115. S. Marsland to Boulton & Watt n.d. Box 3 M; J. Southern to Barnes, 22nd July 1796, Foundry Letter Book.
116. R. Dearman to Boulton & Watt 5th May 1794.
117. J. Petty Dearman to J. Watt 5th May 1794.
118. S. Pollard, *The Genesis of Modern Management*, p.248.
119. N. McKendrick, 'Josiah Wedgwood and Cost Accounting in the Industrial Revolution,' *Econ. Hist. Rev.* 23, 1970.
120. G.A. Lee to J. Watt jun. 11th March 1797.
121. Ibid.
122. Draft memo M. Boulton to S. Wyatt, Albion Mill Box, A.O.
123. M. Boulton to J. Watt 15th March 1786.
124. Draft memo M. Boulton to S. Wyatt op.cit.

John Wood, Bradford, 1833
A worsted spinning and weaving factory. The New Mill was designed by Fairbairn. It was fireproof and had the pilasters at each corner so favoured by him.
Portf: 508. Scale approx 1in : 24ft.

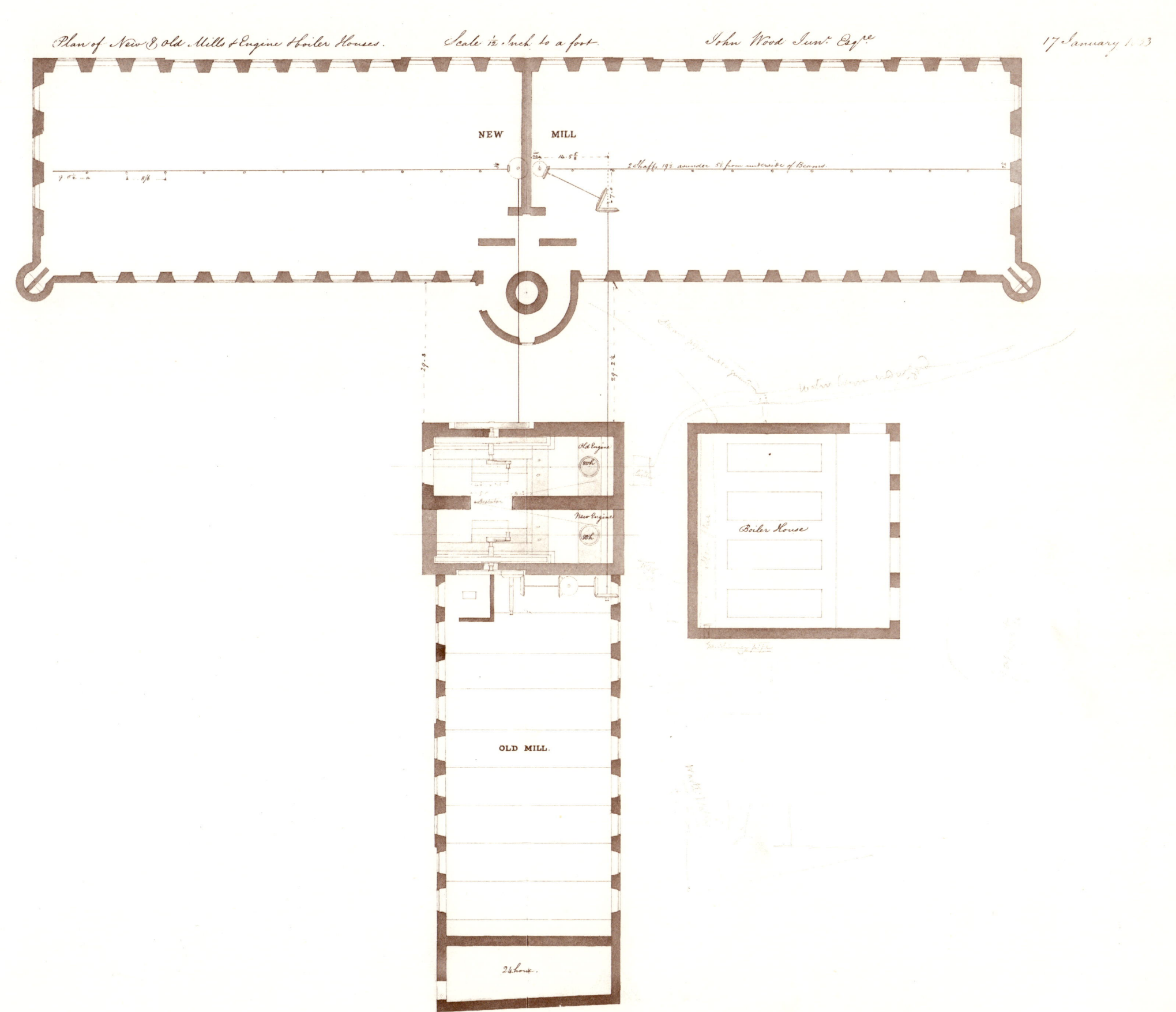

John Wood, Bradford, 1833.

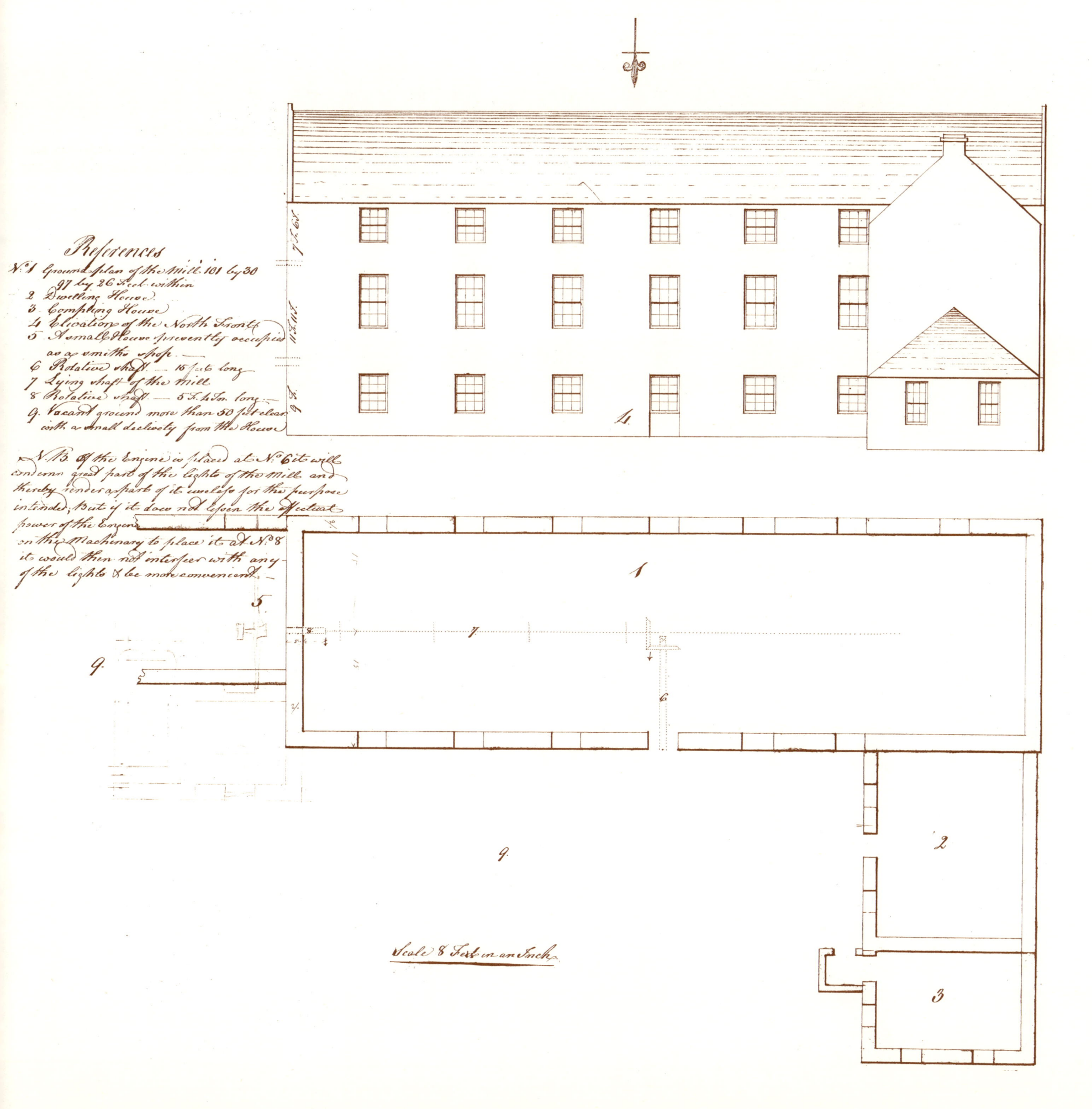

Flax
In 1787 Kendrew and Porthouse of Darlington obtained a patent for a flax spinning frame.
George Wilkie, Dundee, 1799
The drawing shows a mill built on the scale of an Arkwright cotton factory approximately 96 ft by 30. The main shaft went along the centre of the mill with frames placed along either side of it.
Portf: 187 . Scale approx 1in : 16ft.

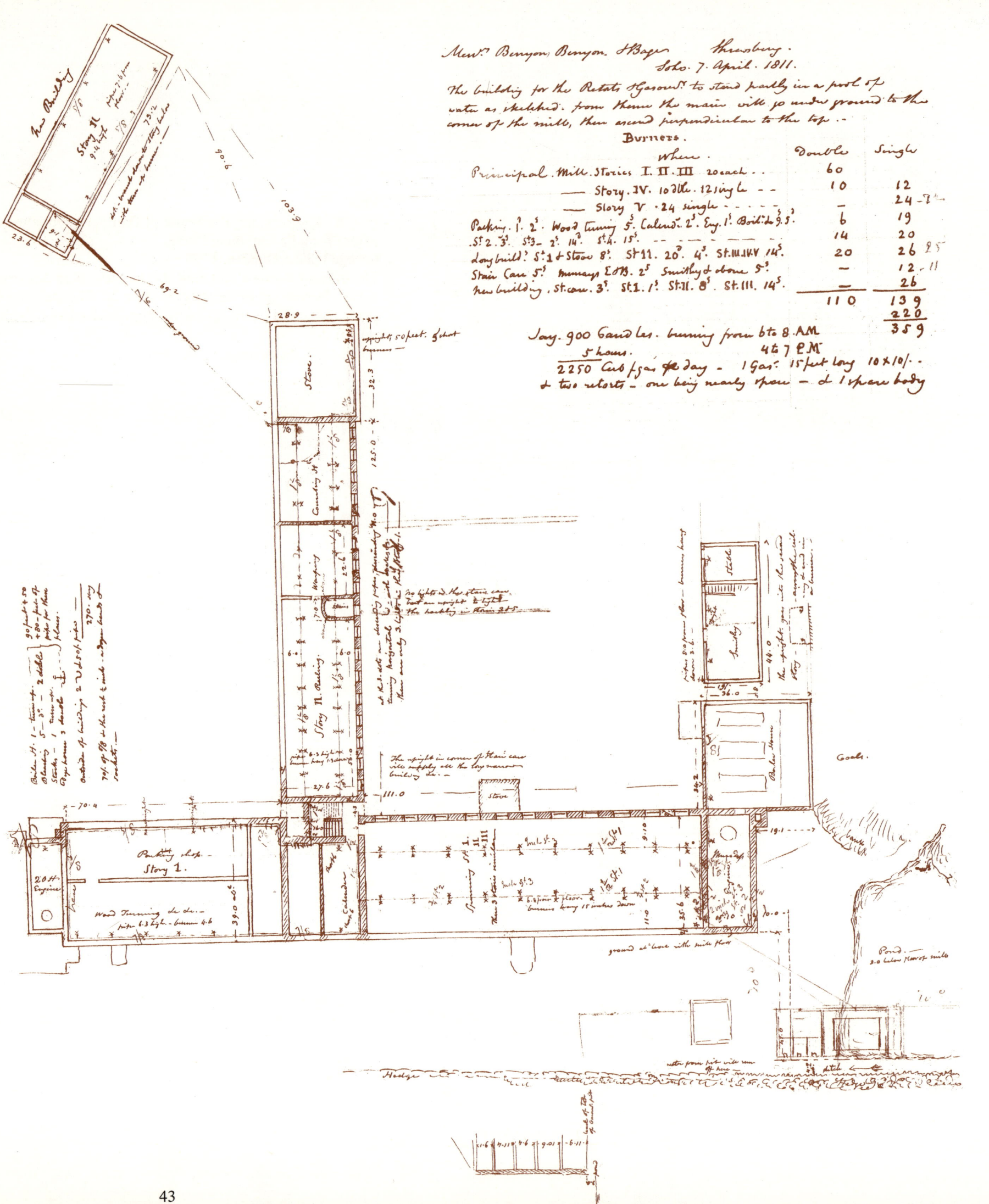

Benyon, Benyon, & Bage, Shrewsbury, 1811
When the partnership between Marshall and the Benyons was dissolved, the Benyons with Bage built flax mills at Leeds and at Shrewsbury.
Portf: 805

Marshall, Hives & Co, Shrewsbury, 1811
John Marshall took the Benyon brothers into partnership and the Benyons in partnership with Bage built this flax spinning mill in 1796-7. It is the first iron framed building in the world. The partnership was dissolved and Marshall took other partners. This drawing was made for the purpose of installing gas lighting.
Portf: 811

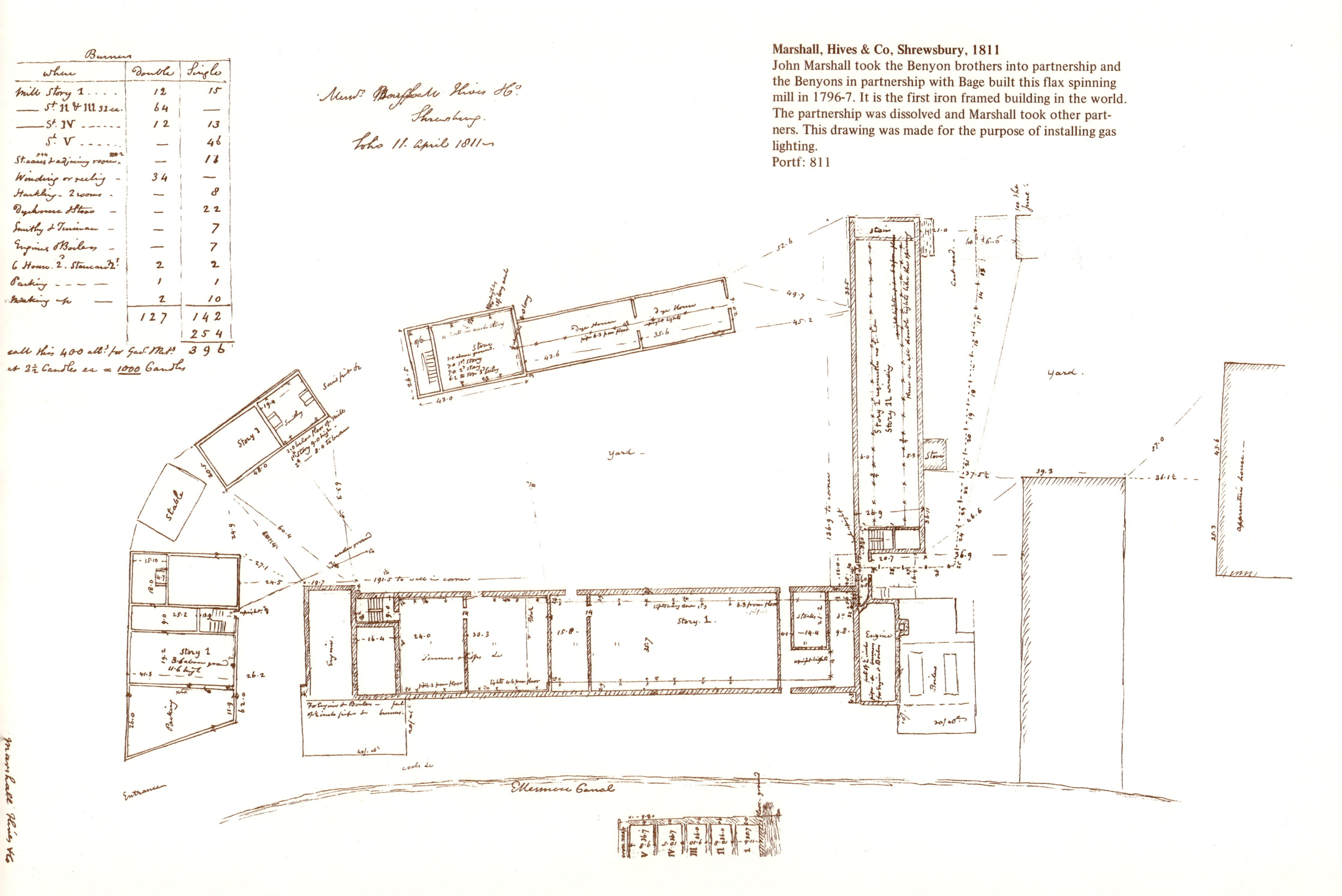

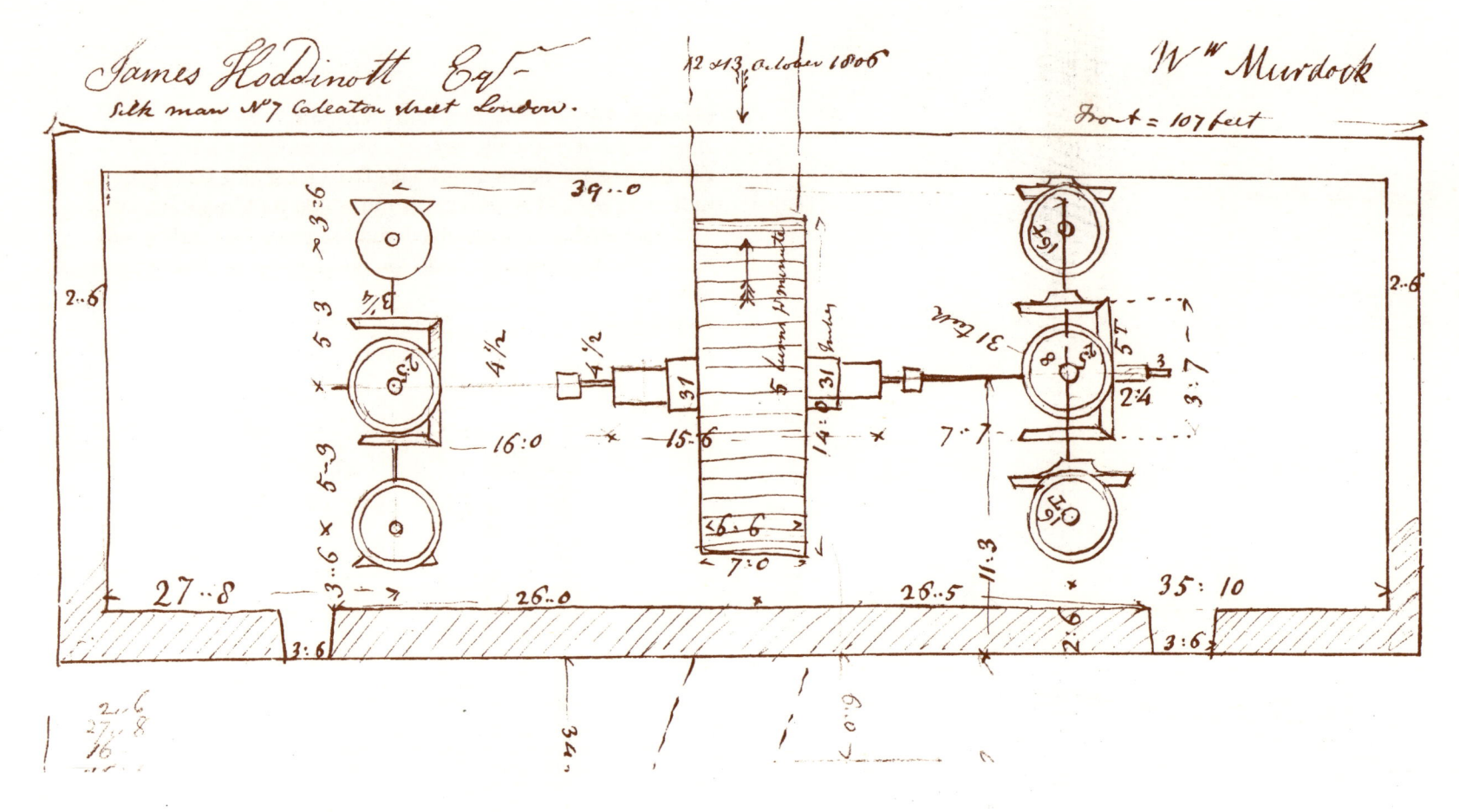

Silk
Silk throwing by power had been firmly established in the early 18th century by Lombe at Derby. After the expiry of Lombe's patent his method of silk throwing was extensively copied.
James Hoddinot, nr Shepton Mallet, Somerset, 1806
This plan drawn by Murdock shows the main drive to the throwing frames.
Portf: 757-766

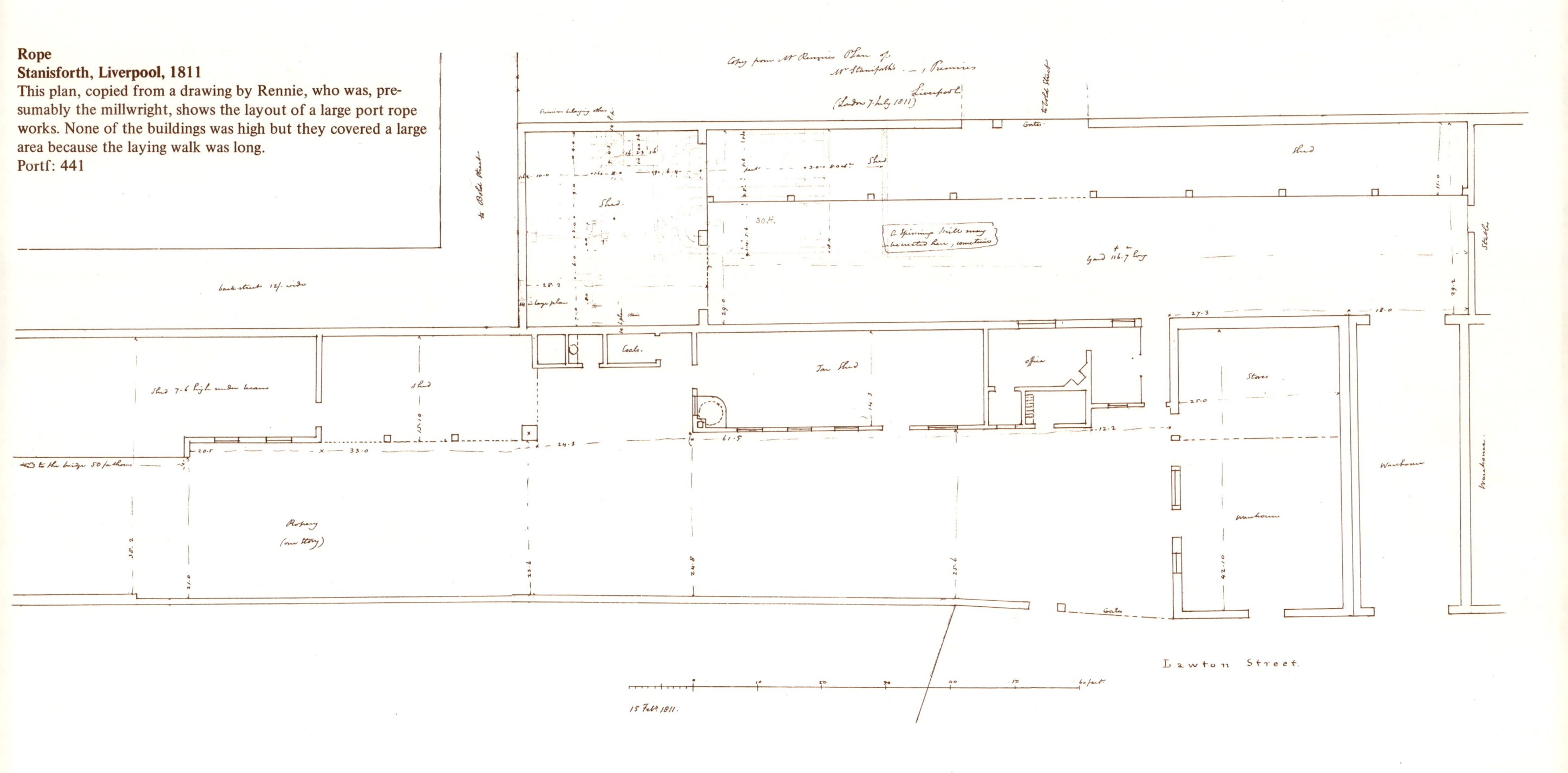

Rope
Stanisforth, Liverpool, 1811
This plan, copied from a drawing by Rennie, who was, presumably the millwright, shows the layout of a large port rope works. None of the buildings was high but they covered a large area because the laying walk was long.
Portf: 441

3. Animal Power in the Factory

1. D. Liptrap to Boulton & Watt 2nd December 1785.
2. e.g. Hull Anti Mill & Hull Subscription Mill.
3. Smeaton designed a number of industrial wind mills. *A Catalogue of the Civil and Mechanical Engineering Designs of John Seaton*, p.29, *Reports of the Late John Smeaton FRS*, 2nd Ed, 1837, vol. 2 p.423.
4. J. Cooper to Boulton & Watt 7th April 1787.
5. M. Boulton to T. Gilbert 2nd February 1771, Letter Book, A.O.
6. M. Boulton, Notebook 13, p.166, A.O.
7. Peter Mathias, *The Brewing Industry in England, 1700-1830*, p.79.
8. Portf 4.
9. Portf 27.
10. Portf 10.
11. Portf 398.
12. Portf 2, portf Misc Mills.
13. *Reports of the Late John Smeaton* op.cit. vol 2, plate VII.
14. Portf 767.
15. Peter Mathias, op.cit. pp.79-80.
16. Gordon, Notes on Sugar Mills, June 1790, Box 4.
17. J. Rennie to M. Boulton, 28th August 1789, A.O.
18. A.P. Wadsworth & J. de Lacy Mann, *The Cotton Trade and Industrial Lancashire*, p.431.
19. Portf 56.
20. T. Harris to Boulton & Watt, June 16th 1784, 4th December 1793, Person & Grimshaw to Boulton & Watt 12th April 1786, 18th January 1792.
21. I am greatly indebted to Dr. S.D. Chapman for showing me his forthcoming paper 'James Longsdon 1745-1821, Farmer and Fustian Manufacturer', *Textile History* vol. 3 1970.
22. Edwin Butterworth, *Historical Sketches of Oldham*, 1856, p.116.
23. J. Aikin, *A Description of the Country from thirty to forty miles round Manchester*, 1795, pp.238, 480.
24. J. Rimmer to Boulton & Watt 14 Dec. 1798.
25. E. Butterworth, op. cit. p. 116.
26. J. Sutcliffe, *Treatise on Canals and Reservoirs...* 1816, pp. 62-3.

At the beginning of the industrial revolution the only sources of motive power available were water, wind or animal power. Animal or wind power was best suited to the small enterprise such as a small brewery, a carding mill or bleachworks. But they frequently had to be employed in larger factories where there was no good water power available. Manufacturers in London, Hull, Liverpool and other towns poorly supplied with water power were forced to use distant water mills or to find alternative sources of power. Some, such as D. Liptrap, a London distiller, had their grain ground at a water mill 'a considerable distance from my manufactory'.[1] Large industrial wind mills were built on the outskirts of London, Hull and Liverpool to grind corn,[2] oil seeds, snuff, flint, and to saw timber.[3] But industrial wind mills represented a considerable capital investment for a factory dependent on the caprices of the weather and many manufacturers turned to animal power.

Apart from operating winding gear at mines, horses or donkeys had been used from at least the 17th century to operate pumping machinery and to turn crushing and grinding machinery of various sorts such as cider mills, bark mills, dyewood mills, bone mills and corn mills. With the increased mechanisation of industry during the 18th century other, new uses were found for animal power and, far from decreasing, the use of animals, if anything, increased during the first phases of the industrial revolution.

Cotes and Co of Hull used a horse wheel turned by three horses to operate edge stones and stampers in their oil mill. 'The velocity of the horses is ten times round in one minute with all the works in motion', the millwright noted in 1784 (page 56). One London manufacturer used eight horses to work a sheet lead works 'and [in addition] six are at work at a time in the white lead works.'[4] Matthew Boulton used four horses to turn his machinery when water was low at Soho Manufactory in 1770.[5] Horse power was used at the Ewell Powder Mills, Surrey in 1776 and Boulton noted some particulars: 'Mr. Taylor thinks that 8 horses will do the work of 2 mills but that the revolutions of the stones when working with horses will be to the revolutions when working with water as 4 is to 7.... A horse will draw against a weight of 180 pounds at the rate of 1 mile and ¾ per hour.'[6]

One of the most widespread uses of animal power was in the brewing industry. In the 18th century breweries required power for only a few operations, chiefly pumping and malt milling. Thus even the large London breweries had need for only a few horses at a time. Truman had five in 1766-7 but this number rose to 20-22 in the period 1793-1808.[7] Whitbread used six horses around a 21 ft 6 in diameter wheel making 4½ revolutions per minute in 1784. Matthew Boulton noted on the plan of his brewery (page 100)[8] '6 horses at present which grind 14 qr per hour with 2 pr of stones but he wishes for 3 pr of stones.

3 do. for 1 pr more stones

2 do. for the liquor and wirt pumps

<u>1 do. extra</u>

12 Horses'

Gyffords had at least two horse wheels one of which worked four pairs of grinding stones, a liquor pump and sack tackle.[9] Barclay and Perkins had a 25ft 2½in horse wheel which, making 3½ turns per minute worked two pairs of stones, wort, liquor and cleansing pumps[10] (page 106). The horse mill at Constitution Brewery worked three pairs of stones[11] (page 104), Goodwyn and Co had two horse wheels, one of which worked two pairs of stones and wort pumps and the other of which worked the liquor and cleansing pumps[12] (page 98). John Smeaton designed a horse wheel to work the pumping machinery at Weevil Brewery, Gosport[13] and Madder of Dublin had such a small two storey brewhouse that most of the ground floor space was taken up by his horse wheel (page 104).[14]

The brewery mill horse was a poor beast, often blind, with a working life of ten or twelve years doomed to walking round in a circle. Compared with the magnificant dray horses they were a different race. Dray horses were valued at £16 each in the mid 18th century rising to £40 each after 1800 whereas mill horses were rarely valued at more than £5 each and Truman valued his entire set of five horses at £7 10s and £12 10s in 1766-7. The mill horse was cheap to buy but it nevertheless required feeding, stabling and men to look after it. Whitbread paid £200 in farriers, smiths, wheelwrights and collarmakers bills and £200 for hay, oats, straw, beans and bran in 1748-9 for his mill and dray horses. Joseph Delafield recalled that it cost £40 a year, many times its capital value, to 'keep up and feed' a mill horse.[15] The death rate of mules in West Indies sugar mills seems to have been particularly high. Thirty six mules at £25 each were said to be sufficient to work 300 hogshead of sugar in five months. In costing a sugar mill £225 was allowed for the replacement of 25% of the animals per year. Oats for feeding cost £150 per year.[16]

Horse wheels were employed extensively for pumping water. Usually such a small amount of water was required that replacement of the horse wheel by a steam engine was uneconomic and horse wheels survived longest performing this function. John Rennie, apart from being a skilled millwright in the newer fields of engineering, was asked to install several horse wheels for pumping including one for Bristol hot wells.[17]

But it was in the textile industries that animal power was most widely used. Paul and Wyatt's spinning and carding machines erected in a Birmingham warehouse were powered by 'two asses walking round an axis'.[18] Paul and Wyatt's ventures were a failure in economic terms but the precedent for power carding and spinning had been set. With Arkwright's spinning and carding inventions the use of animal power spread. When he built his first Nottingham mill in 1769, a four storey building 117 ft long by 27 ft wide, a 27 ft horse wheel was installed in a 30 ft square wheel house attached to the end of the mill. A rough plan amongst the Boulton and Watt papers records that '6 strong horses [work] at a time — Mr. Arkwright said 9 horses'.[19] T. Harris and Pearson and Grimshaw also of Nottingham used horses to drive spinning machinery on Arkwright's plan.[20]

The horse wheel was of greatest significance in the lower echelons of the cotton industry where with small capital the entrepreneur could set up a carding shop and, by ploughing back profits, build up his fixed capital investment until he was in a position to consider the use of a waterwheel or a steam engine to drive warp spinning machinery. The horse wheel was well suited to the transition from hand to power operated machinery. Such an entrepreneur was James Longsdon, a farmer and fustian manufacturer of Longstone, Derbyshire.[21] He built a mill and warehouse costing about £650 which contained a horse wheel to drive carding machines. There were also nine jennies for hand spinning in the building. Longsdon put little value on his mill horse for his farm accounts of 1795 record 'stock on hand, one blind mare for engine £5 14s 6d'. Between 1776 and 1778 six small mills had been built in Oldham, three of which were horse driven. In 1772 John Lees of Oldham invented a feeder for the carding machine and established a small carding factory worked by horses.[22] There were four horse driven carding mills in Royton and 'a great number' of small horse and water driven carding mills at Mottram in the 1790s.[23] John Rimmer of Lymm, Cheshire, had a horse driven factory in 1798.[24] Butterworth comments that 'the greater part of the earliest cotton mills were moved by horse power.'[25] and these details accord with the millwright John Sutcliffe's ideas on the optimum factory size for someone beginning in the trade:[26]

'My opinion is, that the most simple, the least expensive, and most useful, would be that which is worked by the hand; except that in some instances the carding might be performed by a horse.... A horse would find employment for four spinners and a room six yards square would be sufficient for this card room; another room sixty yards square would be sufficient for the [spinning] wheels. Here would be a factory large enough to employ a moderate family and three or four additional hands; and how much better would it be for the children to be employed under their parents than in a large factory.'

Bleaching, Dyeing, Printing
Ground plan of a bleaching mill, n.d.
This plan, possibly in the hand of Rennie, is undated and no name or location is given but it is a good illustration of the small factory that developed and prospered in the first phases of the expansion of the cotton industry. Motive power was provided by horses which operated dash (wash) wheels and calenders.
Portf: Misc. Mills

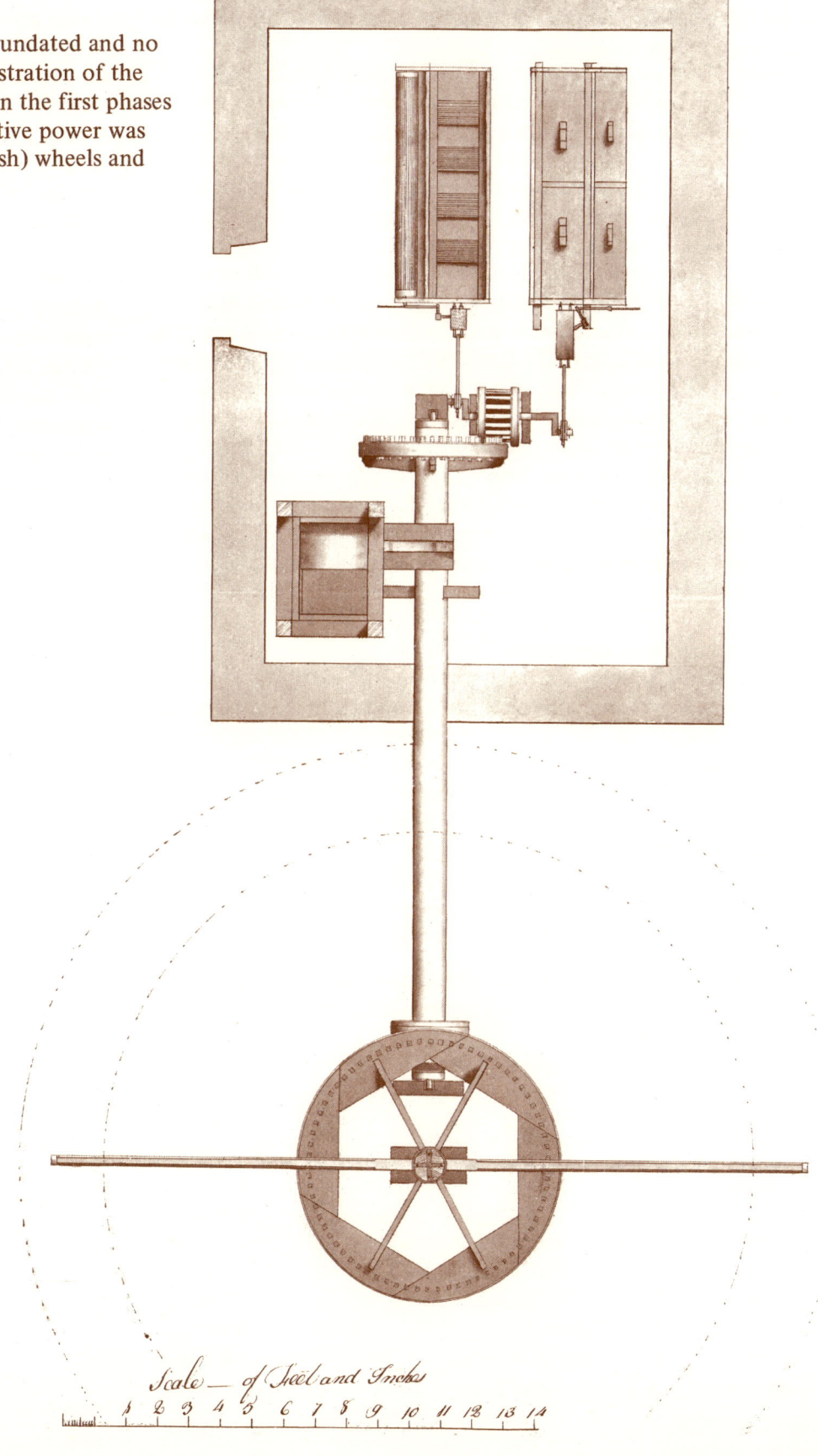

27. F. Naish to Boulton & Watt 26 Aug. 1796.
28. P.P. 1802-3, VII, p.312.
29. F. Hyett, *Glimpses at the History of Paiswick*, p. 78; Glos. City Lib, RX 319.I(10).
30. *A Catalogue of the Civil and Mechanical Engineering Designs* ...., p.29.
31. Box 3.
32. J. Cartwright to Boulton & Watt 15 Aug. 1788.
33. Enid Gauldie, 'Mechanical Aids to Linen Bleaching in Scotland', *Textile History*, Vol. 1 No. 2.
34. P. Ewart to Boulton & Watt 14 Mar. 1791.

Wool carding and scribbling machines were occasionally driven by horses in Yorkshire and the West of England although these regions were on the whole well endowed with water power. F. Naish of Trowbridge enquired after a steam engine in 1796[27] for scribbling and carding wool and for cleaning and thickening the cloth 'which is now done by water and where water cannot be had we substitute horses'. Horses were used for scribbling and carding wool by Daniel Marklove of Berkeley Gloucestershire because water resources were poor.[28] Two Painswick Gloucestershire clothiers used ox power in the 1820s and a sale advertisement dated 1814 described a building at Uley Gloucestershire as 'newly erected ... with a shed for a horse wheel ... well deserving the attention of persons wanting a clothing manufactory.'[29] Both Painswick and Uley possessed good water power resources but such was the overcrowding of water mills in the valleys that an animal wheel may have been more reliable. Fulling was not a suitable process for horse power; the hammers were heavy and the process had to continue uninterrupted for about twelve hours. Daniel Marklove of Berkeley, noted above, used horses for preparing his wool but he sent his cloth to the nearest water mill to be fulled. But John Smeaton designed a fulling mill for Mr. Hill of Colchester in 1760 in which the stocks could be driven by either a waterwheel or a horse wheel.[30]

Crompton's mule, in the days before William Kelly of New Lanark successfully harnessed it to a power source, was a hand machine. But from 1790 there are occasional references to spinning mules being driven by animal power. The fullest information is for Scotland where McTaggart drew up a list of manufacturers who used horse gins in the Glasgow and Paisley districts in about 1797:[31]

'Horse gins used by

Glasgow	Chadwick and Whyte, Grahams Square	24 mules
Glasgow	McKerle and McTaggart, Grahams Square	14 mules
Glasgow	John McTaggart, Ruthglen	14 mules
Glasgow	James Monach, Lawson Street	common (jenny) and mule
Glasgow	Peter Buchanan, George Street	common (jenny)
Paisley	Walter Corse and Co, Newtown	20 mules
Paisley	Daniel Finlay and Co	14 mules
Paisley	William Clarke	20 mules
Paisley	John Davidson	14 mules
Paisley	John Dalyed and Co	14 mules
Paisley	A. Pollock and Co	24 mules'

It was not only in the carding and spinning of textiles that animal power was widely used. Edmund Cartwright experimented on his early power loom at Doncaster 'where some of the weaving machinery is already at work by the power of horses.'[32] But power loom weaving was slow to be adopted. Animal power was frequently used in the bleaching industry (page 50). Several processes had been crudely mechanised before the introduction of chlorine bleaching[33] but each machine required only a small quantity of power and the layout of the bleachcroft was generally ill-suited to the establishment of a central power source. Where possible it was preferable to install separate prime movers for individual machines and groups of machines. At some sites it was possible to direct separate water courses through the buildings to waterwheels which powered the dash wheels, calenders and beetling machines. At others animal power was used. A large horse wheel turned Samuel Oldknow's bleaching and printing works at Stockport in the early 1790s.[34] The Lancashire expression 'as blind as a calenderer's horse' is a legacy of this phase of the industrial revolution.

Joseph Baker & Co, Raikes, Bolton, Lancashire, 1790
A plan of the dash wheel house.
Portf: 53. Reverse.

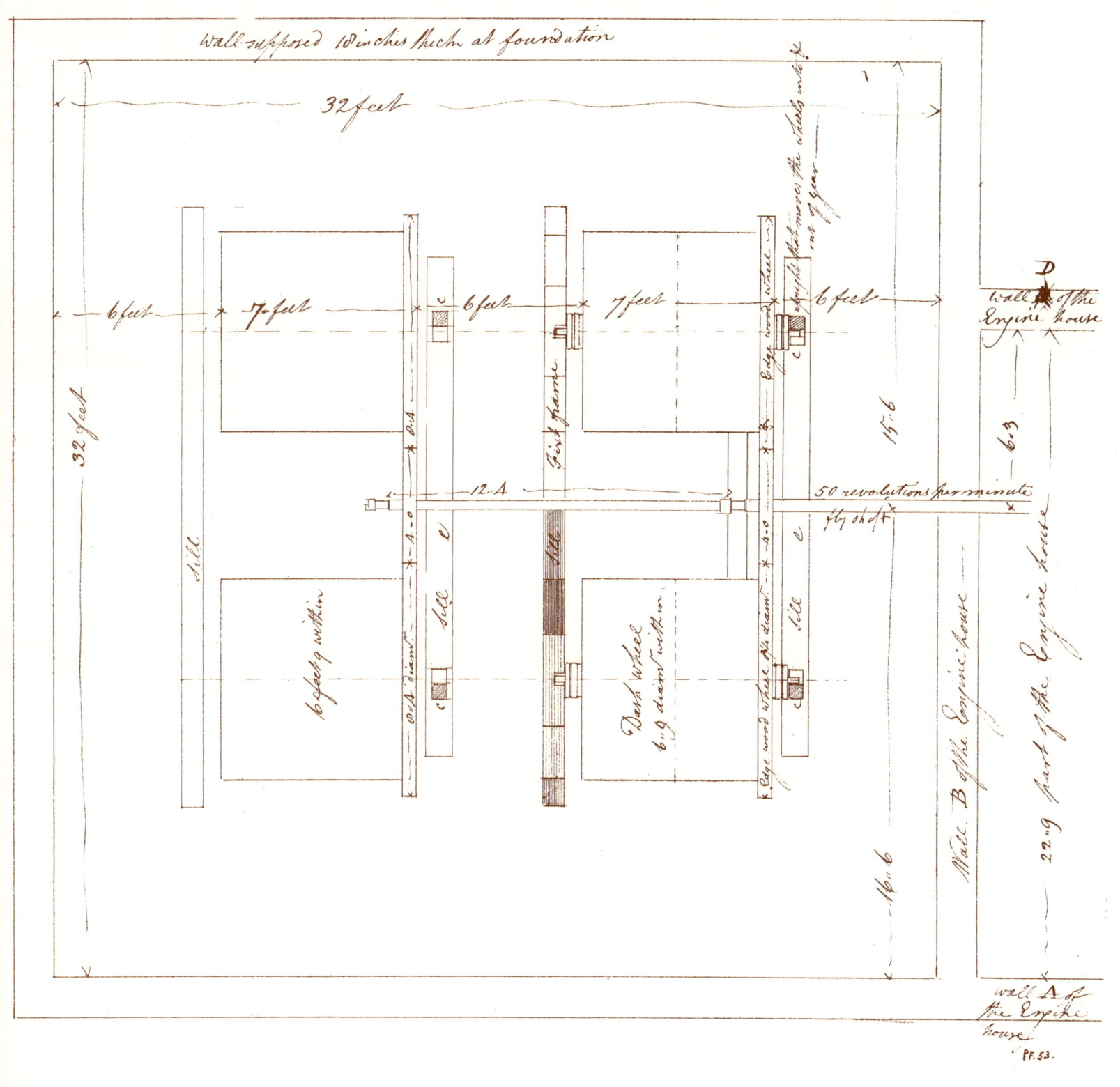

51

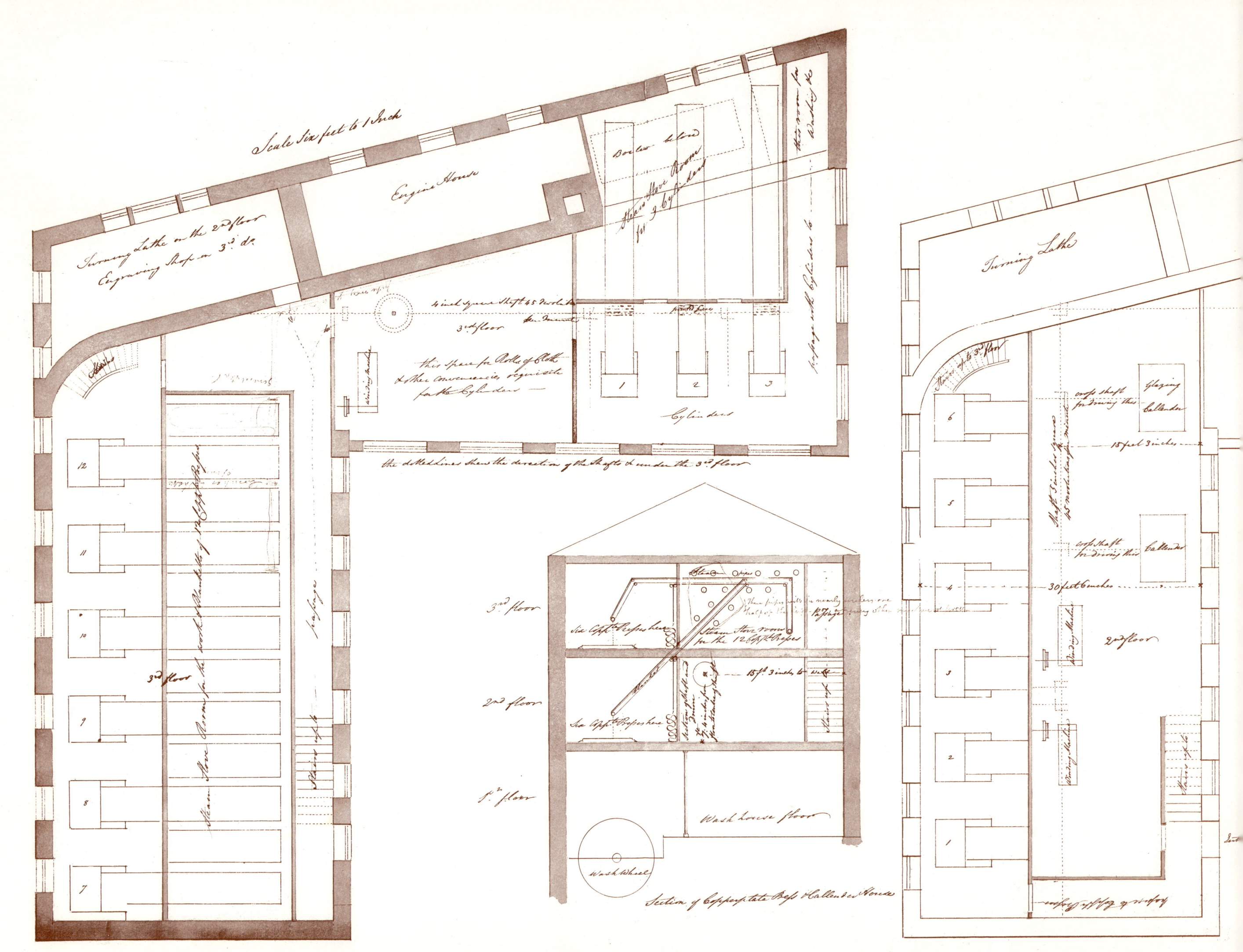

Duffy, Byrne and Hamill, Donnybrook and Ballsbridge, Dublin, 1809
This firm owned extensive bleach and printworks in Ireland and developed the use of steam for heating liquids and for drying. This drawing shows part of the bleach works with steam heated presses, cylinders and stove rooms.
Portf: 394

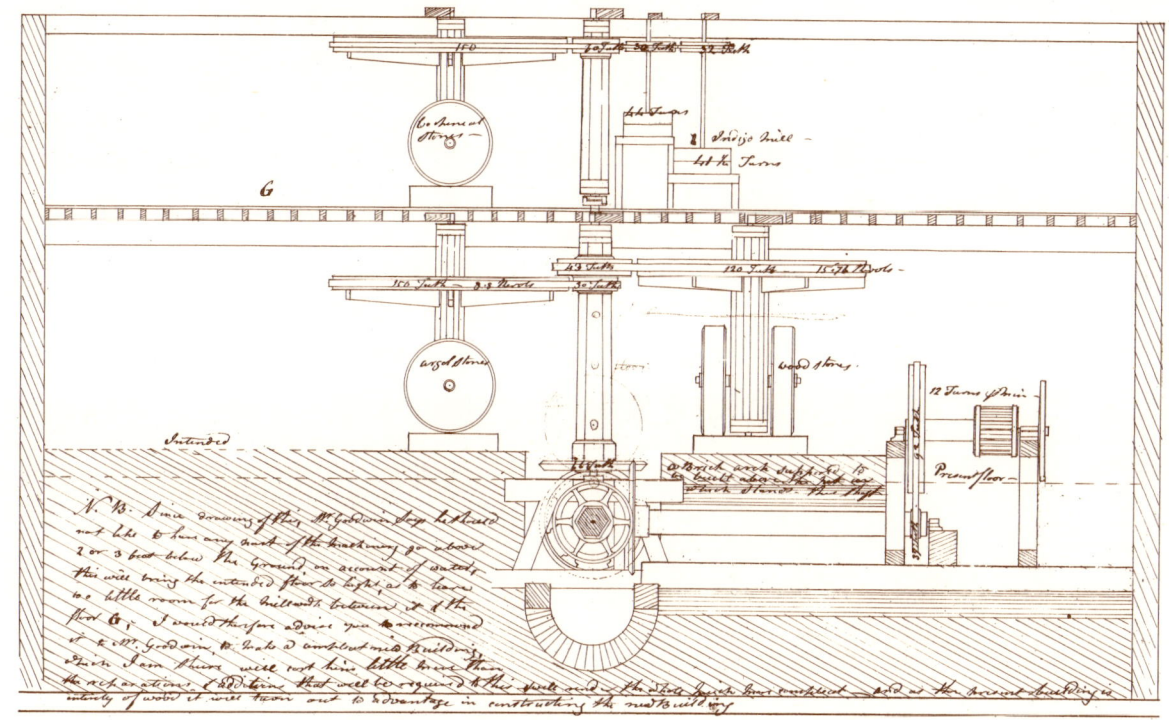

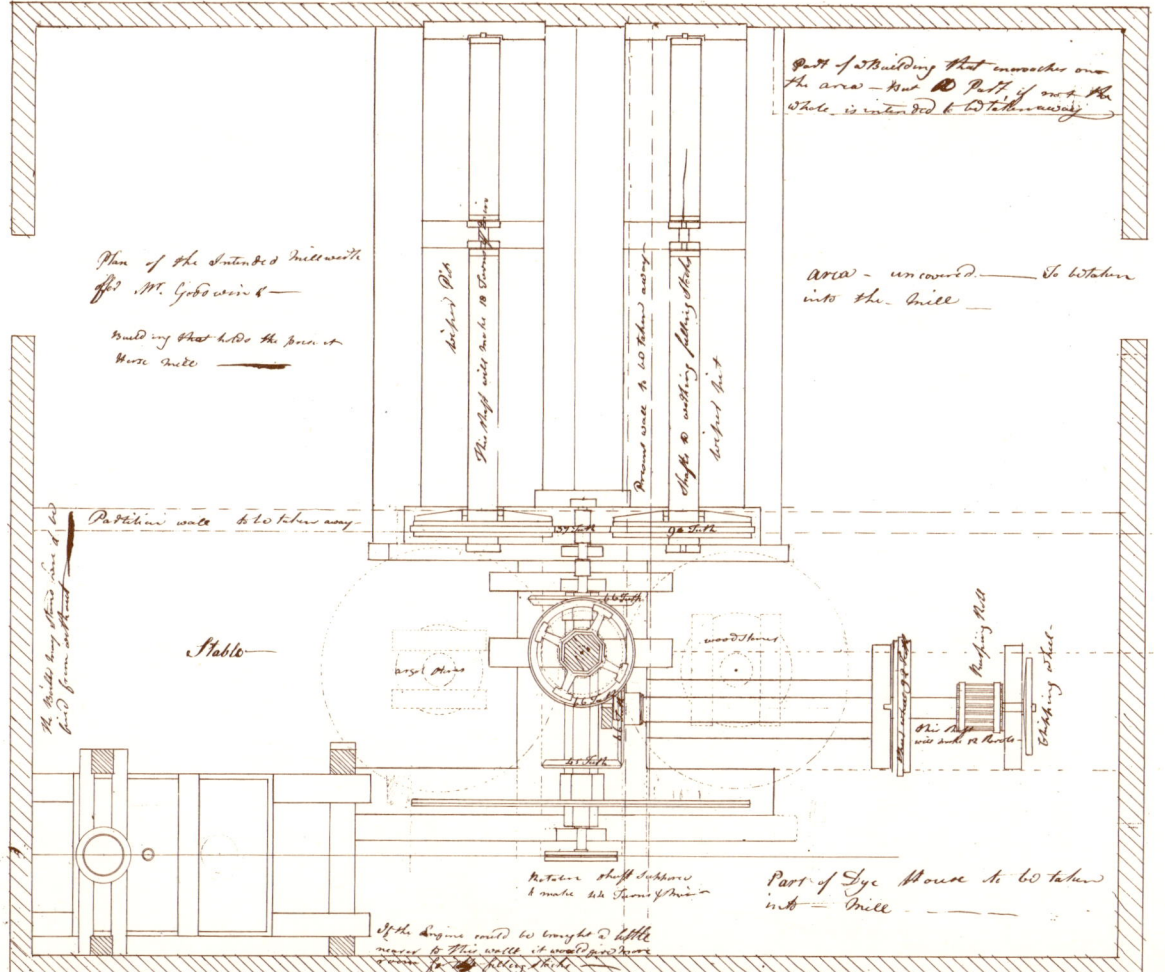

Goodwin, Platt, Goodwin & Co, Southwark, London, *c.* 1790
The dyegrinding mill which adjoined an extensive dyeworks at which Bancroft witnessed some experiments. Horse power was originally used but a steam engine was installed in 1786. Horizontal and vertical stones ground and crushed a variety of dyestuffs and dyewoods.
Portf: 14. Scale approx 1in : 7ft 6in.

Daintry, Ryle & Co, Manchester, 1805
This was a large printing and bleach works. Steam was conducted in pipes along the dyehouse floor to each boiler and to the kier in the bleach house. Boulton and Watt were in correspondence with several dyers about the best methods of using steam heat in the dyehouse and they probably advised over the design of the boilers.
Portf: 363

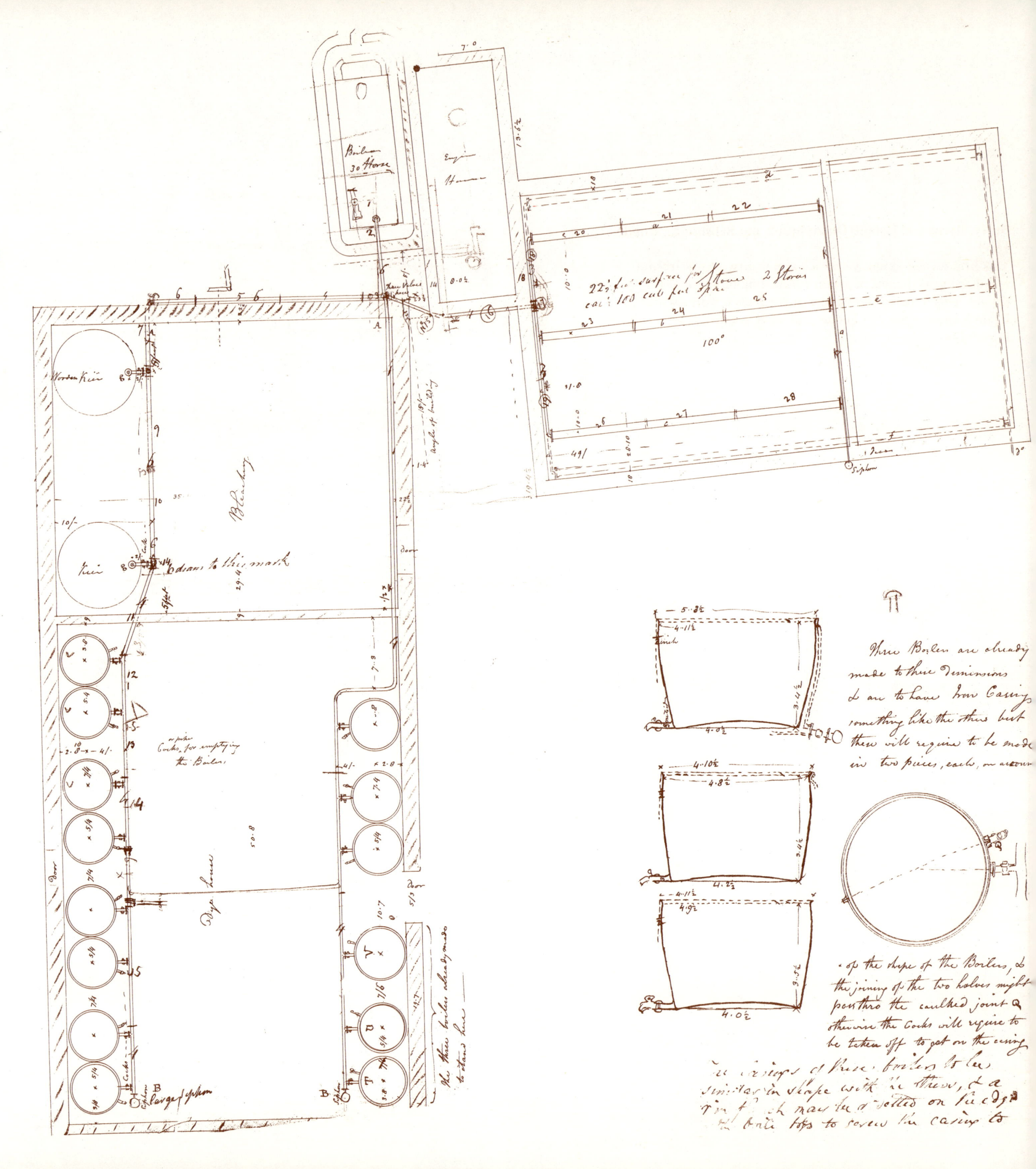

Daintry, Ryle & Co, Manchester, 1805
Part of the printworks with the steam pipes leading to colour
pans, presses and cylinder machines.
Portf: 363

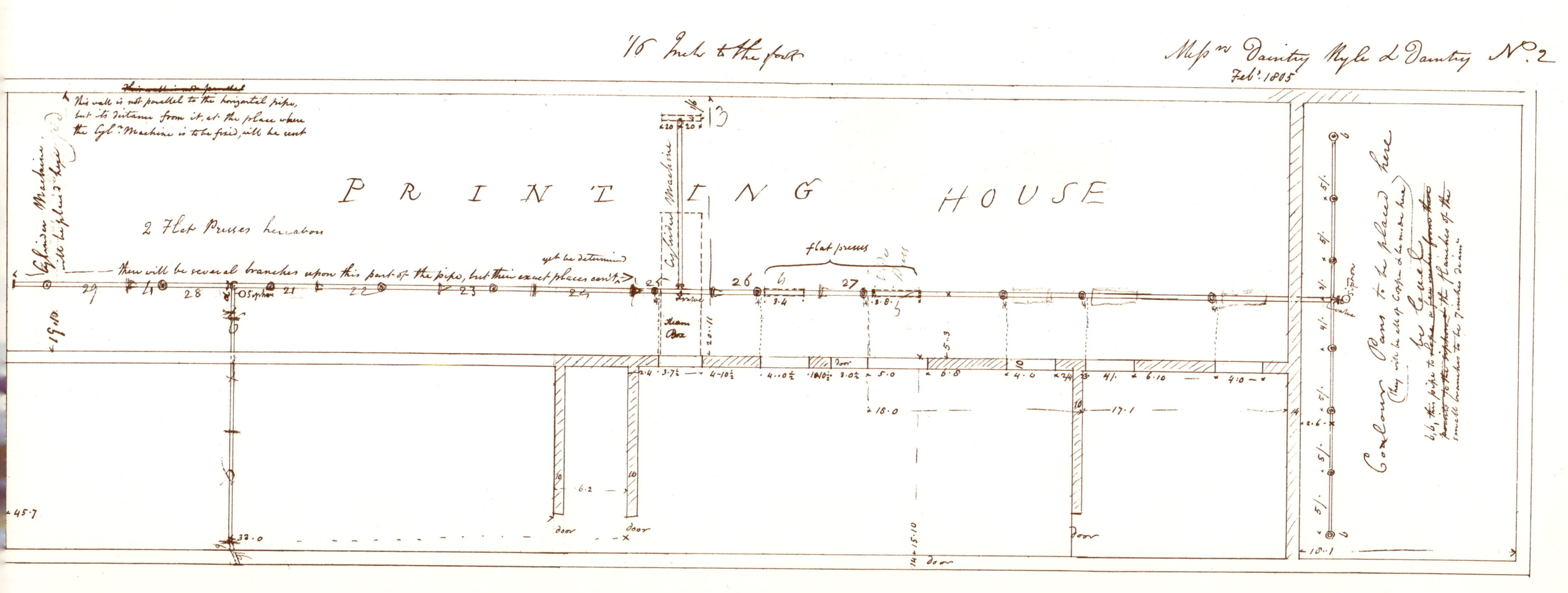

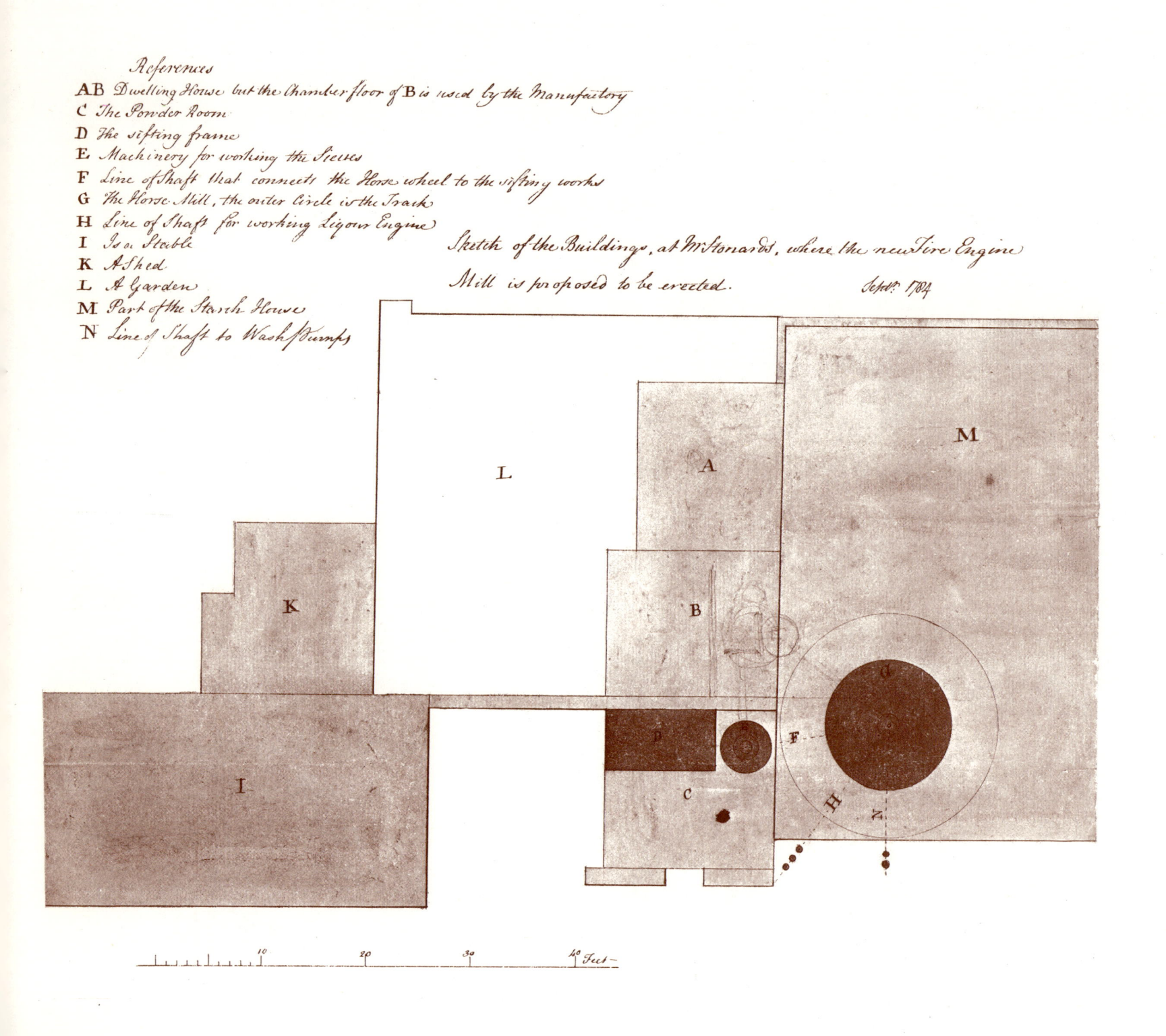

References
AB Dwelling House but the Chamber floor of B is used by the Manufactory
C The Powder Room
D The sifting frame
E Machinery for working the Sieves
F Line of Shaft that connects the Horse wheel to the sifting works
G The Horse Mill, the outer Circle is the Track
H Line of Shaft for working Liquor Engine
I Is a Stable
K A Shed
L A Garden
M Part of the Starch House
N Line of Shaft to Wash Pumps

Sketch of the Buildings at Mr Stonards, where the new Sieve Engine Mill is proposed to be erected. Sept: 1784.

Starch
Stonard & Curtis, London, 1784
This is an attractive small factory building with pediment and lunette. The horse wheel shown on the plan worked powder sifters, a pump and liquor engine.
A wind mill is also mentioned.
Portf: 3

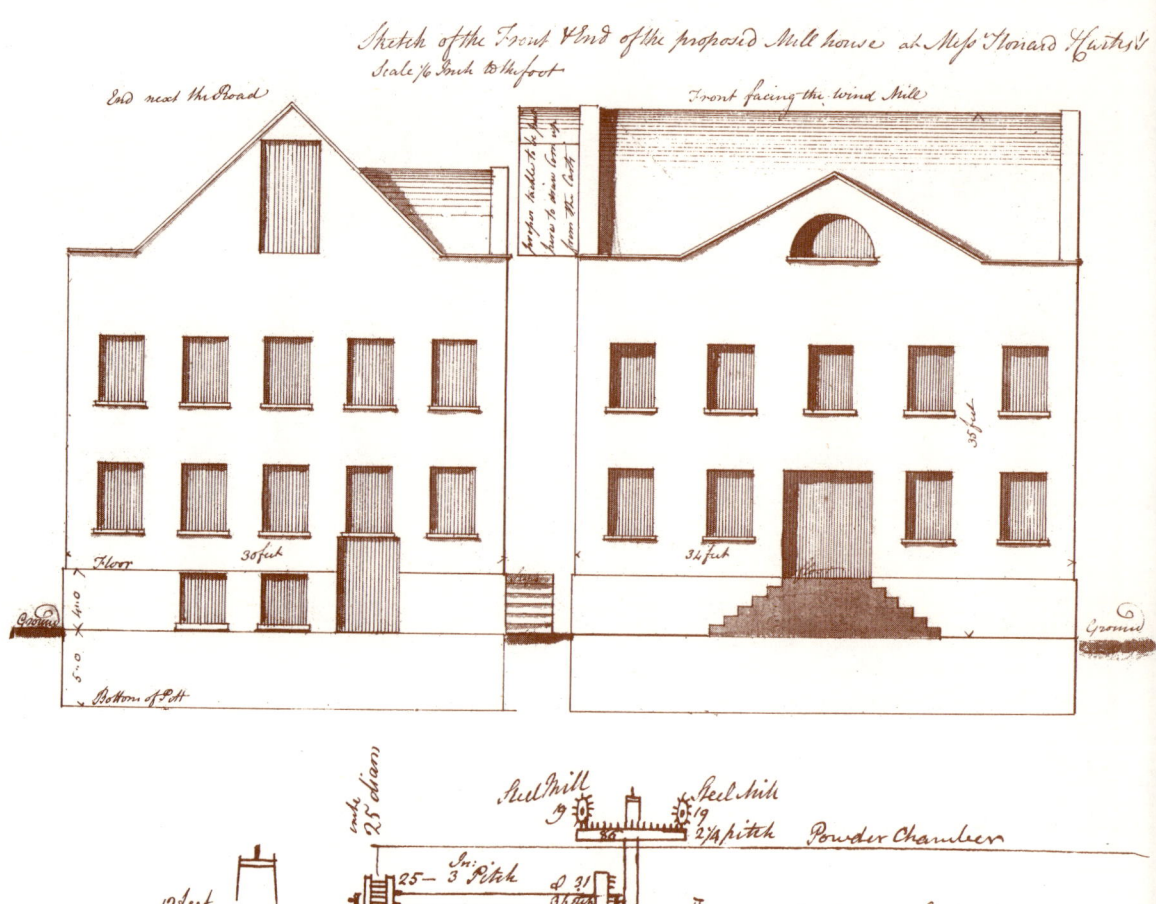

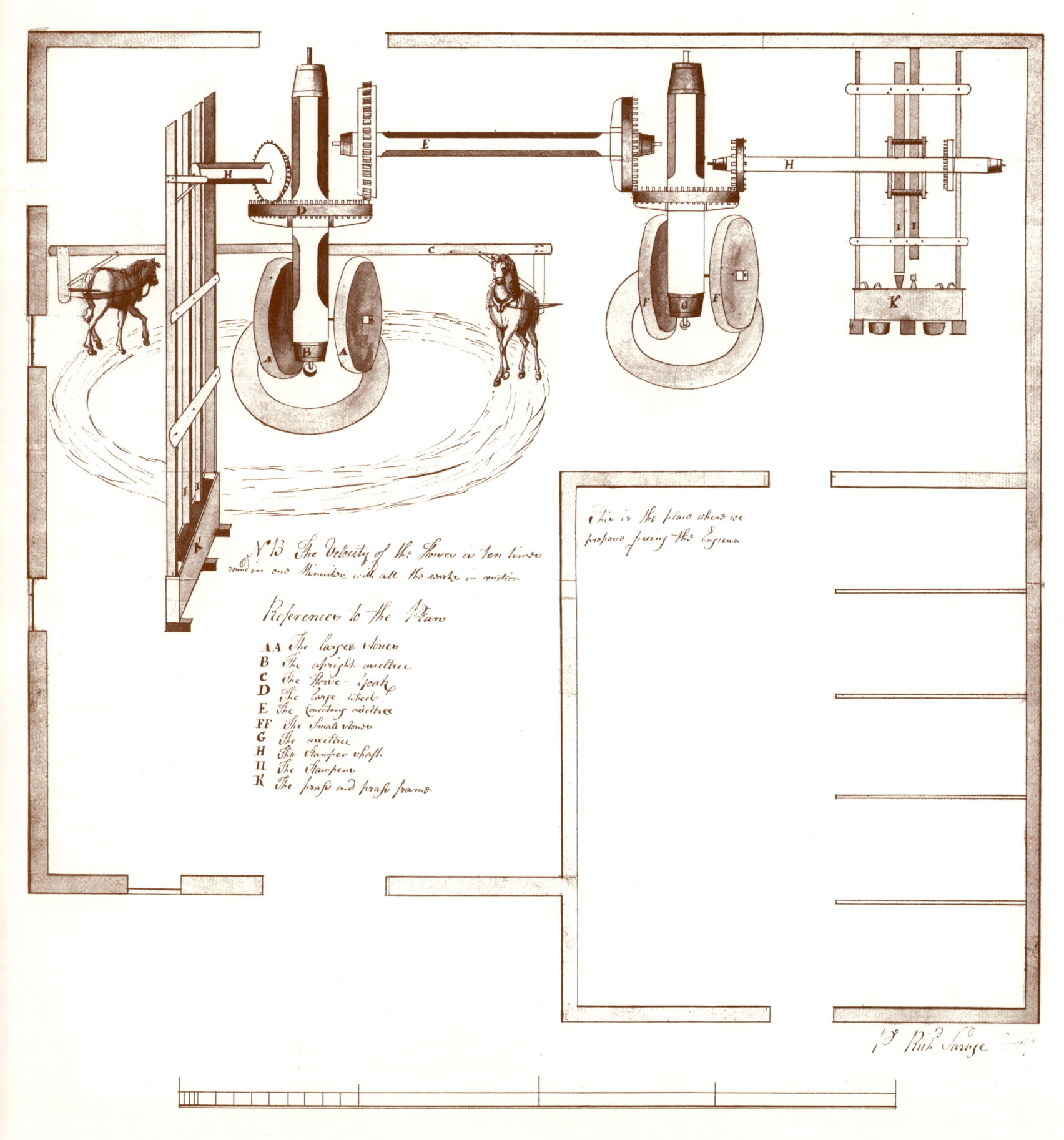

Oil Seed Crushing
Cotes & Co, Hull, *c.* 1784
Oil Seed crushing was concentrated in and around Hull and Gainsborough in the late 18th century. Hull was poorly supplied with water power and the millers had to depend on animal or wind power. In this drawing horses are shown working edge runners and stampers.
Portf: 9

Brooke & Pease, Hull, 1795
A fine plan and section showing the crushing machinery, boilers and storage cisterns. Scale approx 1in : 7ft.

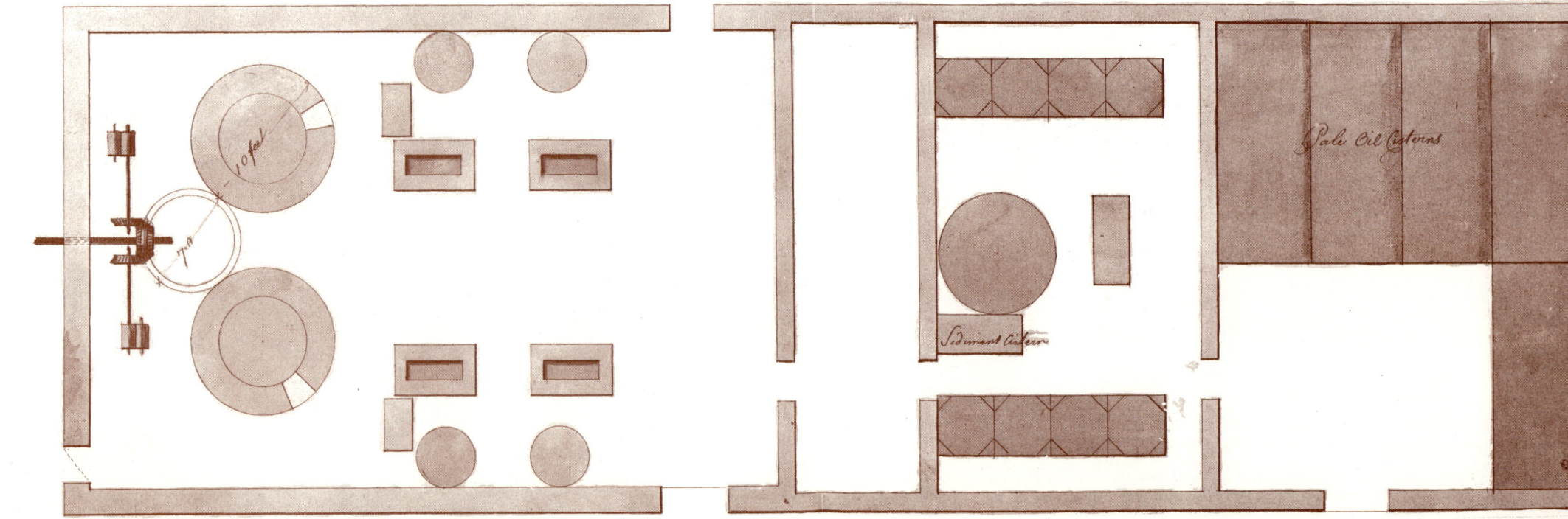

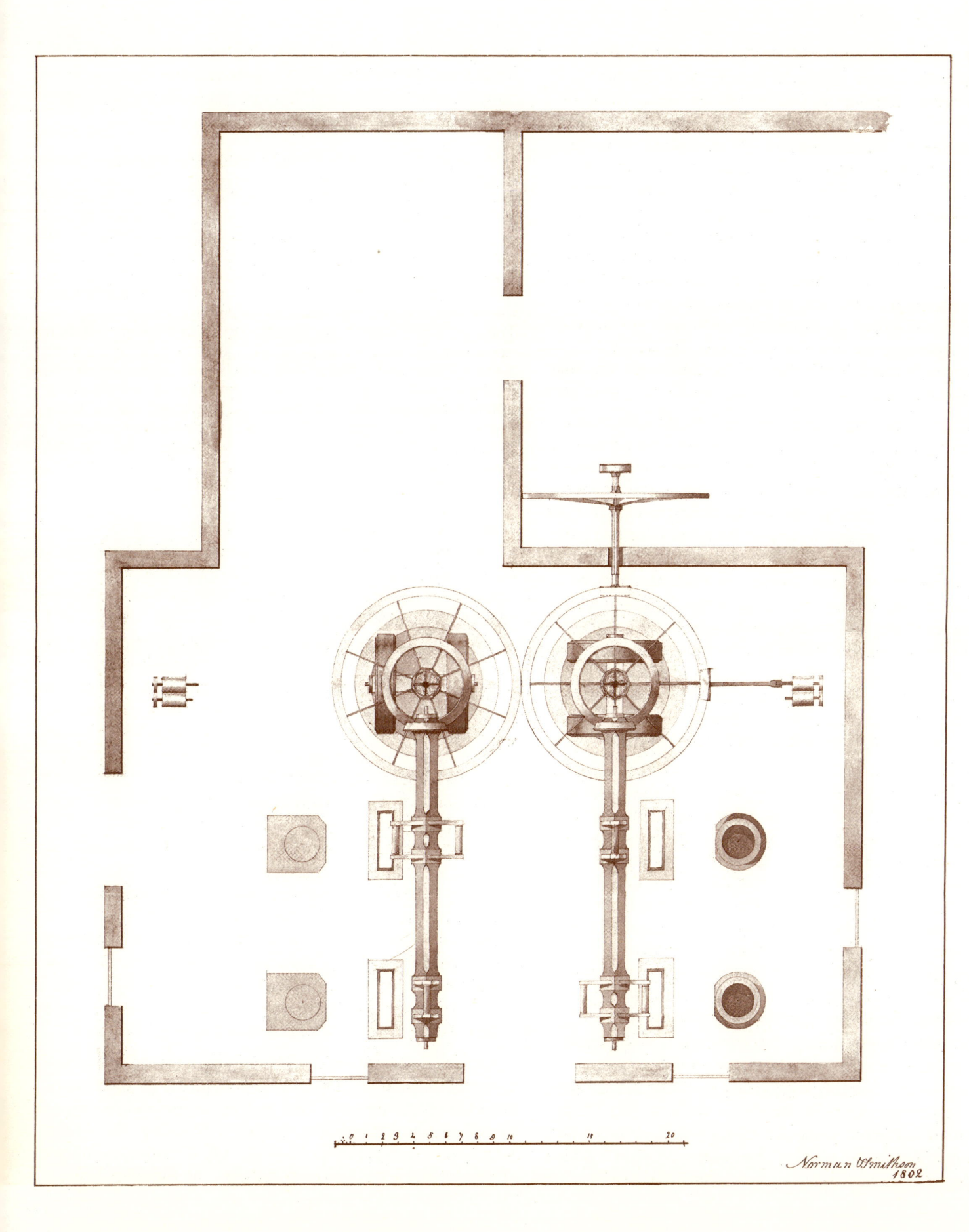

Cotes and Co, Hull, 1802
A plan of the millstone floor showing the edge runners. Norman and Smithson were probably the millwrights.
Portf: 1

4. Water Power in the Factory

1. Much of this and the following paragraph are based on Owen Ashmore, *Industrial Archaeology of Lancashire*, pp 40-7.
2. M. Hartley & J. Ingilby, *The Old Hand Knitters of the Yorkshire Dales*; Tuke, *General View of the Agriculture of the North Riding*, p 312; J. H. Priestly, 'The Black Brook...', *Halifax Antiquarian Society*, Jan. 1963. J. Hodgson, *Textile Manufacture and other industries in Keighley*, 1879.
3. S. D. Chapman, *The Early Factory Masters*, pp 62-76.
4. John Butt, *Industrial Archaeology of Scotland*, pp 64-75.
5. Michael Davies-Shiel & J. D. Marshall, *Industrial Archaeology of the Lake Counties*, pp 92, 96-98.
6. Edward J. Foulkes, 'The Cotton Spinning Factories of Flintshire, 1777-1866', *Flintshire Historical Society Publications*, XXI, 1964.
7. S. D. Chapman, op cit pp 62-76.
8. Ibid, pp 101-124.
9. J. Chapman & P. Andre, *Map of Essex 1772-4*. Whereas a medium sized woollen mill containing 4 prs of stocks and carding machines required 5 hp, a cotton spinning factory on the Arkwright pattern containing 1,000 spindles required 10 hp.
10. Jennifer Tann, *Gloucestershire Woollen Mills*, p 47.
11. B.L.C. Johnson, 'The Foley Partnerships: The Iron Industry at the end of the Charcoal Era,' *Economic History Review 4*, 1951-2.
12. British Association for the Advancement of Science, 'Birmingham in its Regional Setting.'
13. Misc Box, Mills etc.
14. Robin Chaplin, 'A Forgotten Industrial Valley,' *Shropshire News Letter* (Shropshire Archaeological Society) 36, June 1969. I am indebted to Mr. Chaplin for information from the Attingham Papers, Salop Record Office.

For a crucial period of thirteen years between Arkwright's roller spinning patent of 1769 and Watt's development of rotary power from the steam engine in 1782 the British cotton industry had to depend upon natural sources of power; and in practise the industry was dependent on water power for a good deal longer. Arkwright used a horse wheel in his first factory at Nottingham, but by 1771 he had built a water powered factory and he did not return to the use of horse wheels. The horse wheel was best suited to the small entrepreneur entering the trade perhaps from agriculture or as a skilled artisan. The power of the horse wheel was limited. It was to water power therefore that the larger cotton spinners and carders turned in the interval between Arkwright's first patent and Watt's double acting rotary engine patent. The important first stages of factory development during the industrial revolution were achieved largely by water power.

One of the consequences of this was that the cotton industry was scattered to areas where water power facilities were good, and to some where they were not. In Lancashire and Cheshire town sites were quickly developed and the Pennine valleys of the Irk, Irwell, Tame, Roch, Calder and Darwen became crowded with water powered factories. As pressure on resources grew remoter sites were developed. Many were difficult of access being at heights of up to 1000 ft above sea level.[1] In the Shuttleworth valley north of Bury there were at least ten mills within the space of one mile. This was made possible by the fact that the stream falls about 500 ft in that distance. In Rossendale water powered factories were built not only along the main valleys of the Irwell, Whitewell Brook and Limey Water but also along many of their small tributaries.

Cotton mills sprang up north of the Ribble too in the hinterland of Lancaster. The Calder and Lune valleys and their tributaries contained many mills — there were five at Caton for example. From the valleys on the Lancashire side of the Pennines the cotton spinning industry spread over to Yorkshire to remote areas in Wensleydale such as Aysgarth and to the Aire and Calder valleys and their tributaries. A number of the woollen and worsted factories in the Calder valley and around Keighley were originally built as cotton spinning mills.[2]

In the Midlands cotton carding and spinning spread far beyond Nottingham where Arkwright and Hargreaves established their first factories. Arkwright and his partners developed water powered sites at Cromford, Belper, Matlock Bath, Bakewell, Wirksworth, Cressbrook, Rocester, and possibly Ashbourne and other spinners built factories on the Arkwright pattern in Nottinghamshire, Derbyshire, Warwickshire and Staffordshire.[3]

Cotton spinning spread rapidly in Scotland, particularly around the Clyde;[4] into Cumberland, Westmorland and Furness where there were around fifty mills;[5] into north Wales where there was an important development at Holywell;[6] to Warwickshire, Leicestershire and Northamptonshire.[7] By the 1790s there were cotton mills in Bristol, London, Exeter, Norwich, Carlisle, Darlington, Dublin and the Lagan Valley of Ulster.

The wool textile industry was equally dependent on water power but in the first phases of the industrial revolution the power requirements were less. The spinning machines of Hargreaves, Arkwright and Crompton were designed for, and first used in, the cotton industry. Early attempts were made to adapt these machines to the spinning of woollen and worsted yarns. The jenny was suited to woollen yarn but experiments showed that it was unsuited to worsted. Arkwright's machine on the other hand was ill-suited to spinning woollen yarn but could be adapted to spinning worsted. The mule was only slowly adopted by the woollen industry in the late 1820s, thus power spinning was introduced into the woollen industry during the age of steam rather than in the heyday of water power. On the other hand water driven worsted spinning mills on the Arkwright principle were built in Yorkshire and the Midlands — there were thirteen such mills in the Midlands before 1800.[8] Thus in the woollen industry power was required only for scribbling and carding and for fulling. In Yorkshire small water driven scribbling and carding mills were built in many remote valleys to serve the country clothiers. Sometimes, but not always, the machinery was installed in an existing fulling mill. In the West of England and East Anglia scribbling and carding machinery was generally installed in existing fulling mills but much of the wool preparation seems to have continued to be done by hand. Fulling mills could be built in situations where spinning mills could not, and a number of these mills were erected on the sluggish streams around Colchester[9] and in the Severn valley in Gloucestershire.[10]

Water power, or the lack of it, exerted a strong influence on the development of the iron industry for, where there was an abundance as in the Forest of Dean, numbers of furnaces (with water-operated bellows), forges and slitting mills were built.[11] The development of primary industry in the centre of the Black Country was hindered by poor water power and the growth of the heavy iron industry in this area dates from Wilkinson's introduction of the steam blowing cylinder at Bradley. But radial streams flowing from the edge of the Black Country and the Birmingham plateau were powerful enough to supply forges and slitting mills and the Stour had at least eleven forges and four slitting mills along it in the early 18th century.[12]

When Matthew Boulton went into partnership with James Watt he conducted a survey of some of the nearby water-powered ironworks, possibly with a view to going into the iron trade as an extension of the steam engine business. In 1775 he visited Cradley and West Bromwich forges making detailed notes about the quality and quantity of their output. Knight's rolling and slitting mill was visited and scale drawings made of his works and machinery (page 72). An observer was sent to Birch's rolling and slitting mill at Halesowen and the report is worth quoting in detail because it demonstrates the precision that could be achieved in a water powered factory.[13]

'The head of water of Mr. Birch's mill is 33 feet when the pool is full . . . the diameters of the wheels are 22 feet and the width 4 feet. When the head of water is 33 feet the mill can roll and slit 1 heat (that is about 14 or 15 cwt) in 16 minutes the water wheels making about 24 not less revolutions in a minute. When the water is lower the mill rather comes short of its quantity of work, that is it does not do work in the ratio of the falls of water owing probably to the sluices not being drawn higher in the proportion that the head of water gets less.
The two water wheels are exactly equal and the apertures for the water are always regulated with great exactness each wheel turns 1 upper and 1 lower roll, that is, one of the wheels turns the upper roll and the lower slitter and the other turns the upper slitter and the lower roll . . . . The water wheels went round with great steadiness and did not seem to accelerate in the least during the little time between the putting the iron into the slitters after it had past thro' the rolls . . . . As to the work of the mill it seems to be exceedingly well Executed.'

Many areas that had a thriving industry in the age of the water wheel failed to keep pace during the age of steam and the industrial sites fell into disuse. It is these areas that have on the whole been neglected by historians. A good example is provided by a remote Shropshire tributary of the Severn, the Tern and its tributary the Roden.[14] Historians of Shropshire's industrial history have tended to concentrate their attention on those areas where heavy industry has survived to the present day — Coalbrookdale for example — yet Tern Forge within the grounds of Attingham Park was at least as important in the early to mid 18th century as the better known Shropshire works. Abraham Darby I, although not a partner, was closely concerned with his relative Thomas Harvey at Tern from the outset. The fixed capital investment amounted to £10,000, there was a labour force of between eighty and ninety men and the site, including housing and a series of locks and wharfs leading from the Severn spread over about two miles of the Tern valley. Another site farther up the valley (Moreton Corbet Forge) was leased by John Wilkinson and Edward Blakeway before they made Willey the centre of their Shropshire activities. Hazledine owned land near the forge and his younger son, William, managed and later leased Upton Forge lower down the valley. This example is by no means unique.

Similarly, students of the British silk industry have tended to concentrate their attention on developments in Spitalfields, Macclesfield, Congleton, Leed and Stockport and to neglect the developments at Sherborne, Dorset or Blockley, Worcestershire which were important nuclei of the silk throwing industry in the era of the water powered factory. The Blockley mills

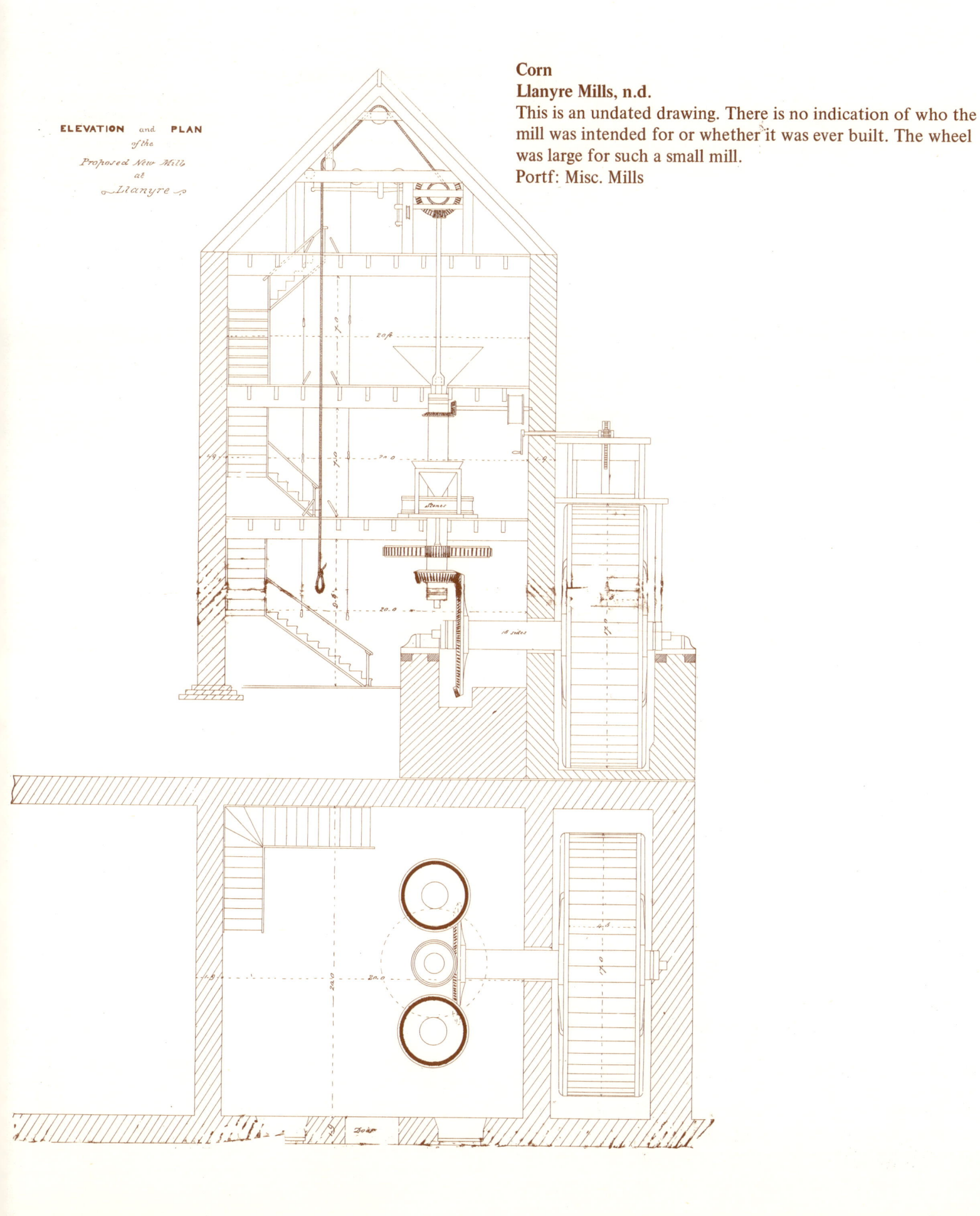

ELEVATION and PLAN
of the
Proposed New Mill
at
Llanyre

Corn
Llanyre Mills, n.d.
This is an undated drawing. There is no indication of who the mill was intended for or whether it was ever built. The wheel was large for such a small mill.
Portf: Misc. Mills

supplied the Coventry ribbon weavers.

Technical development

Between 1700 and 1825 the waterwheel was, in theory, transformed from being the product of the village craftsman to being the product of the engineer. Water power, it has been said, had been converted into a scientific technology.[15] Discussion on the mechanics of waterwheels was opened by the Frenchman, Antoine Parent, in 1704. He produced calculations on the efficiency of undershot wheels which were accepted by many later engineers. Deparcieux in France and Smeaton in England found that, in practice, undershot wheels with paddles could yield more power than the limit suggested by Parent. Smeaton, by constructing models of overshot wheels and varying the height at which water hit the wheel, showed that, all other factors being equal, the same wheel was twice as efficient overshot as undershot.[16] An undershot wheel worked chiefly by the impulse of the water and a good deal of the potential 'power' of the water was lost in collision. An overshot wheel worked chiefly by the weight of the water which was retained in buckets. Smeaton showed that 'fall' was important but 'the higher the wheel is in proportion to the whole descent the greater will be the effect.' If the fall alone was increased without the diameter of the wheel being increased the power would not be increased by more than one seventh of the increase in height. 'Therefore it depends less upon the impulse of the head and more upon the gravity of the water in the buckets: and if we consider how obliquely the water issuing from the head must strike the buckets we shall not be at a loss to account for the little advantage that arises from the impulse thereof.' Smeaton showed that the wheel should have a slower velocity than the velocity of the water hitting it for the more slowly the wheel moved the greater the quantity of water each bucket would contain 'so what is lost in speed is gained by the pressure of a greater quantity of water acting in the buckets at once.' He suggested that 20 rpm would achieve the greatest effect but added that experience showed that with large wheels a greater deviation from the rule was possible before any power was lost.

Breast wheels worked both by the impulse and weight of the water as Smeaton in his inimitable manner explains: 'the effect of such a wheel will be equal to the effect of an undershot whose head is equal to the difference of level between the surface of the water in the reservoir and the point where it strikes the wheel, added to that of an overshot whose height is equal to the difference of level between the point where it strikes the wheel and the level of the tail water.'

Smeaton was responsible for the introduction of new materials in the construction of waterwheels. In 1769 he introduced cast iron wheelshafts and in 1780 wrought iron buckets were added.[17] Cast iron arms gradually replaced the wooden ones.

15. D.S.L. Cardwell, 'Power Technologies and the Advance of Science 1700-1825', *Technology and Culture* 6, 1965.'
16. J. Smeaton, *Experimental Enquiry Concerning the Natural Powers of Wind and Water to turn Mills and other Machines depending on a Circular Motion*, 1796.
17. Paul N. Wilson, 'Water Power and the Industrial Revolution', *Water Power* August 1954; Paul N. Wilson, 'The Water Wheels of John Smeaton', *Trans Newcomen Soc.* 30, 1955-7.

18. Science Museum, Goodrich Papers, Journal & Memoranda Book 66, 13 March, 1830.
19. W. Fairbairn *Mills and Millwork*, 4th Ed., 1878 p.115.
20. P.N. Wilson, 'Water Power and the Industrial Revolution' op.cit. p.312.
21. D.S.L. Cardwell op.cit. p.193.
22. Jennifer Tann, 'Some Problems of Water Power...' *Trans Bristol & Glos Arch Soc.*, 84, 1965 pp.56-66.
23. David Smith, *Industrial Archaeology of the East Midlands*, pp.80-84.
24. W. Fairbairn op.cit. p.87.
25. Ibid pp. 83-5.
26. D.S.L. Cardwell op.cit. p.194.
27. See eg. A.E. Musson & C. Robinson, *Science & Technology in the Industrial Revolution.*

This solved the problem of 'wet and dry' which occurred with wooden wheels standing in water. An important development took place at the end of the century with the introduction of the rim gear wheel. This transmitted power to a pinion running at a much higher speed. Apart from relieving the waterwheel shaft of strain, higher speeds could be achieved within the factory and it became possible to transmit power along shafting for greater distances. When Goodrich visited Strutt's factory at Milford in 1830 he noted[18] that power from a wheel moving at 3 to 3½ ft per second was taken 110 ft to a mill and from this mill the shafting was continued another 375 ft. The waterwheel worked pumps and rasping machinery and ten or twelve pairs of bleaching stocks.

An important development in waterwheel technology took place during the millwrighting career of Thomas Hewes.[19] Hewes was responsible for the installation of waterwheels at some of the major factories in the early 19th century including Belper, Styal and Tutbury mills. He replaced the heavy cast iron arms of the wheel by light wrought iron cross rods which were supported in suspension, giving the wheel the appearance of a giant bicycle wheel. This development paved the way for the large industrial wheels of the 19th century, some of them over 60 ft diameter, such as those installed at Catrine by Fairbairn.

Rennie made an important contribution to the efficiency of waterwheels by designing the sliding hatch in 1783.[20] This admitted water to the wheel at as high a point as possible and was particularly valuable at sites with an irregular supply of water. By this means the maximum head of water was maintained and there was a reduction in airlocking the buckets. The problem of airlocking was solved by the ventilated bucket popularised by Fairbairn. By a series of slits air trapped in the buckets by the water could escape and turbulence was markedly reduced.

In 1824 Poncelet designed an undershot wheel with curved blades into which the water entered without shock and left practically without velocity. It was claimed that an efficiency of up to 60% could be achieved and Poncelet pointed out that in many situations undershot wheels were cheaper to install and operate.[21]

Just as there was the most appropriate type of wheel for a particular site so there was the most suitable type of watercourse. There were two main alternatives.[22] The factory builder could throw a dam across the stream creating a fall and a pond for storing water. The factory could then be built either across the stream course below the dam or on the bank alongside. This type of site was best suited to a hilly region where the gradient of the stream was steep and the valley narrow so that a good fall could be obtained without large areas of the valley having to be banked up to prevent flooding. If the factory was to be built in a wide flat valley the best method of obtaining a fall safely was to build a low weir across the river and divert some of the water into a narrow embanked channel or leat which led, without losing much height, to the factory which was built perhaps half a mile downstream from the weir. Sometimes the water was conveyed in a wooden trough carried on stilts. A good example of a leated factory site was one of the Robinson's factory complexes at Papplewick, Nottinghamshire. Grange mills were driven by a mill leat which began as an overflow from the pond of the mills above. The leat continued for over a mile beyond Grange Mills to Middle Mill lower down the stream.[23] Fairbairn conveyed the water to Catrine Mills in a 100 yd long tunnel.[24]

Weir design had been improved by the early 19th century.[25] The standard weir was a diagonal one which was used where there was a large supply of water as at Carron. This was often associated with a leat. V-shaped weirs were built if the current was particularly swift as the resisting powers were increased. Sometimes they were stepped in profile to prevent erosion of the foundations. 'The most perfect weirs however, are formed of stone, built of solid ashlar, and usually forming part of the segment of a circle' as at Belper, Milford and Masson Mills on the Derwent.

Innovation and technical problems

In this examination of water power technology an important question arises. We have discussed the improvements and devments propounded by millwrights and engineers but comparatively little is known about who adopted these improvements. Presumably the millwright engineers used their own improved systems in the works they carried out, but we know of only a handful of examples of the work of Hewes and Fairbairn. Did other millwrights use the improved waterwheel or make a scientific appraisal of the proposed site and design the most appropriate system of weirs, reservoirs and leats? D.S.L. Cardwell writes 'It is said that undershot wheels were rarely erected in England after the publication of Smeaton's paper; practically all those to be seen had been put up before 1759. It seems clear that the union of greater scientific understanding and enhanced technical skills resulted in progressively larger and more powerful waterwheels and, in a country of limited water power like England the gain in efficiency may have been crucial in expediting the Industrial Revolution.'[26] There has been recent discussion of the growing scientific approach of the entrepreneur during the industrial revolution[27] but there is evidence to show that in water power technology at least and, one suspects in other fields too, much of the so-called scientific approach was pure empiricism or a blind faith in 'scientific gentlemen' some of whom were little more than quacks. Field work does not tend to support the claim that few post-Smeaton wheels were undershot. A number of undershot wheels survive. Where they are wooden it is highly unlikely that they are pre-1759 in date and, as many undershot wheels contain iron parts, this puts

Thompson & Baxter, Hull, 1788
This mill, built in 1787, was one of the earliest steam corn mills in England. The elegant proportions of the mill are emphasised by the pediment and string courses. The mill initially contained four pairs of stones and was obviously not working to capacity, even allowing for a large amount of storage space.
Portf: 29

No. 8
Front View of the
Mill house —
Mess Thompson & Baxter
Jan.y 1788

1/3 Inch to the foot

28. Reports of the Late John Smeaton, 2nd Ed., Vol.2 p.118.
29. R.S. Fitton & A.P. Wadsworth, *The Strutts and the Arkwrights*, pp.210-211.
30. F.J. Glover, 'Dewsbury Mills, A History of Messrs. Wormalds & Walker Ltd., Blanket Manufacturers, Dewsbury,' Ph.D. Thesis Leeds, 1959-60, p.455.
31. Leonard A. Jones, 'The Development of Water Power in Wales', MSc thesis, Univ. Wales, Cardiff, 1965, p.200.
32. F.J. Glover thesis op.cit. pp.455-456.
33. J. Sutcliffe, *Treatise on Canals and Reservoirs...1816*, pp.67-69. Sutcliffe was not immune to misjudging the work of others for he recommended that B. Gott should buy a steam engine from Bowling Iron Works. Gott opted for a Boulton and Watt one.
34. Ibid p. 274.
35. Jennifer Tann, *Gloucestershire Woollen Mills* p.47.
36. Jennifer Tann, *Some Problems of Water Power*, op.cit. p.62.
37. PP 1834, XX pp.254-255, PP 1833 XX pp.944.
38. P. Ewart to J. Watt 3 July, 1972.
39. P. Ewart to J. Watt 6 Jan., 1792.
40. M. Boulton to T. Gilbert, 2. Feb., 1771, Letter Book. A.O.

their date at the end of the Smeaton mill-building era or after.

Wooden parts remained common in waterwheels well into the 19th century. Smeaton himself suggests the reason for this; for when reporting on a gun factory he wrote 'If the practice of making iron axes for the waterwheels be found to recommend itself so far as to be *equally cheap with wood* they may also be used here.'[28] Strutts, innovators in many aspects of factory building, installed a waterwheel at their West Mill, Belper in 1795-7. This was made of the traditional material — oak, the only concession to modern materials being the use of iron hoops in bands around the shaft.[29] Wooden wheels were still working in part of Dewsbury Mills in the 1840s.[30]

The rim gear wheel was well-known yet it is a fairly rare survival, and Hewes' suspension wheel was developed too late to be widely adopted. These developments made larger wheels, and a greater horse power, possible but they were introduced at a time when many new entrants to industry were building steam mills and, with some exceptions, were probably only used at already established water power sites which were being brought up to date or extended. Ventilated buckets seem to have been fairly widely adopted but Cardigan Foundry was still making unventilated buckets in the 20th century.[31] Hagues, Cook and Wormald had some unfortunate experiences with their new iron wheels and one wonders whether enough notice was taken of the vagaries of the river or whether the mid 19th century millwright/engineer felt that water power could be reduced to a series of abstractions. 'We are tired of writing to you on this matter of the wheel ... your theories think nothing of our observation — we would not set down our judgement against practical men like you but Fairbairn was just as pertinacious as you are, until he got all his stays — or four fifths of them smashed to pieces ... although they were made of beaten iron.'[32] John Sutcliffe, the Halifax millwright, was sceptical of the work of some of the men of science who went about the country recommending new power systems:[33]

'For the last twenty years gentlemen have gone through the nation giving lectures upon the application of water, and also upon mechanics, philosophy, astronomy etc., who have exhibited small models of water wheels and steam engines, by which they have pretended to shew the best mode of applying the powers of those two great agents of nature. By these paltry models and the representations given of them the public have been greatly mislead; and large sums of money have been expended to no purpose in making water wheels after such imperfect models. I know of a number of good water wheels, judiciously constructed, well speeded, and the water systematically applied, that some of these ingenious gentlemen have persuaded their owners were unfit for use, because in their opinion they moved too quick and consumed too much water; which they were prevailed upon to reject, and make others according to their models.'

Sutcliffe gave as an example one of his friends who had removed a good water wheel and replaced it by a wheel of a new design suggested by one of these gentlemen. He had to reject it and substitute for it a wheel similar to the original one, wasting £400 in the process. Sutcliffe was well aware of the element of chance involved in the progress of mechanical improvements.[34] 'We see the most trifling and insignificant improvements, if such they should be called, make rapid strides, and, for a time, arrest the attention and mislead men of sound judgement, witness the numberless plans that have been designed for steam engines, and for the make and movement of water wheels, which are vanished like a dream.'

When it came to the engineering works associated with the water powered factory, even if the builder or millwright was aware of the ideal solution, there was a strong possibility that he would be unable to achieve it. Overcrowding created havoc in some river valleys. By the early 19th century there were twenty six factories along 5½ miles of the Stroud valley where there was only a moderate gradient.[35] Leated sites were best suited to this gentle gradient but the overcrowding was so intense that it became impossible to construct elaborate leats and factory owners had to make do with less fall. Because of the little 'fall', low breast or undershot wheels had to be used instead of the more efficient high breast or overshot wheels.[36]

Apart from the fact that the ideal layout was frequently unattainable there were other problems. Many mill owners complained of a severely diminished supply of water in summer. At Ebley Mills, one of the largest Gloucestershire woollen mills, there was sometimes no water until noon in summer and Charles Hooper, owner of the lowest factory in the Stroud valley often saw no water until 1 pm.[37] Peter Ewart described the problems at the water driven cotton and woollen factory of Mr. Lodge at Oakenrod near Rochdale;[38] the cotton mill contained 804 spindles with carding engines and the fulling mill contained stocks, carding and other machinery. 'I was informed that in a dry season there is not water enough to turn the four stocks even when the cotton mills is stopped.' Lawrence and Yates, 'extensive manufacturers and merchants', had a large water mill containing about 2,500 spindles 'in wet times they have plenty of water to turn them ... but in dry weather they have not water enough to turn four or five hundred.'[39]

Matthew Boulton suffered both from summer shortage and from the proprietors of the Birmingham Canal taking water from his stream:[40] 'In all the dry part of last summer I was obliged to stop half the Movements in my Mill ... for want of water although there was last summer double the usual quantity of rain. ... Let Mr. Smeaton or Brindley or all the Engineers upon Earth give what evidence they will before Parliament, I am convinced from last summers experience that if the Proprieters of the Canal continue to take the two streams on which my mill depends

Unknown mill, n.d.
This is a fine undated unnamed drawing of a corn mill containing three pairs of stones. The millwrights shop is an indication that the mill may have been designed for a country estate.
Portf: Misc. Mills. No Scale.

Union Company, 1789
This drawing shows the use of cast iron in an intermediate stage of factory building when wood beams, and floors were used but cast iron columns supported the floors. The window frames were also probably of cast iron.
Portf: 122

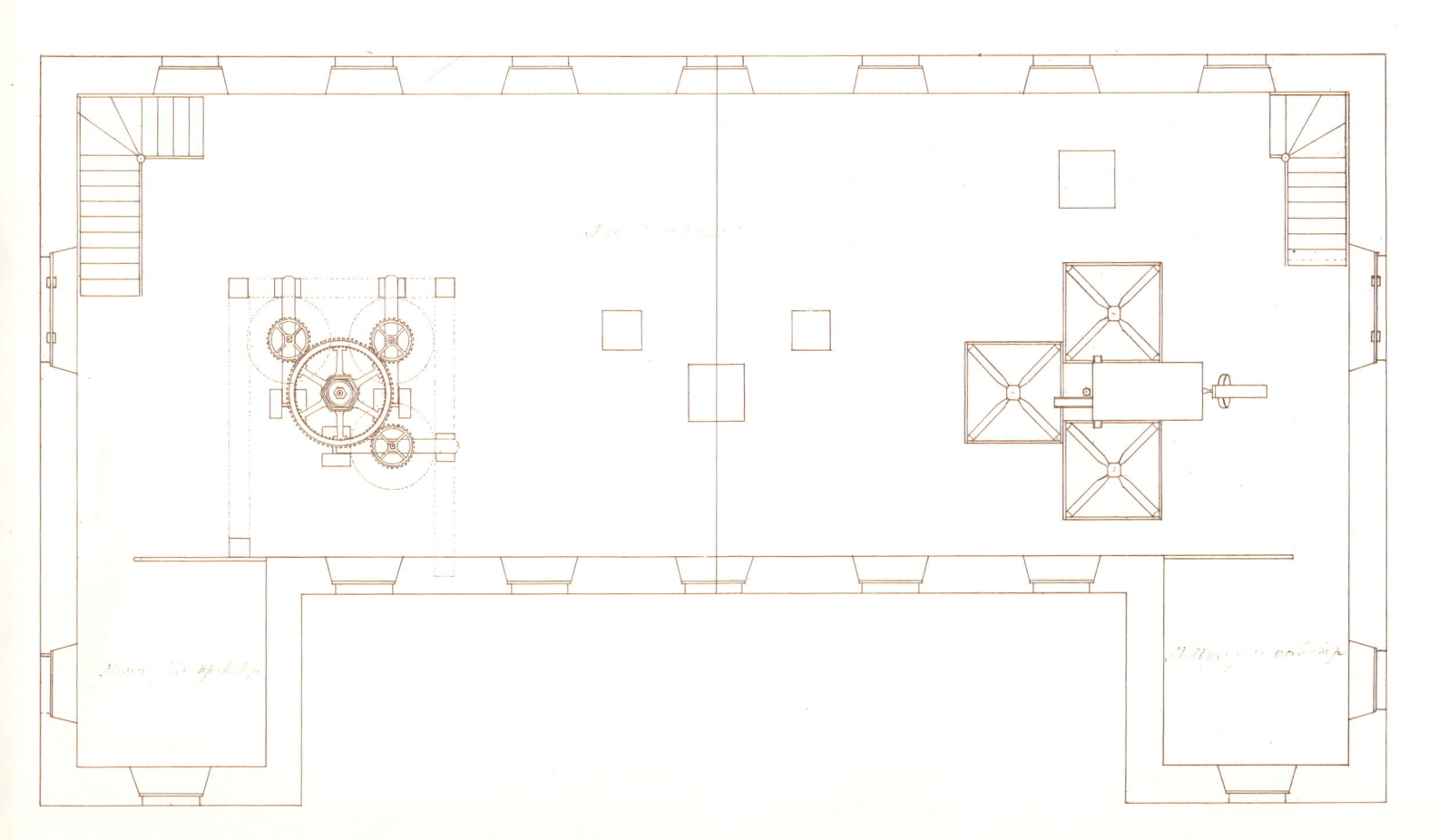

Cross Section of Mill House &c.
Union Company. 3 Inch to the foot   23rd March 1789.

41. J. Kellet to M. Boulton, 9 March, 1797, A.O.
42. Gloucester City Library Box 69.
43. D. Harrison to Boulton & Watt 30 May, 1785.
44. Gloucestershire Record Office D 654T/II/4/T1.
45. Ibid D 186V/33.
46. Ibid D 947/M2.
47. John Graham, 'History of Printworks in the Manchester District from 1760 to 1846',Ms, 1847, Manchester City Library.
48. A. Neilson to J. Watt 22 Feb., 1788.
49. S.D. Chapman, op.cit. p.67.
50. M. Boulton to T. Gilbert 10 Feb., 1771, Letter Book, A.O.
51. Water wheel at Macclesfield £40, Sun Insurance 87/840981.
52. R.S. Fitton & A.P. Wadsworth op.cit. pp.210-211.
53. David Smith op.cit. pp.83.
54. M. M. Edwards, *The Growth of the British Cotton Trade, 1780-1815*, pp.209.
55. *Reports of the Late John Smeaton* Vol 1, pp.170, vol 2 pp.392.

it is ruin'd. I might as well have built it upon the Summit of the Hill.'

Many water powered factories were shared. Aston rolling mill, near Birmingham, contained two water wheels one for grinding corn and the other for rolling metal. A clause in the lease stipulated that the lessee of the rolling mill should allow water to be drawn at both gates at the same time.[41] Bowbridge woollen mills, Gloucestershire, were shared in 1810 and a clause was inserted in the agreement to the effect that when water was short both parties should cease work until there was sufficient water for them both to start work again.[42]

One of the most involved problems of water power was the height of water between factories. If the owner of factory B wished to raise the head of water at his factory he frequently affected the water power at the site below him C and above him A. The owner of factory C might find like Harrison of Doveridge, Staffordshire that 'the millers above keep back the water to replenish their dams.'[43] Humphrey Austin, a Gloucestershire woollen manufacturer when leasing a small mill upstream of his own took care to see that the lessee would not 'dam or pen up the water . . . or do any other act which may hinder or obstruct or in any wise injure or impede the working of another [Austin's] mill.'[44] Alternatively it was the owner of factory A who was in difficulties. When B raised his dam the water level was likely to rise upstream and flood the tail race and wheel pits at the factory above. An overshot wheel was less able to work in tail water than either a breast or undershot wheel – 'an undershot wheel will work in tail water better than any other.'[45] The raising of a mill dam in Dursley, Gloucestershire, caused a chain reaction up the Ewelme valley. The owner of the uppermost mill concerned in the ensuing court case wrote, 'if your water should be penned to the highest height and my wheel should be going, I should be deprived of a very considerable part of the fall attached to my mill.'[46] In some crowded valleys it may have been preferable to install a breast or undershot wheel, even when there was sufficient fall for an overshot wheel, in order to avoid the difficulties of an overshot wheel working in tail water.

Finally there was the insuperable problem that good streams did not necessarily flow in the most convenient places. Some of the Lancashire and Cheshire print works were situated in remote valleys where transport was a great problem. Crag Print Works near Macclesfield was said to lie 'quite in the mountains as if at the far end of the world'.[47] Scottish entrepreneurs faced a similar problem. Archibald Neilson of Dundee enquired after a steam engine[48] 'not, as you know because we want water in general in Scotland, but it is often so far from the towns in which our Manufacturers are, as to be unserviceable for that purpose; at this very town for example, we have no stream nearer than two or three miles from us in which moreover there is too little water for our purposes.' Arkwright's motives in choosing Cromford as the centre of his empire were complicated; he appears to have chosen not only because he thought that it had good water power but because there was a good chance that he would be able to satisfy his social ambitions there.[49] The Bonsall Brook on which he built his first mill seems to have provided insufficient water power and when Arkwright attempted to augment it from other sources he involved himself in a legal dispute which dragged on for five years.

Perhaps Matthew Boulton should have the last word on the hazards of water power.[50] He suggested that there was only one solution to his problem 'and that is by erecting a fire engine at each [mill] . . . and I suppose the Company would rather be inclined to let the canal go dry than practise that expedient.'

The cost of water power

In considering the capital costs of water power installations one must distinguish between the water power installations of the period c. 1750-1800 and those of the period after 1800. The crude wooden paddle wheels of many of the smaller water powered factories were cheap. Insurance valuations in the region of £40 to £60 were common for this type of wheel. Larger wheels were valued at about £100.[51] Maintenance work on such wheels could be done cheaply by the carpenters and clockmakers frequently employed to make machinery but depreciation would have been higher than on an iron wheel. Larger, more sophisticated wooden wheels, were being built by the end of the century. The oak wheel installed in Belper West Mill cost £629. 19. 4¾d.[52] but even this was cheaper than a steam engine of equivalent power.

However, the wheel itself was by no means all. If the mill was at a new water power site, watercourses had to be constructed. If an old site was being developed the existing watercourses sometimes required reconstruction. In estimating the cost of water power one must bear in mind the advantage of a transfer of capital in the form of existing watercourses. Robinson of Papplewick, for example, developed a site known as Forge Mill and there are many instances of cotton spinning gaining sites at the expense of the older established woollen industry.[53] The cost of building water courses for a Lancashire spinning mill came to £223. 3. 0d. in 1786.[54] Smeaton estimated the cost of a two mile leat through hard rock at £400 and he prepared a detailed estimate for a new dam head and watercourse for Milltown Mills, New Tarbet in about 1771.[55]

	£.	s.	d.
Timber at 4s. per ft. and workmanship	45.	4.	0.
Iron work	3.	0.	0.
Carpentry total ...	48.	4.	0.
Casting foundations	3.	0.	0.
Dam wall 4 ft. thick, 8 ft. high, 96 ft. long			
Flue walls in sluice, 2 ft. thick, 4 ft. high, 70 ft. long	31.	12.	6.
Rubble masonry at £5. 10s. per rood 3d. per ft. allowance for larger stones	14.	8.	0.
Freestone for surface of the dam ...	42.	15.	0.
72 cubic yds. quarry rubble for filling dam at 1/6d.	5.	8.	0.
Total of masonry ...	97.	8.	6.

	£.	s.	d.
Cutting a new aquaduct 3 ft. deep, 555 yds. long at 6d.	13.	17.	6.
Banking from dam's end wall to high ground, 27 yds. at 4d.	3.	18.	0.
Total of spade work	17.	15.	6.

Abstract

Carpentry	48.	4.	0.
Masonry	97.	3.	6.
Spade Work	17.	15.	6.
Neat estimate ...	163.	3.	0.
Contingencies at 10%	16.	6.	4.
	179.	9.	4.

Deptford Corn Mill, 1825
This was designed by Rennie's sons for the victualling yard at Deptford.
Portf: 499

66

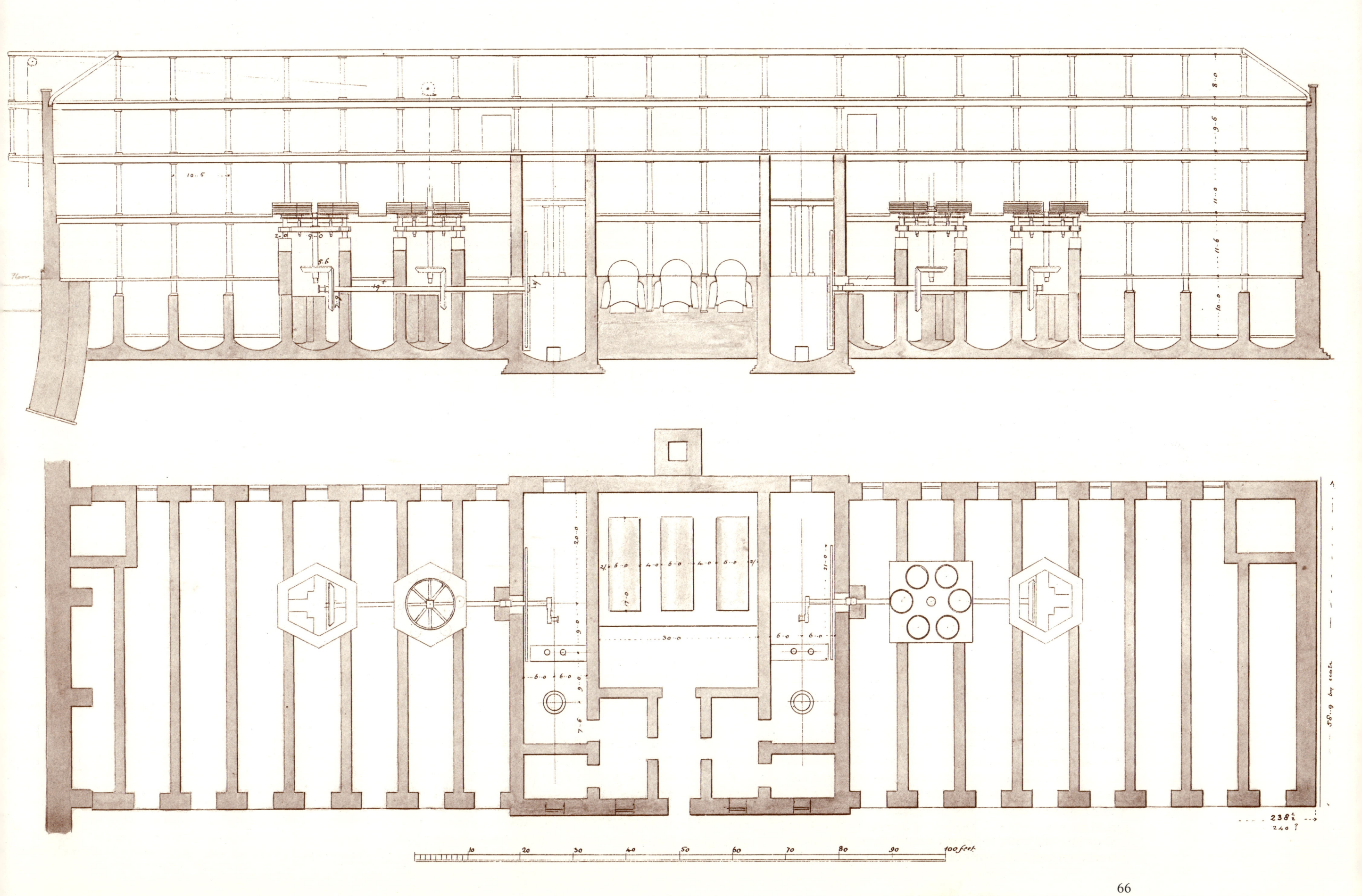

56. R.S. Fitton & A.P. Wadsworth op.cit. pp.210-11.
57. *200th Anniversary, A History of Playne of Longfords Mills*, pp.40.
58. W. Fairbairn, *Mills & Millwork* 1878, pp.87.
59. W. Pole ed., *The Life of Sir William Fairbairn*, 1877, pp.147.
60. F.J. Glover, Thesis op.cit. pp.400, 455.
61. J. Rennie to M. Boulton, 5 July, 1788. A.O.
62. W.T. Miller, *The Water Mills of Sheffield*, 1949.
63. T. Fox to Boulton & Watt 21 April, 1791.
64. Jennifer Tann, 'Some Problems of Water Power', op.cit. pp.76-77.
65. Engine Book; S.D. Chapman, 'Fixed Capital Formation in the British Cotton Industry 1770-c. 1815', *Economic History Review 23*, 1970. I am grateful to Dr. Chapman for showing me his typescript before publication.
66. J. Southern to Gardiner, 27 Feb., 1792.
67. W.T. Lawson to Boulton & Watt, 5 Dec., 1802.
68. Portf. 9.
69. Peter Temin, 'Steam and Water Power in the Early 19th Century', *Journ. Economic History*, XXVI, 1966.
70. P. Marsland to Boulton & Watt n.d. Box 3.
71. Brotherton Library, Marshall MSS 63, p.32.
72. Alex J. Robertson, 'Robert Owen and the Campbell Debt. 1810-1822', *Business History* XI, 1969. I am indebted to Dr. Butt for information on the rate of interest.

The watercourse reconstruction at Belper West Mill cost £1142. 2. 1½d. including the cost of the wheel house.[56] In none of these examples is the cost of land or rent of water rights included. In 1806 one of the largest mill reservoirs in Gloucestershire was constructed at Longfords Mill, Avening.[57] This was an old water power site and the Playne family had owned the fulling mill for several generations. With the mechanisation of more of the cloth manufacturing processes a greater quantity of power was required. A 150 yd 30 ft high dam was built across the valley forming a lake of 15 acres. The necessary land was purchased for £400 with a peppercorn rent of 2/6d. per year reserved. The dam cost £945 to construct excluding the leat which supplied the wheels and for which there is no recorded cost. By the end of the 18th century the capital costs of water power, including water rights, the construction of dams and leats as well as the wheel itself were higher than for steam power.

This gap widened during the early part of the 19th century with the introduction of large iron wheels built by Hewes, Fairbairn and Lillie and others. Fairbairn was asked to reconstruct the power system at Catrine cotton mills in 1824. He installed two 50 ft diameter, 10 ft 6 in wide wheels which generated 120 hp, each, leaving space for two more should they be required. The watercourses were altered to re-unite a fall of 47 ft and to provide reservoirs covering 120 Scotch acres. The total cost was £18,000, £75 per hp — small wonder that the other two wheels appear not to have been installed.[58]

	£.
Water privileges and land	4000
Cost of weir	1000
Head race, tunnel, leat	3000
Archways, cisterns, sluices	1000
Wheel house, foundations	1500
Tail race	1500
Water wheels etc.	4500
Contingencies	1500
	18000

The annual cost of this power system for interest on capital at 7% was £1260. Shortly after completing the Catrine works Fairbairn installed water wheels at Deanston; at his own works on the site of a dyeworks and cotton mill he installed a 62 ft diameter, 130 hp wheel.[59] In 1827-8 a Fairbairn and Lillie wheel was installed at Hagues, Cook and Wormald's Dewsbury Mills. The total cost, including the new wheel house and improvements to the leat was the more modest sum of £2899. Whither of Leeds erected a new iron breast wheel at another part of the works in 1844. The cost of this was £1366. 3. 10d.[60]

Once installed water power was cheaper to run than steam power. Until 1800 the owner of a Boulton and Watt engine had to pay a premium for the use of the patent rights and coal costs rose enormously away from the coal fields. Many an owner of a water powered factory must have been deterred from installing a steam engine. Even Matthew Boulton preferred water power for his rolling mill in 1788.[61] An engine required a man to tend it constantly and steam engines depreciated rapidly. Waterwheels depreciated much more slowly and a smaller depreciation charge was almost certainly added to the annual costs. The low running cost of water power was undoubtedly one of the main factors which led to the retention of the waterwheel for so long in the Sheffield area, where every available stream was crowded with wheels in the 19th century.[62] This same factor applied to the West of England woollen area in the late 18th and early 19th centuries. Thomas Fox of Wellington, Somerset[63] gave as his reasons for not ordering a small engine 'the expense of a small engine as well as the consumption of coal and water being much greater than I apprehended would be required. For our work it seems more adviseable to place our machines on a stream of water about a mile from our house.' Fox highlights another factor operating in favour of water power — the relative expense of a small engine. This was something that Boulton and Watt pointed out to enquirers after small engines but where coal was costly the lower relative cost of a larger engine did not compensate for the fact that running costs would be considerably higher. The woollen manufacturer only required a small amount of power for fulling and preparing wool; similarly the Sheffield blade grinder or forge owner needed only 4 to 8 hp.

The West of England clothier was conservative in his attitude to power and steam power was only very slowly introduced. During the major factory building phase of 1813 to 1825 when power spinning was introduced, only one factory was built away from the stream side, to be driven wholly by steam; all the others were built or re-built alongside the rivers with provisions made for water power.[64] This is not to suggest that a conservative attitude was the only reason for the West of England manufacturer clinging to water power but it may have contributed.

In view of the fact that the capital costs of water power were higher than for steam power by the end of the 18th century, why did water power survive for so long? — why indeed did some entrepreneurs apparently prefer it to steam power? The question is an involved one on which all too little work has been done. At this point all that can be done is to suggest some possible answers.

By the 1790s, a water wheel could generate more power than a single steam engine could. With the era of Hewes and Fairbairn 100 hp and more could be generated from one water wheel. From the 1790s to about 1825-30 the owner of a factory requiring large quantities of power could find that a water wheel or wheels served his purpose better. The Robinsons of Papplewick, Nottinghamshire were the first cotton spinners to install a Boulton and Watt rotative engine (they bought two). With a fixed capital investment of something in the region of £21,000[65] it is unlikely that they could not afford two small steam engines. But Southern noted in 1792[66] that John Pappworth of Papplewick, probably Robinsons' manager, wished to dispose of an engine which had only been erected 'in case they should lose their water course which at the time a trail with Ld. Byron made them afraid of but in which they were successful.' In 1802[67] Lawson wrote that 'Mr. James Robinson ... has not worked his engine for near three years past — till about a fortnight since — and it was again stopt some days ago from late rains having given them water enough.' One of Robinsons' water wheels was 30 ft diameter and 6 ft 11 in wide and another was 42 ft diameter and 6 ft wide.[68] The power that these wheels could generate would have been greater than the power of two steam engines whose combined horse power was about 22 hp. The Arkwright method of cotton spinning required more power per spindle than the mule. By using water power spinners like the Robinsons could increase their means of production beyond what could be achieved by using steam, and could survive, for a while, the competition of the mule spinners.

Peter Temin has suggested[69] in a study of steam and water power in America that the choice between them at a given location was affected by the interest rate. At a 6% rate, 60% of the costs of water power were capital costs while they comprised only 20% of the costs of using steam in 1840. When the interest rate was higher the relative cost of water power increased. There is too little data available to show whether there were significant regional variations in the interest rate on private loans in Britain but in a period when many entrepreneurs were finding difficulty in meeting their fixed capital requirements the rate of interest would have been an important consideration. Robert Owen, for example, having managed the first steam powered cotton factory in Manchester went into partnership with some merchants at a site adjoining Marsland's factory. Owen and Co's shortage of capital was apparently so great that Samuel Marsland had to lend them half the cost of their steam engine.[70] Yet Owen left this steam driven factory for a water powered factory complex in Scotland — New Lanark. No records survive which indicate the rate of interest that Owen and Co paid to Marsland but 10% was not uncommon on private loans; nor are there sufficient records of the New Lanark partnership. However, Owen's father-in-law, David Dale was a respected banker and Marshall noted that Owen spent much of his time at the bank.[71] Owen had little capital of his own and his shares in the New Lanark Twist Company were somewhat shadily purchased with capital formerly entrusted to David Dale on which he paid 5% interest.[72] Owen's move to Scotland was probably prompted partly by the availability of capital for long term investment and by a lower interest rate as well as by the superior natural advantages for water power which would have enabled him to obtain more power than one or two steam engines could have provided.

The factory movement was well under way in Britain before the development of satisfactory rotary power from the steam engine. The first generation of larger factory masters had no alternative to water power and when rotary power was available, many manufacturers were hesitant to use it. They invested heavily in water powered factory buildings and long-term loans were difficult to raise. It is sometimes imagined that there was an almost unlimited supply of capital available for long-term investment, but there is much evidence to show that the entrepreneur found difficulty in meeting the demands for fixed capital. Boulton and Watt found that 'the all devouring works we are erecting required money faster than we can easily get it.'[73] Once the water powered factory had been built — perhaps in a remote valley — capital was, as it were, locked in the building and the entrepreneur was tied to his water powered site. In other words the extensive fixed capital investment of the first few decades of the industrial revolution served to prolong the use of water power.

The dispersal of industry in search of water power could have far reaching consequences for that industry. During times of machine breaking riots or when the entrepreneur wished to keep a process secret a virtual 'safe-box' could be created by isolation. But a high price was paid in other ways. The factory was sometimes remote from merchants and markets, transport costs could be high. Information about the prices of imported raw materials could take longer to reach the remote factory,[74] and, probably most important of all, the water powered factory was often remote from centres of innovation. Unless industry in these remote areas could develop some internal economies it was almost bound eventually to be overtaken by industry in a better situation. Arkwright's choice of Cromford cut him off from the leadership of the cotton spinning industry by isolating him from the major centres of the cotton industry, Manchester and Nottingham. Lancashire, in particular, developed important external economies which partly account for the dominant position it acquired and maintained in the cotton industry. Arkwright's sons contracted their business after their father's death. In Shropshire the iron industry failed to develop important external economies. There was a migration of capital to Staffordshire and although even in the 19th century few manufacturers doubted the superior quality of Shropshire iron this industry, located at water powered sites, failed to keep pace with the south Staffordshire iron industry using steam power.

73. J. Southern to Barnes, 22 July, 1796, Foundry Letter Book.
74. But M. M. Edwards has shown that the smaller remote firm could survive by producing grades of goods which the larger firm could not or did not wish to produce, so that in some respects the small and the large firm were complementary in the late 18th and early 19th century cotton industry. M. M. Edwards, *The Growth of the British Cotton Trade*, p.131.

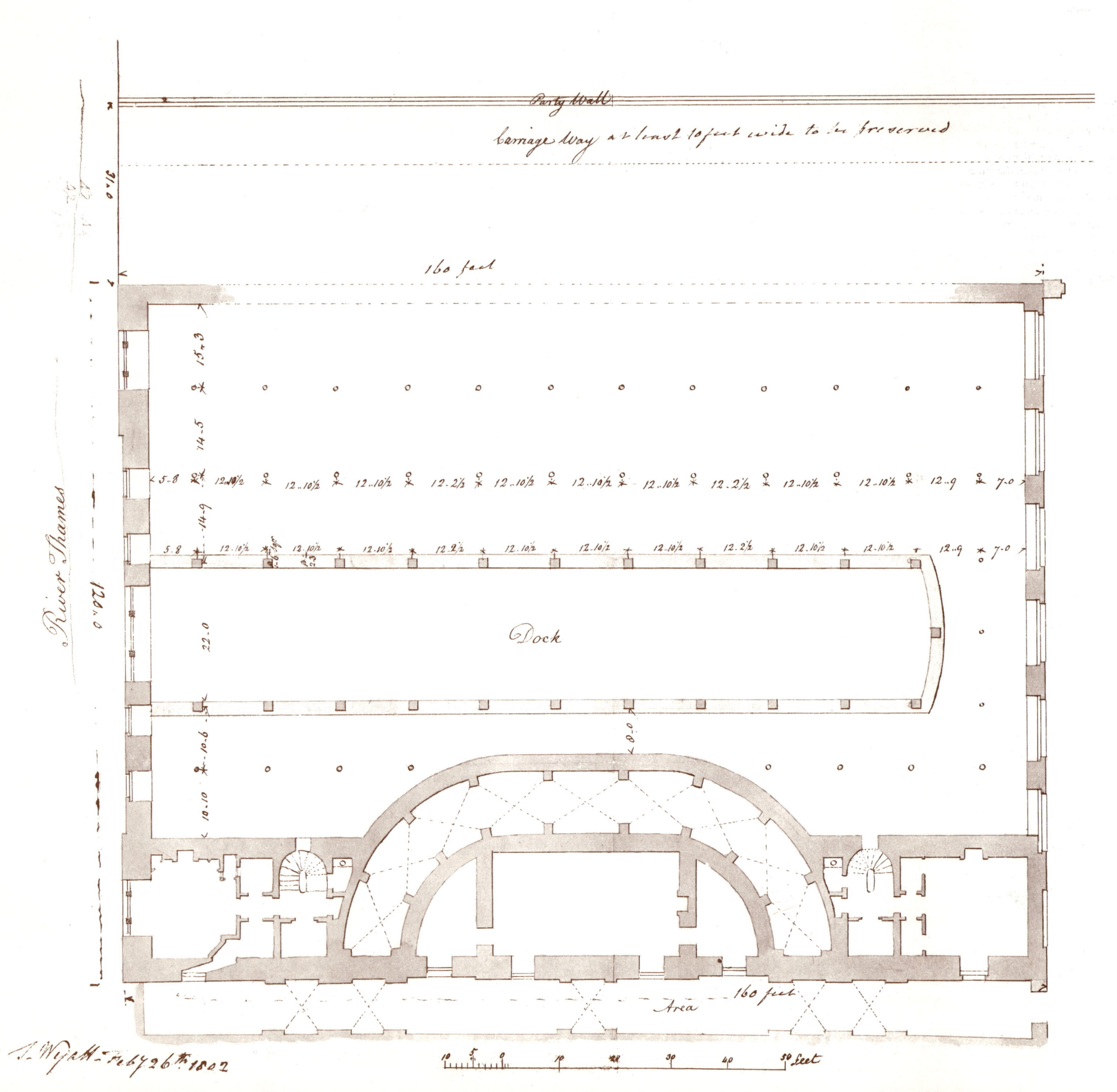

Albion Mills, London, *c.* 1794, 1802
Albion Mills were built in 1784 and the plan has special interest in being signed by the architect Samuel Wyatt. Boulton and Watt and Wyatt, amongst others, were partners in the concern which was a financial failure. But the mills were one of the wonders of London and provided a fine advertisement for Boulton and Watt's engines. Rennie was millwright and responsible for the introduction of cast iron gearings in the mill.
Some of the subsidiary operations in corn milling were mechanised for the first time at Albion Mills.
When completed the mill would have contained thirty pairs of stones but it never reached capacity. It was burnt down in 1791.
Portf: 152

70

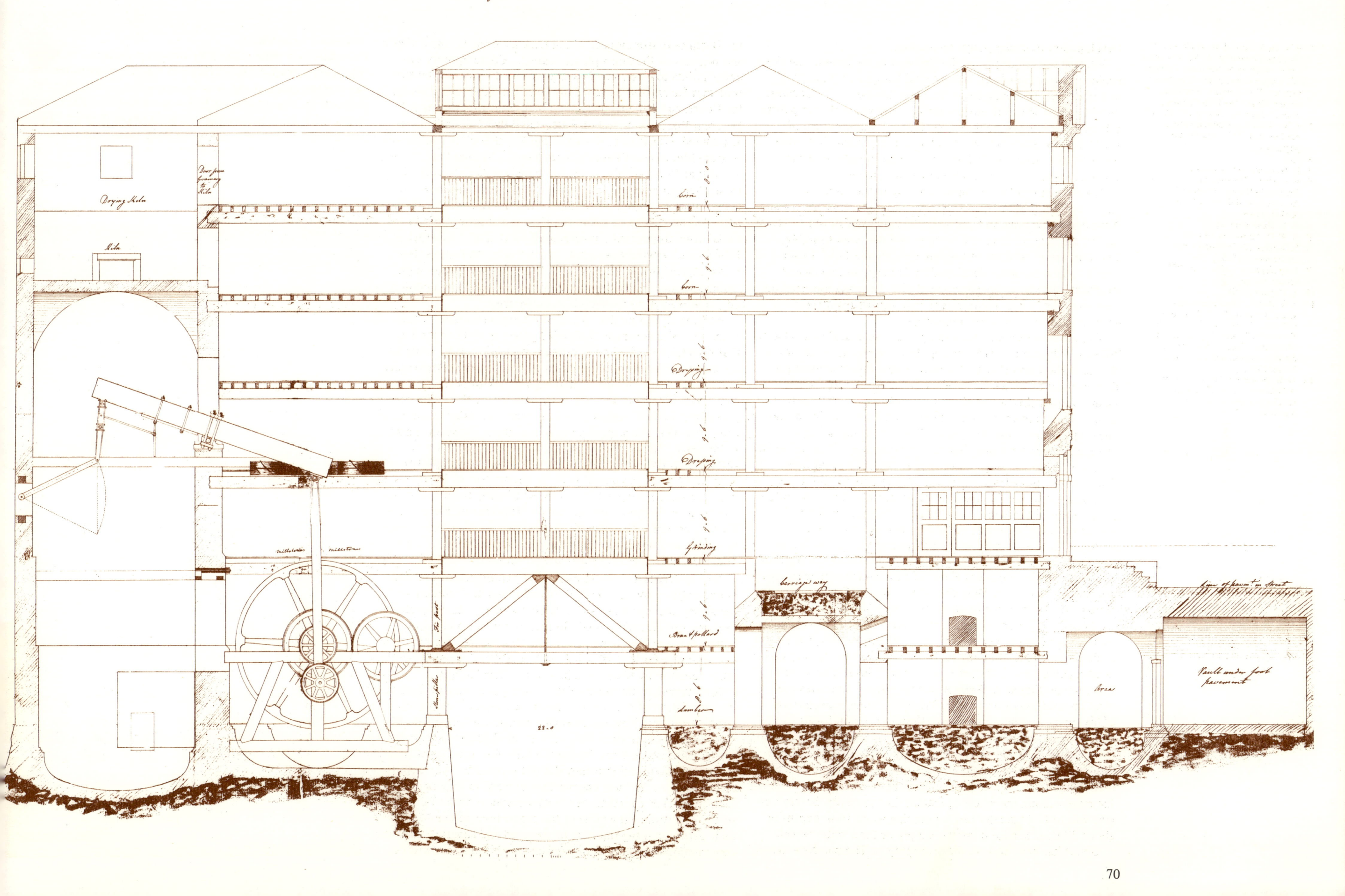

5. Steam Power in the Factory

1. H.W. Dickinson, *A Short History of the Steam Engine*, p. 15-16. For other people working on the idea of steam as a motive power in the 17th century see Rhys Jenkins, 'The Heat Engine Idea in the 17th Century,' *Trans. Newcomen Soc.* XVII, 1936-7.
2. A. E. Musson, introduction to 2nd edition of H. W. Dickinson, no pagination.
3. Ibid.
4. H. W. Dickinson op cit p.20.
5. E. Hughes, 'The First Steam Engines in Durham Coalfield', *Archaeologia Aeliana*, 4th ser. XXVII, 1949.
6. A. E. Musson op cit.
7. H. W. Dickinson op cit p.29.
8. S. Switzer, *Hydrostaticks and Hydraulicks*, 1729, p.342.
9. *Encyclopaedia Britannica* 1797, article 'Steam'.
10. E. Hughes op cit.
11. J. R. Harris, 'The Employment of steam power in the 18th century,' *History*, 52, 1967, p.132; Marie B. Rowlands, 'Stonier Parrott and the Newcomen Engine,' *Newcomen Society* unpublished paper, read 1 Jan. 1969.
12. H. W. Dickinson, op cit. p.61-2; *A Catalogue of the Civil and Mechanical Engineering Designs of John Smeaton*, p.57-85.

Development of steam power technology

Mystery surrounds one of the first people to develop the steam engine. Edward Somerset, later second Marquis of Worcester, described a steam apparatus in his book *Century of Inventions*, published in 1663. But the description is so vague that it is difficult to assess what he achieved. Dickinson[1] was sceptical of the Marquis's claim to have invented a steam engine but other writers[2] have agreed with James Watt who thought that the evidence, as far as it went, was convincing enough.

The Frenchmen Denys Papin, working in the 1690s, was theoretically far ahead of his English contemporaries Savery and Newcomen. He experimented with the expansive power of steam and its application to rotary motion, and was even thinking, like Watt later, of ways of keeping the cylinder hot.[3] Papin aimed to have his ideas developed by industry but he lacked practical business and engineering ability. He was supported by pensions from governments rather than profits from the commercial exploitation of his discoveries.

Thomas Savery is also something of an enigma. He was elected FRS in 1705 and in 1714 was appointed Surveyor of the Water Works at Hampton Court. Dickinson called him 'the most prolific inventor of his day'[4] but since Dickinson wrote Savery's claim to originality in inventing the steam engine has received several severe blows. A contemporary of Savery's claimed that Savery had copied the Marquis of Worcester's engine[5] and Desaguliers was informed that he had bought up all the available copies of the Marquis of Worcester's book and burned them. The first mentioned authority was Stonier Parrott and it has been suggested that Desaguiler's informant was Henry Beighton FRS.[6] Both of these men were interested in getting Savery's patent repealed, nevertheless their statements seem to be accurate. In 1698 Savery took out a patent for 'a new Invention for Raising of Water and occasioning Motion to all Sorts of Mill Work by the Impellent Force of Fire, which will be of great use and Advantage for Drayning Mines, Serving Towns with water and for the Working of all Sorts of Mills where they have not the benefit of Water nor constant Windes.' The grant was for the usual term of fourteen years but in 1699 Savery obtained an Act of Parliament extending the term of the patent for twenty one years making a total of thirty five. Savery's book *The Miners' Friend or an Engine to Raise Water by Fire*... indicates the most obvious use of his engine. Savery had high expectations of his engine and he set up a workshop in London for its manufacture. But the engine could only satisfactorily raise water from between 20 and 50 ft. Since many mines were 50 or 60 fathoms deep the only way in which the engine could have been used would have been to place engines at intervals of 50 ft. down the mine making the whole operation far too costly. Savery's engine was to be of greater use in pumping water from depths of about 20 ft. either for domestic supply or for a water wheel.

Doubt has been cast on the originality of Thomas Newcomen, the Dartmouth ironmonger whom Dickinson called 'the inventor and begetter of the atmospheric engine'.[7] Contemporary evidence points to Newcomen having produced his engine by the time that Savery had designed his 'only the latter, being nearer the Court, had obtained his Patent before the other knew it.'[8] Dr. John Robison, friend of Watt, stated that Newcomen was acquainted with Robert Hooke of the Royal Society and obtained from him notes of Papin's experiments.[9] Stonier Parrott believed that Newcomen's engine was based on Papin's work.[10] Desaguliers commented that Newcomen and his associates not being 'either Philosophers to understand the Reasons, or Mathematicians enough to calculate the Powers and proportion of the parts' often discovered improvements by chance. Nevertheless Newcomen's engine was a great improvement on Papin's. His knowledge of the ironmongery trade and his association with 'admirable and ingenious workmen' near Birmingham undoubtedly contributed to his success in engineering.

Newcomen was unable to patent his engine because of the comprehensive coverage of Savery's patent in spite of the fact that his engine worked on a different principle and he made an agreement with Savery. This was not, however, the disadvantage that it at first appears for by making a deal with Savery, Newcomen was able to protect his engine for longer than if he had patented it himself, unless a Parliamentary extension could have been obtained. The wide distribution of the Newcomen engine in the early 18th century owed much to the partnership of two energetic colliery lessees Stonier Parrott and George Sparrow.[11] They came to an agreement with the proprietors of Savery's Patent, Savery having died in 1715, which gave them licence to erect engines in their own coal works. This venture being successful the partners acting sometimes together, at other times separately, obtained licences to erect engines in other parts of the Midlands, in Lancashire, Wales and Northumberland.

New materials were introduced in the construction of atmospheric engines. By the 1730s the Coalbrookdale Company was supplying cast iron pipes and cylinders which gradually replaced the brass ones formerly used. Stonier Parrott introduced iron boilers made of hammered plates like the Cheshire salt pans. The size of atmospheric engines increased and 70 in cylinders became quite common by the second half of the 18th century.

John Smeaton, improver of the water wheel, was involved in the installation of a number of pumping engines.[12] Being dissatisfied with the Newcomen engine he made an experimental model which broke away from the standard design in several fundamental ways. When he put the ideas into practice at New River Head he found that they did not succeed but he began to collect data on the large engines then at work in the country and then conducted a further set of experiments at his home in Leeds. In 1772 he compiled a table of the proportions of engines and their parts and using this data he designed and superintended the construction of a number of engines. By his improvements he

Iron and Engineering

Comparatively little machinery was to be found at an 18th century ironworks. The slitting mill was often separated from the forge because of the demands for water power; there were many of these mills around Birmingham and the Black Country to supply the nailers and chain makers of the district. The hammer was the important tool at the forge or smithery.

72

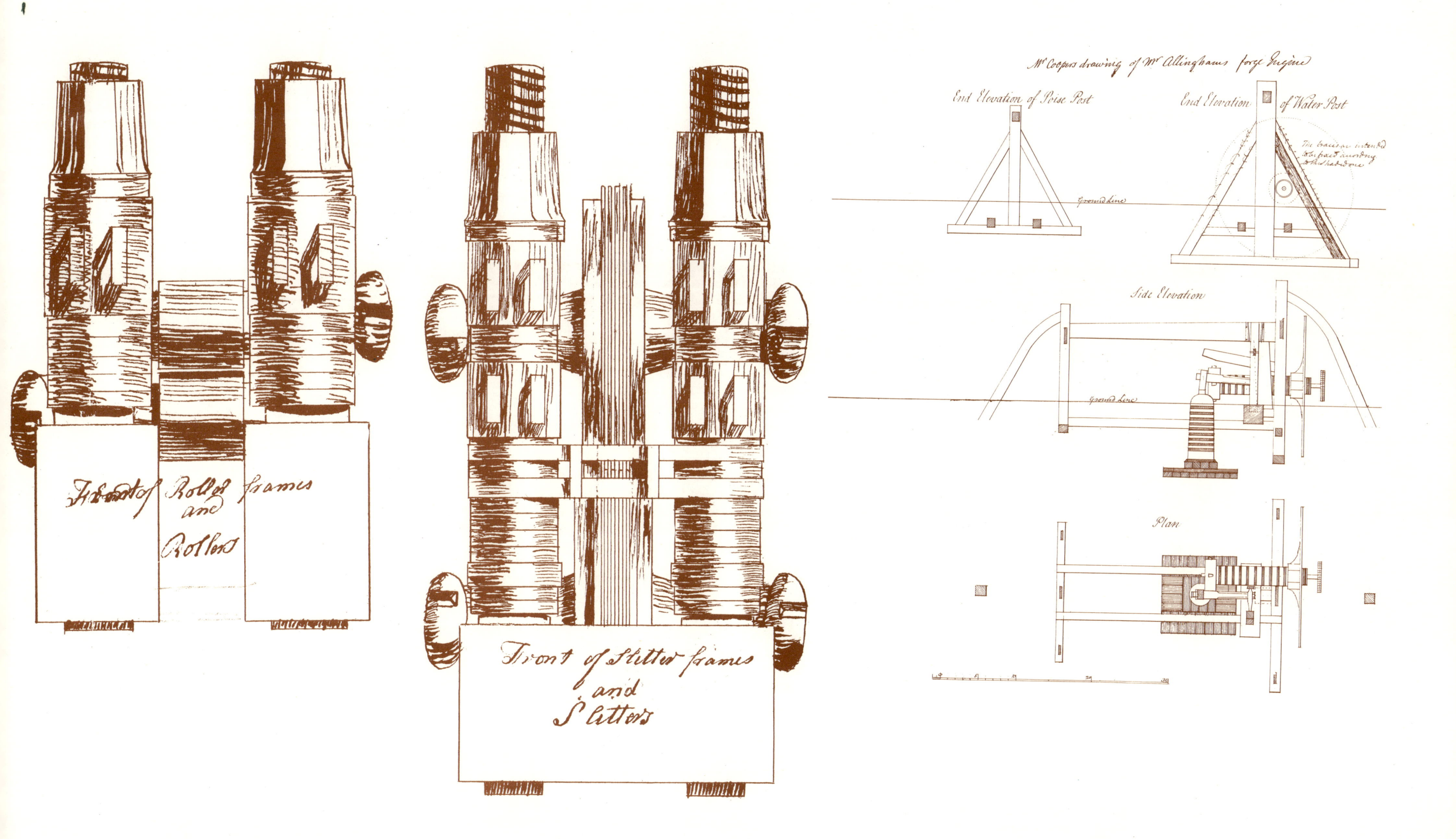

13. For a general account of Watt's inventions see Eric Robinson and A. E. Musson, *James Watt and the Steam Revolutions*; H. W. Dickinson, op cit; H. W. Dickinson and Rhys Jenkins, *James Watt and the Steam Engine*.
14. Eric Robinson and A. E. Musson op cit p.8.
15. J. Watt jun. to M. R. Boulton 12 June, 1802.
16. ibid.
17. J. Watt jun. to M. R. Boulton 17 June, 1802.
18. A. E. Musson and E. Robinson, 'The Early Growth of Steam Power', *Science and Technology in the Industrial Revolution*.
19. J. R. Harris, 'The Employment of Steam Power in the 18th Century', *History*, 52, 1967.
20. B. & W. Sandford to Boulton & Watt, 5 Feb., 1796.
21. A. E. Musson and E. Robinson op cit p.400.
22. M. R. Boulton to J. Southern 19 May, 1796.
23. J. Kennedy, 'A Brief Memoir of Samuel Crompton', *Misc. Papers*, 1849, pp 71-2.
24. J. Watt to T. Jordain 26 Nov. 1784 Office Letter Book; A. E. Musson and E. Robinson, op cit p.402.
25. A. E. Musson and E. Robinson op cit p.403, quoting J. Watt jun. to J. Watt 13 March, 1791.
26. P. Vaughan to Boulton & Watt 18 May, 1798.
27. S. Unwin to Boulton & Watt 30 Jan. 1798.
28. Brotherton Library, Leeds, Marshall MSS 57, p.2.

doubled the performance of the atmospheric engine.

While Smeaton was improving the atmospheric engine James Watt was in the process of making the most important advance in steam engine design of the 18th century. Realising that the alternative heating and cooling of the cylinder resulted in the loss of a great deal of heat and a wastage of fuel Watt constructed a separate vessel in which the steam was condensed. He made several models and was helped in his experiments by a loan from Dr. Joseph Black of Glasgow University. In 1769 Watt patented the separate condenser and five years later he moved to Birmingham to join Matthew Boulton. In the following year Parliament extended Watt's patent for twenty five years (to 1800) and Boulton and Watt's partnership commenced in the same year. Watt's first engine was a single acting pumping engine but the idea of a rotative engine was considered as early as 1774, although it was not developed until 1781. By that time other engineers had been considering the question of rotary power. In 1779 Matthew Wasborough of Bristol developed rotary power from a common engine using a ratchet and pawl mechanism and a year later James Pickard of Birmingham replaced the ratchet and pawl by a crank and connecting rod. Pickard's patent prevented Watt from using the crank until 1794 and he devised a substitute which he patented in 1781 — the sun and planet gear. In 1782 the double acting engine was patented and two years later Watt's parallel motion overcame the problem of an inflexible connection of the piston rod to the beam. The centrifugal governor which ensured steady motion even when the load on the engine varied was introduced in 1787 by which time Watt's rotative engine was standardised.[13]

Wasborough and Pickard were by no means the only people apart from Watt to patent steam engines in the 18th century. In 1781 Hornblower patented his engine. This was followed by Cameron's in 1784, Symington's in 1787, Heslop's in 1790 and Francis Thompson's in 1792. There were others besides and Boulton and Watt listed thirty four patents other than Watt's between 1698 and 1793. Many of these men had access to Watt's inventions either directly by spying or by enticing his employees away or indirectly through Wilkinson. A few of these engines, notably those by Francis Thompson, were adopted by the early factory masters but on the whole their practical application was limited.[14]

A far more widespread rival to the Boulton and Watt engine in the factory was the modified Savery or Newcomen engine. The Savery engine or sometimes the Newcomen engine could be used to raise water for a waterwheel; alternatively the Newcomen engine could also be used with a connecting rod, crank and flywheel to provide rotary power. These engines were made by Joshua Wrigley, M. Murray, J. Wilkinson, J. Young, J. Bateman and W. Sherratt, Ebenezer Smith, Alexander Brodie, and Robert Lindsay and Co. Some of these engineers, Wilkinson, Murray, and Sherratt and Bateman for example, in addition to making atmospheric engines produced 'pirate' engines embodying some of Watt's patent devices.

In 1799 Matthew Murray, described by Watt jun. as 'a very able mechanic',[15] took out his first steam engine patent. This was followed by others in 1801 and 1802 and Murray became a formidable rival to Boulton and Watt in the early 19th century: 'I have been surveying the environs of this rival establishment and making enquiries respecting the property and tenure of the neighbouring lands with a view to seeing whether we could purchase any thing under their very nose that might materially annoy them and eventually benefit ourselves.'[16] So wrote Watt jun. in 1802 and his fears were partly justified for Murray was acknowledged to use some superior techniques in casting engine parts.[17]

The high pressure steam engine was developed simultaneously in England by Trevithick and in America by Oliver Evans. Trevithick's patent of 1802 covered steam engines and a steam carriage and in the following year Woolf patented his high pressure compound engine. The stage was set for the development of the high pressure steam engines of the mid 19th century multi-storey factories.

Innovation

Dickinson assumed that the Savery engine went out of use early in the 18th century but Musson and Robinson[18] and J. R. Harris[19] have shown that it was improved and was widely used later in the 18th century for pumping water onto waterwheels. Joshua Wrigley and Joseph Young, both of Manchester, were well known manufacturers of Savery engines. Although more is known about their activities in Lancashire than elsewhere Wrigley is known to have built engines for factories in Yorkshire and London as well. B. & W. Sandford of Manchester had a Wrigley engine.[20] In 1784 two of his Savery engines were erected at Thackeray and Whitehead's cotton factory at Garrat, Chorlton-on-Medlock.[21] Nightingale, Harris and Co had 'A Savery [engine] which now turns their works'[22] in 1796 and John Kennedy said that when he improved the mule for fine spinning in *c.* 1793 he used a Savery engine and waterwheel.[23] Watt corresponded with Thoman Jordain of Manchester about the erection of a Boulton and Watt engine but Wrigley obtained the contract instead.[24] Watt jun. noted wryly that Wrigley had 'orders for 13 Engines for this town [Manchester] and neighbourhood all of them intended for working Cotton Machinery of one kind or other by the Medium of a Water Wheel.'[25]

The desire for a small engine to raise water for a wheel was by no means confined to Lancashire, Philip Vaughan of Carmarthen[26] said that he needed an engine 'just big enough to raise water to work a forge hammer to Bloom and draw Bar iron at a place where it will be in use only in dry seasons' in conjunction with a 18 ft overshot water wheel. Samuel Unwin[27] of Sutton in Ashfield, Notts. had a steam engine for returning water to a wheel and Marshall of Leeds noted that 'J. Wrigley says there is nothing gained by a crank instead of a water wheel because of the great weight they are obliged to use at the beam end.'[28]

Knight's slitting mill, Whittington, n.d.
This is one of the earliest factory drawings in the Boulton and Watt Collection. Although undated it was probably made in the period 1770-1780, when Matthew Boulton was collecting data on local ironworks.
Portf: Misc. Mills

House for packing the Slitted Iron.

House for holding the Shears & barrs of Iron

Part of Mill Lead

Crane

Crane

Slitters

Rollers

Furnaces for heating the Iron that is to be Slitted

Plan of Mr Knights Slitting Mill at Whittington

Crane

Part of Mill Pool

Smiths Forge

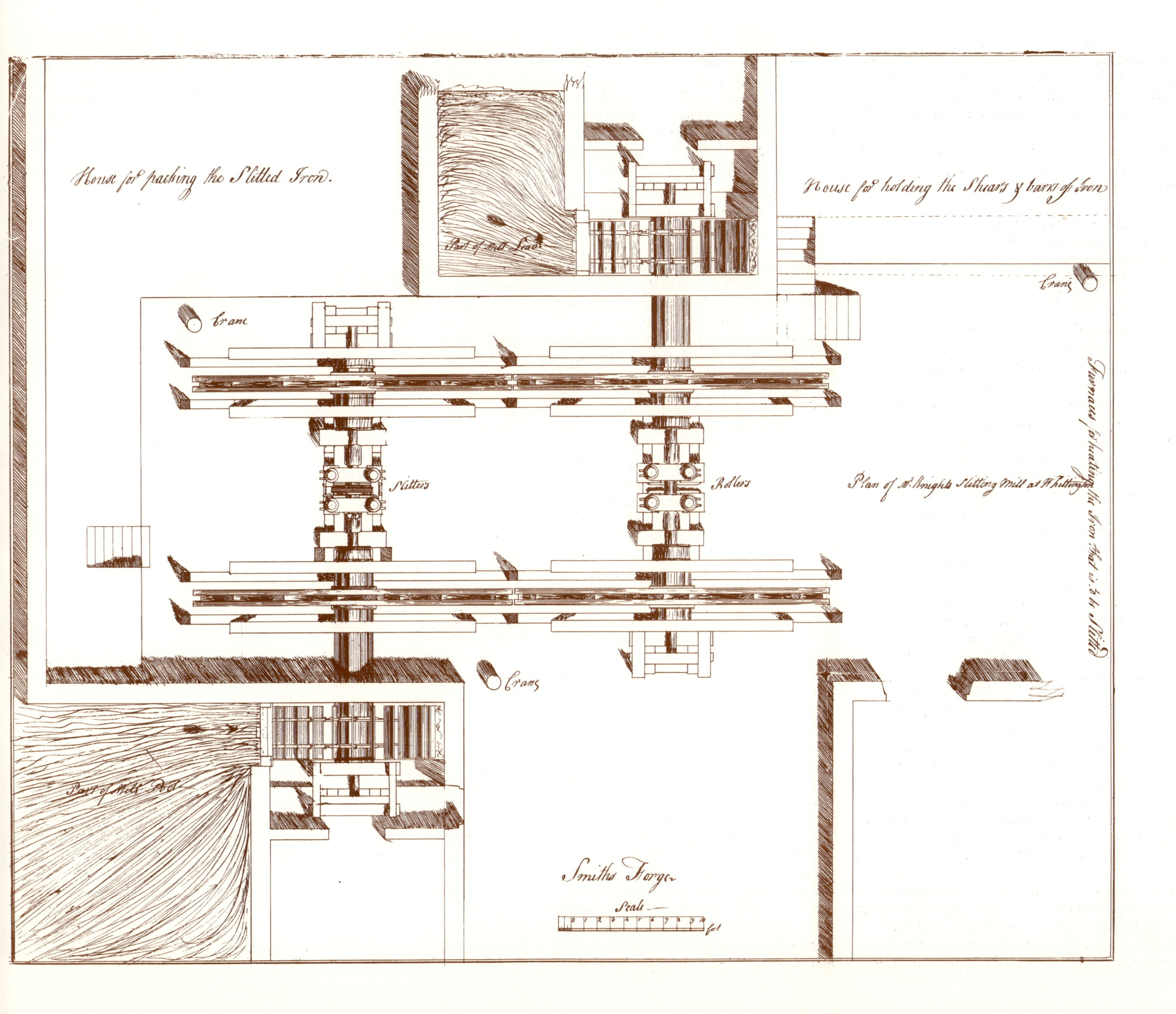

Scale

29. Engine Book.
30. J. Farey, *A Treatise on the Steam Engine*, 1827, p.122-5.
31. A. E. Musson introduction to H. W. Dickinson op cit.
32. R. A. Pelham, 'Corn Milling and the Industrial Revolution in England in the 18th Century,' *University of Birmingham Historical Journal*, 6, 1957-8 p.166.
33. Marshall MSS 57 p.1. He had two 30 in. pumps and a 30 ft. water wheel.
34. J. Bateman to J. Scale 6 July, 1783.
35. (J. Ogden) *A Description of Manchester . . . 1783* p.16, quoted A. E. Musson and E. Robinson p.395. This is also suggested in a letter from Wilkes.
36. P. Drinkwater to Boulton & Watt 3 April, 1789.
37. J. Farey op cit. p.422
38. Ibid.
39. There were 38 in Lancashire before 1800, A. E. Musson and E. Robinson op cit p.423.
40. Guildhall Library, Sun Insurance Registers 7/636089; A. E. Musson and E. Robinson op cit. p.417.
41. W. Joulet to Boulton & Watt, 27 Oct, 1797.
42. A. E. Musson and E. Robinson op cit p.410.
43. W. H. Chaloner, 'The Stockdale Family, the Wilkinson Brothers and the Cotton Mills at Cark in Cartmel'; *Trans Cumberland and Westmorland Antiquarian and Archaeological Soc.* LXIV, 1964, p.364.
44. A. E. Musson and E. Robinson op cit.
45. J. Southern to J. Champion 10 Dec. 1793 Foundry letter book.
46. B. and W. Sandford to Boulton & Watt, 5 Feb. 1796.
47. *A Catalogue of the Civil and Mechanical Engineering Designs of John Smeaton*, p.85.
48. J. Southern to J. Watt 22 April, 1799. Foundry Letter Book.
49. J. Watt jun. to Gregory Watt n.d. 1803
50. P. Ewart to Boulton & Watt, 25 Sept. 1792
51. J. Cartwright to Boulton & Watt 15 Aug. 1788
52. Marshall MSS 57 p.1.
53. J. Cooksen to Boulton and Watt 7 Feb. 1792
54. Marshall MSS 57 p.1.
55. Leeds City Archives DB 233. I am indebted to Professor M. W. Beresford for this reference

Marshall installed a Wrigley engine in conjunction with a water wheel but appears to have been dissatisfied with it for within a year he had replaced it by a Boulton and Watt engine.[29] Farey describes a Wrigley engine which was used to turn lathes at Peter Kier's factory in London.[30]

The Newcomen engine was used for mine pumping during much of the 18th century but it was also used to return water to water wheels from quite early in the century. One was in use at Coalbrookdale in 1742 and there was one at Carron in the 1760s.[31] Smeaton approved of this method of obtaining rotary power and he designed several such engines.[32] But towards the end of the 18th century various attempts were made to obtain rotary power direct from the Newcomen engine. One of the earliest steam engines to be installed at a cotton factory was one at Arkwright, Simpson and Whitenbury's Shudehill mill in Manchester. The engine was used to pump water onto a water wheel latterly[33] but a letter from Bateman in 1783 hints that an unsuccessful attempt may have been made to obtain rotary power direct:[34] 'Mr. A.'s works to go by fire engine are all to pieces.' Ogden[35] also gives the impression that the machinery was being driven direct by the steam engine for he reported that the carding and spinning machines were 'setting to work by a steam engine'. From 1780 onwards a considerable number of common engines were erected in factories. Drinkwater said in 1789[36] that there were 'a great number of the common old smoking engines in and about the town' and Farey said that 'great numbers of atmospheric engines were also made for turning mills, particularly in the districts where coals were cheap.'[37] The atmospheric engine in order to compete with Watt's engine had to be made double acting. Farey noted that 'The principal makers of these engines were Messrs. Bateman and Sherratt of Manchester . . . and . . . Mr. Francis Thompson of Ashover in Derbyshire who made engines for that district and for Sheffield and Leeds.'[38] Musson and Robinson have shown that Bateman and Sherratt made a considerable number of the engines in use in the Manchester district in the 1780s and 1790s.[39] Bateman was given details of the Boulton and Watt engine by Boulton after he had appeared to be interested in buying one and he also employed ex-Soho men as did Matthew Murray of Leeds. Bateman and Sherratt built three engines for J. and S. Horrocks, cotton spinners of Preston,[40] and William Joulet and Son[41] of Salford had a 6 hp Bateman and Sherratt engine in 1797. A Bateman engine was erected at Garrat by Thackery and Whitehead to replace an earlier Wrigley[42] engine and one was installed at Cark, replacing an engine by Rowe who had formerly built one of his own invention at Bersham.[43]

Some of the Bateman and Sherratt engines were improved common engines but others infringed Watt's patent rights by having air pumps and condensers. In view of this Boulton and Watt turned detectives to track down the infringers with Lawson, one of their employees dressed, somewhat inappropriately, as a jockey.[44] He travelled all over Lancashire inspecting suspect engines and sent reports of the infringements back to Boulton and Watt. Occasionally Boulton and Watt allowed an air pump and condenser to be applied to an existing common engine on payment of a premium, provided that the work was done under Boulton and Watt's direction. John Champion of Bristol, a corn miller, was permitted to do this.[45]

Smith and Co of Chesterfield, who made parts for Francis Thompson's engines, installed an engine for Nash and Co and one for Houldsworth of Manchester in 1796.[46] Thompson also made engines for James Kennedy's cotton mill in Manchester, Daintry and Co's Macclesfield cotton mill and Mr. Patten's works at Cornbrook near Manchester 'to grind charcoal and allum and other works to make iron liquor' for calico printing. Thompson, as is to be expected, was active in the Midlands. He built an engine for Davison and Hawkesley, worsted spinners of Arnold near Nottingham and for John Bacon at Sutton in Ashfield, Nottinghamshire.[47] John Wilkes used a common engine to pump water back to his water wheel in his cotton factory at Measham, Leics.[48] but it is not known whether the engine was made locally. Dale Abbey and the Butterley Co also produced for the steam engine market but rather later than Smiths. Josiah Spode was investigating the possibilities of Trevithick's engines in 1803. The agent at Fenton Park Colliery, in which Spode was a partner, was a keen advocate of Trevithick's engines and he was sent by Spode to examine and report on the engines. Trevithick apparently bribed the agent with fifty guineas to persuade Spode to use his engines and Watt jun. said that Spode already had one Trevithick engine and was paying £4 to £5 per horse power premium.[49]

A number of Yorkshire textile factories had non-Boulton and Watt rotative and pumping engines before 1800. Many of these were made by Matthew Murray and by Bowling and Low Moor Ironworks and it is often difficult to discover whether they were Savery or atmospheric engines. Boulton and Watt were less active in tracking down infringers in Yorkshire but Murray was certainly infringing Watt's patents. Coupland and Wilkinson 'very sufficient people'[50] of Leeds turned the machinery in their cotton mill by a water wheel and two common pumping engines. Edmund Cartwright installed a 'steam engine of the old construction' at Doncaster in 1788.[51] Rogerson and Co of Hunslet and Holroyds of Shipscar had 32½in cylinder pumping engines,[52] Markland Cooksen and Fawcett of Leeds had 'an Engine of so inferior a construction'[53] that they bitterly regretted not having seen a Boulton and Watt one first. Claytons of Keighley had a 40 in cylinder engine and a 16 in cylinder pumping engine and Ard Walker of Leeds had a 34 in cylinder pumping engine.[54] Richard Paley of Leeds installed an atmospheric engine from Bowling Ironworks in his new factory in 1799.[55] When Benjamin Gott was planning Bean Ing he asked for estimates of the

Tredegar Ironworks, 1802; Wolford rolling mill, n.d.; Mr Ryelands slitting mill, n.d.
These three drawings illustrate the small amount of machinery required at a forge or rolling and slitting mill. Nevertheless a large amount of power was needed, hence the 52 ft diameter wheel at Tredegar.
Portf: 698; Misc. Mills

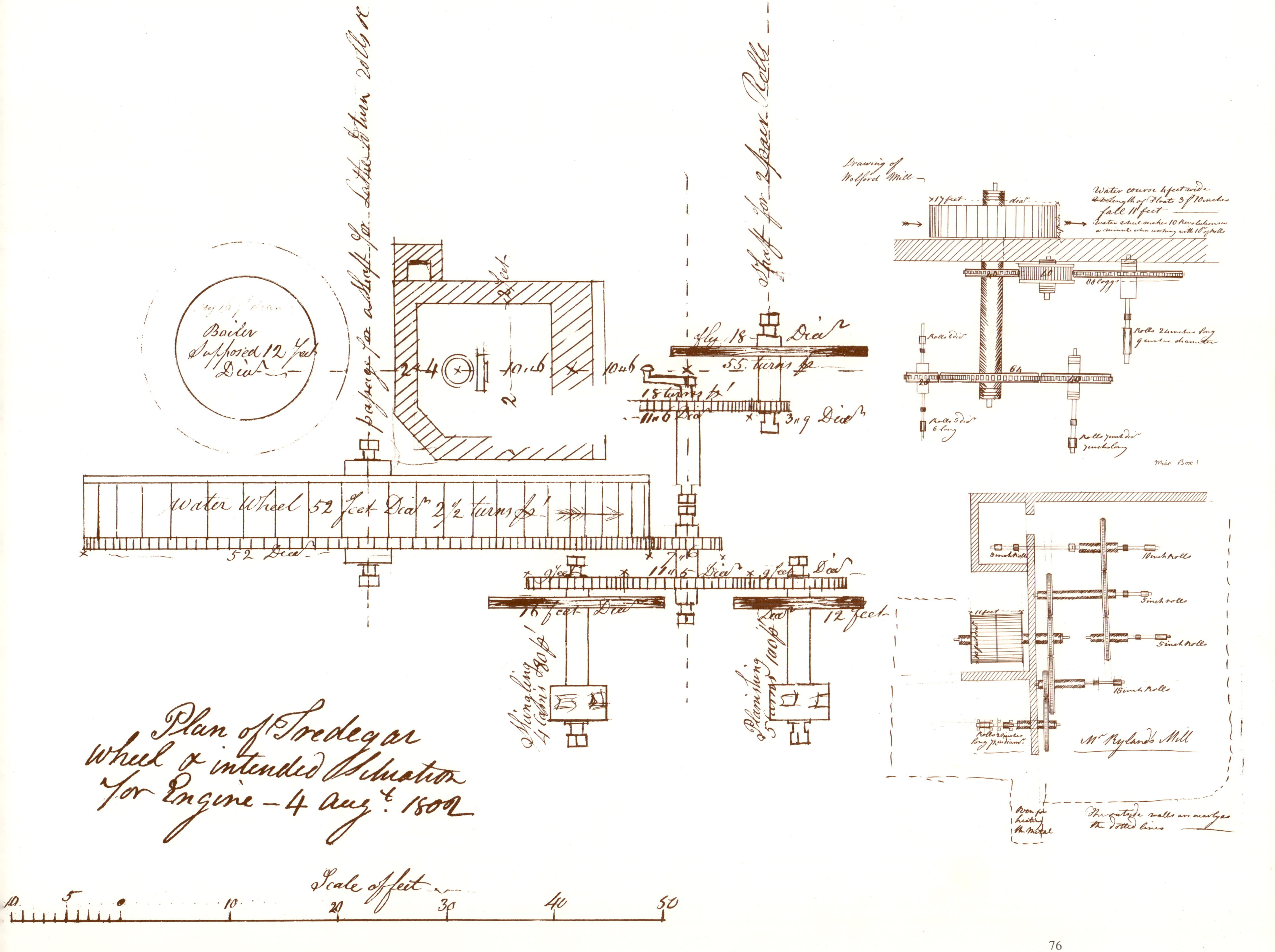

56. Brotherton Library Leeds, Gott MSS; J. Walker to B. Gott 14 May, 1792; Prices of Cast Iron Articles made at the Iron Works at Bowling, Low-Moor and Birkinshaw.
57. Cusworth Hall MSS, Sutcliffe's charges to B. Gott, 22 Dec. 1792
58. Ibid J, Sutcliffe to B. Gott, 10 June, 1792.
59. Ibid. B. Gott to J. Sutcliffe 11 June 1792.
60. Ibid. J. Sutcliffe to B. Gott, 13 Aug. 1792.
61. J. Farey, *Treatise on the Steam Engine* p.422
62. A. E. Musson and E. Robinson op cit p.421
63. R. Hare to Boulton & Watt, 11 Jan. 1785
64. J. Cooper to Boulton & Watt, 7 April 1787.
65. Peter Mathias, *The Brewing Industry in England 1700-1830*, pp. 95-6
66. Steam Engines at present used in Cotton Mills, Box 3.
67. J. Sword to Boulton & Watt, 24 Aug. 1798
68. Marshall MSS 62 p.37
69. Boulton & Watt to J. Rose 18 July, 1800, Foundry Letter Book
70. The author is at present working on this subject
71. J. Wilkes to Boulton & Watt 19 Oct. 1783
72. S. Unwin to Boulton & Watt 30 March, 1798
73. J. Lister to James Watt 14 March, 1785
74. T. Cooper to Boulton & Watt 10 Jan. 1790
75. J. Lister to James Watt 14 March, 1785
76. Hare to Boulton & Watt n.d. 1785 Box 4.
77. Engine Book.
78. J. Rennie to Matthew Boulton 5 Sept, 1785, A.O.
79. Peter Mathias op cit p.80.
80. Engine Book.

cost of engine parts from Joshua Walker and from Bowling and Low Moor Ironworks.[56] He also sent his millwright, John Sutcliffe, to London 'to examine sundry patent Steam Engines.'[57] Sutcliffe favoured Bowling Ironworks and went ahead with plans: 'I shall write to Mr. Sturgess to send his man over to Leeds on Wednesday next and then I will settle with him about the Engine and get it forward as fast as possible.'[58] Gott was not so sure however 'I have to request you will say nothing to Mr. Sturgess related to an engine being of opinion that one of Boulton Watts and Co. will in the end answer better.'[59] Sutcliffe took offence at this 'you... tould Mr. Marshall that I wanted to cram Mr. Sturgises Engine down your throat.'[60]

Farey said that 'some' Bateman and Sherratt engines were sent to London, one or two going to breweries.[61] Miles and Thomas Edwards, cotton spinners of Southwark had one in 1797.[62] Richard Hare, a London brewer confessed to Boulton and Watt in 1785 that he had spoken with Wood of Oxford who wanted to put up one of his patent engines in the brewery.[63] Webster, a London white lead manufacturer was said to be installing a Cameron engine in 1787.[64] Meux Reid's Griffin Brewery had a Hornblower and Maberley engine which was installed in 1797 by Arthur Woolf. Woolf remained as chief engineer at the brewery until 1806. He converted a small Fenton, Murray and Wood engine into a compound engine and it is possible that he erected the first of Trevithick's high pressure engines to be installed outside Cornwall.[65] It is not surprising that a greater variety of makes of engine should be found in London (Wrigley had also sold an engine there) than elsewhere but little detailed work has yet been done on this subject.

Savery and atmospheric engines were installed in a number of factories in Scotland. McTaggart drew up a list of the steam engines used in cotton mills in Glasgow and Paisley in c. 1797,[66] nine were mentioned of which only one was a Boulton and Watt engine. Horse power varied from the 3 hp engine of Scott Stevenson and Co at Glasgow to John Twigg's 24 hp engine at Paisley. John Pattison, a large firm, had an 8 hp engine but in 1799 this was replaced by a 32 hp Boulton and Watt engine. A Mr. Robertson was making an engine to work a boring and grinding mill for James Sword of Glasgow in 1798.[67] When Marshall of Leeds made his first Scottish tour in 1800 he noted 'the steam engines at Glasgow are in general very bad ones, only 2 or 3 of Boulton and Watts.'[68] A list of Lancashire, Yorkshire and Scottish Savery and atmospheric factory engines are in the appendix at the end of this chapter.

Boulton and Watt engines

Boulton and Watt did their best to dissuade prospective customers from buying pumping engines for water powered factories claiming that power was lost and that a rotary engine would be far more efficient.[69] On the whole Boulton and Watt lost these customers who seem to have preferred to buy a Savery or Newcomen pumping engine. Until an analysis is made of the people who enquired after a Boulton and Watt steam engine, distinguishing those who actually bought one it will not be possible to estimate the percentage of successful orders.[70] Wilkes of Measham, Leicester, enquired after an engine in 1783.[71] 'We find by our servant Tho. Jewsbury that you can erect an Ingin to turn any mill with a very small stream' but they did not order an engine for their cotton mill until some years later. Unwin[72] mulled over the matter for some time saying that he was 'at a loss to discover how any advantage could arise to us from one of your engines' and eventually retained his atmospheric engine. John Lister[73] enquired about the cost of a steam engine to raise water from a well to work a water wheel driving three pairs of stones in 1785. T. Cooper of Manchester[74] held out against a rotative engine and decided to use a pumping engine. 'The additional expense from the loss of force by the indirect application of it is no object of consequence compared with the convenience of working where we please by the means of water.'

It is clear that a great number of enquirers after a Boulton and Watt rotative engine did not follow up with an order. In many cases their power requirements were too small to justify expenditure on an engine. Mr. Benet[75] of Dewsbury enquired the cost of a rotative engine to work two or three frizing boards and a scribbling machine but he did not order an engine. Richard Hare[76], a London brewer, discovered that Clowes had calculated the advantage of employing steam power in a brewery and had found that with only nine horses employed there was no saving to be made by using steam.

Brewers were, nevertheless, amongst Boulton and Watt's earliest customers. H. Goodwyn's 4 hp engine was ordered in May 1784[77] and this was followed shortly afterwards by S. Whitbread's of 10 hp. Whitbread took a great delight in the business and much to Rennie's annoyance 'would not allow his engine to be started but in his presence'.[78] The large breweries were already highly mechanised and the machinery was well suited to being driven by a rotative engine. The engine was required to perform 'robust, simple operations', there was no need for meticulous precision.[79] There were few of the mechanical problems such as those that delayed the application of steam power to the textile industries. In many cases the steam engine was harnessed directly to the horse wheel involving, no doubt, a certain loss of power but at the same time cutting down on the costs of installation and avoiding the unknown problems of a direct drive to the machinery. The majority of brewery engines were installed in the large London breweries such as those of Goodwyn, Whitbread, Thrale, Gyfford, Charrington. Several engines were installed in the larger provincial breweries such as Castle and Ames of Bristol (1793), J. Taylor of Liverpool (1795) and Green of Nottingham (1793)[80] but the country breweries which were generally smaller than those in London

Soho Manufactory, Birmingham, 1788
Matthew Boulton's factory for toys and plated ware was begun in 1764 and finished in 1765 at a cost of £9000. The engraving shows the main part of the factory which surrounded the Great Court. Later developments show an intermingling of workshops and industrial dwellings built around smaller courts.

Plan of the Buildings belonging to the Works at the Soho (except the Mint & Building adjoining) standing on the Land now under Lease to Mr Bolton taken in June 1788.

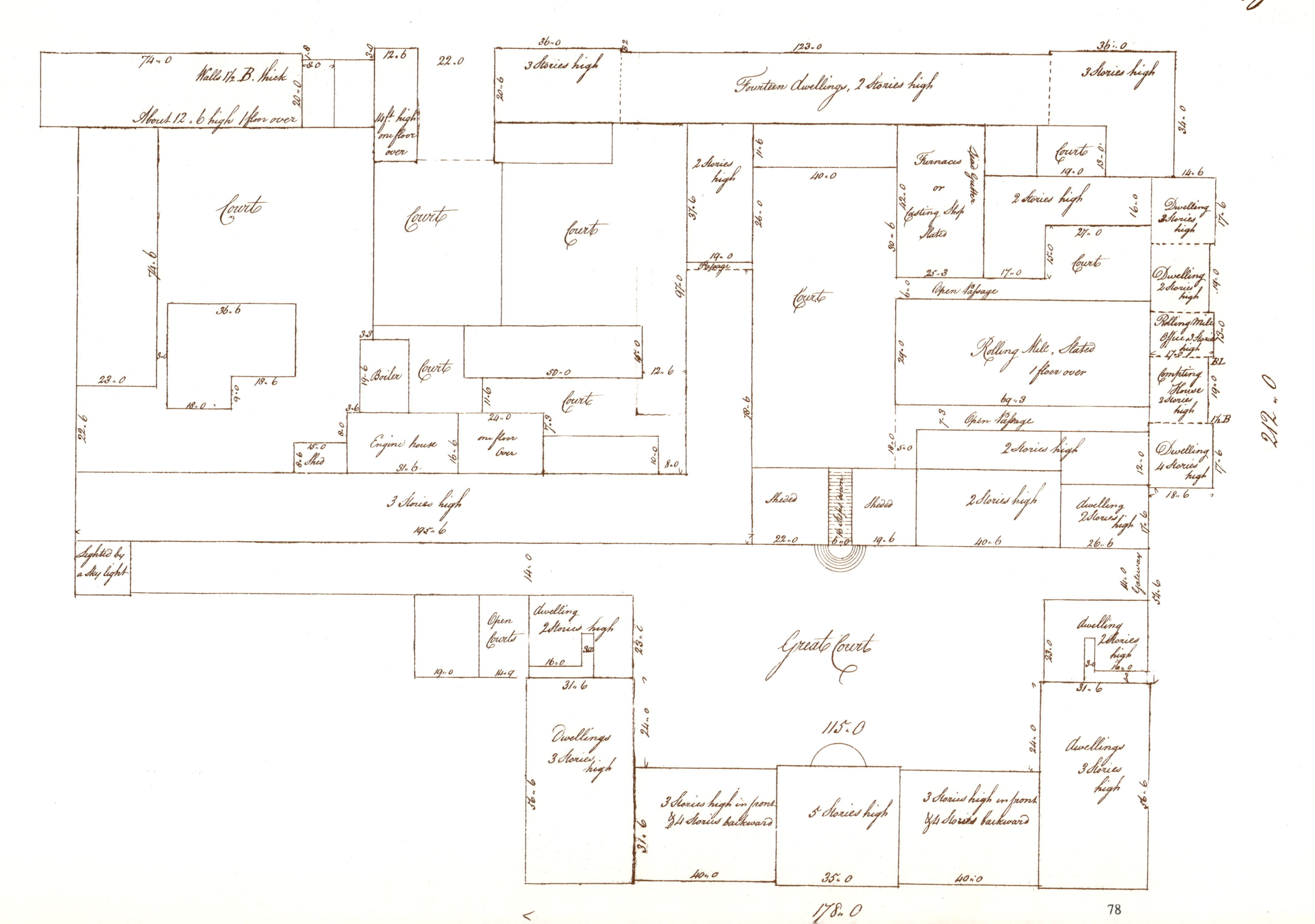

By J.A. Smith.

[81] R. A. Pelham op cit p.164-168.
[82] R. Davison to Boulton & Watt 17 Jan, 1798, 1 June, 1798.
[83] The phrase is Musson and Robinson's.

retained the horse wheel.

The other major user of Boulton and Watt rotative engines in the 18th and early 19th centuries was the textile industry. It is noticeable that many of the purchasers of the first Boulton and Watt engines for the textile industry had already been using some other form of power in their factories, either a horse wheel, water wheel or another kind of steam engine. Comparatively few of the purchasers of early rotative engines equipped their factories with Boulton and Watt rotative engines from the outset. The bulk of the sun and planet engines for the cotton industry went to Lancashire and Cheshire (sixty one in all), nineteen went to the Midlands and seven to Yorkshire and Durham. At this stage few engines were employed in the woollen and worsted industries — only ten altogether.

Boulton and Watt sun and planet engines

	Scotland	Lancs, Chesh.	Yorks, Durham	Midlands	London	W. England
Cotton	9	61	7	19	4	—
Wool/Worsted	—	2	6	3	—	1

Source: Engine Book

Most of the engines employed in the cotton industry were between 8 and 14 hp. Only two of the Scottish factories had sun and planet engines of more than 20 hp (David Dale at Catrine and J. Pattison at Glasgow), fourteen Lancashire factories had engines of 10 hp or less, twenty two had engines of 16 hp or over J. & S. Simpson of Manchester had a 40 hp engine and P. Atherton of Liverpool had one of 45 hp.

Horse power of sun and planet engines in the textile industries

hp	4–6	8–10	12–14	16–18	20–22	24–26
Cotton (No. of engines)	6	30	22	9	8	4
Wool/Worsted	—	—	2	1	6	—

	28–30	32–34	36–38	38–40	over 40
Cotton	12	5	2	1	1
Wool/Worsted	1	—	—	1	—

Source: Engine Book

In the wool textile industry there were no sun and planet engines of less than 12 hp but only two of more than 20 hp John Cartwright ordered a 30 hp engine for his Retford factory and Benjamin Gott ordered a 40 hp engine for Bean Ing, Leeds.

A third developing steam powered industry was corn milling. Several flour mills had been directly driven by steam by Wasborough/Pickard engines[81] before Boulton and Watt supplied any rotary steam engines to flour mills. Although the burning of Albion Mill was a severe loss to Boulton and Watt who had shares in the concern, the mill was at work long enough to impress a number of prospective customers. A direct result was the proposal to set up a factory on the same lines in France and scaled-down versions of the Albion Mill were built in Hull, Chester, Birmingham, Oxford and elsewhere. Flour milling was becoming a port industry.

Early Rotative steam engines in flour mills

Town	Owner	Maker	Date
Bristol	E. Young & Co.	W/P	1779-81
Bristol	?	W/P	?
Southampton	Potter	W/P	before 1783
Birmingham	Pickard	W/P	1784
London	Albion Mill Co. (1)	B/W	1784
Chester	Walkers and Ley	B/W	1785
Measham	Wilkes	B/W	1786
London	Albion Mill (2)	B/W	1787
Hull	Thompson & Baxter	B/W	1788
Banbury	J. Brockles	B/W	1788
—	Union Co.	B/W	1789
London	J. Dunkin	B/W	1789
Sunderland	Ransom & Ellerby	B/W	1790
Falkirk	Dyker Smith	B/W	1800
Glasgow	Corporation of Bakers	B/W	1800
Cork	I. Morgan	B/W	1800
Hereford	Subscription Flour Co.	B/W	1801 (Countermanded)
Snaith (Yorks)	H. Mitton	B/W	1802

Source: Engine Book; R. A. Pelham, op.cit.
W/P = Wasborough/Pickard
B/W = Boulton/Watt sun and planet engines

Boulton and Watt were slow to introduce the crank and flywheel after Pickard's patent had expired. The cost of the engine increased because the castings for a crank were heavier than the sun and planet gears and this may have deterred them in view of the keen competition in Lancashire. But in at least one case they insisted on using a sun and planet gear even when requested by the factory owner to use a crank.[82] Watt's stubborness is noticeable in connection with Murdock's inventions; the apparent unwillingness to use the crank may be another instance. Nevertheless once introduced the crank engine was bought in great numbers in Lancashire and Scotland, far exceeding the number of sun and planet engines bought there. Horsepower increased, Houldsworths of Glasgow ordered a 45 hp engine, and the demand rose for the more high-powered engines.

It is noticeable that in the second stage of the 'steam revolution'[83] steam power was adopted by a wide range of industries such as saw milling, bleaching and printing, paper making, distilling, minting, oil milling and iron manufacture besides textiles, brewing and corn milling. During the early 19th century Boulton and Watt began to make smaller engines of the side lever (boat) type and the independent type. These engines were commonly

Soho Foundry, Birmingham, c. 1796

When Boulton and Watt began to manufacture steam engi[nes] they made few of the parts themselves. This made it diffic[ult] to maintain a high quality of castings and delays were inev[i]table. In 1795 Boulton and Watt decided to build their ow[n] foundry. Stebbing Shaw described the layout: 'The place [of] this work being well digested and settled previous to layin[g the] first stone, the whole is thereby rendered more complete t[han] such works as generally arise gradually from disjointed ide[as.] Scale approx 1in : 36ft.

80

84. Engine Book. There were no 18th century steam engines in the Glos. or Wilts. textile areas. Steam power was still considered a supplement to water power in the 1830s.
85. W. Harper to Boulton & Watt 15 Nov, 1786.
86. Marshall Fenton and Co. to Boulton & Watt 25 Feb, 1789.
87. J. Wilkes to Boulton & Watt 19 Oct, 1783.
88. *Smeaton's Reports* 1797, II, pp. 378-9.
89. 'But the worst of all is we cannot raise steam sufficient to work the engine when there is 6 pr of stones on without the utmost exertion of the fire.' A. Mitchell to Boulton & Watt 30 April, 1786.
90. E. Aldred to Boulton & Watt 3 Nov, 1784.
91. P. Ewart to Boulton & Watt 7 Dec, 1791.
92. A. E. Musson and E. Robinson op cit p. 403.
93. P. Atherton to Boulton & Watt 17 April, 1791.
94. S. D. Chapman *The Early Factory Masters* p. 155.
95. Engine Book; R. Dayus to Boulton & Watt 21 Feb, 1786.
96. Marshall, Fenton and Co. to Boulton & Watt, 25 Feb, 1789.
97. Aitchison and Brown to Boulton & Watt 10 Dec, 1787.
98. Aitchison and Brown to Boulton & Watt 21 Dec, 1789.
99. Lawson to Boulton and Watt 1 Dec, 1802; J. Watt jun. to Gregory Watt 12 Oct, 1803.
100. Gorton and Son to Boulton & Watt, 10 April, 1788.

adopted by manufactures in areas where the water powered factory still predominated such as the West of England textile area.[84]

Technical problems

It has already been noticed that one of the factors operating in favour of the adoption of steam power by the brewing industry was the simplicity and stoutness of the machinery that the engine had to drive. The problems in the textile industry were infinitely more complex. It was essential that the motive power for spinning cotton, worsted (and later woollen) and silk yarn should be smooth running. William Harper of Macclesfield, owner of a silk mill told Boulton and Watt that 'a regular motion is necessary'.[85] It was also a problem raised by Marshall,[86] the Leeds flax spinner. Wilkes[87] of Measham asked 'We beg to be informed if you are certain [the engine] will work as steadily as water as I find Mr. Arkwright was obliged to alter one he erected at Manchester to a water wheel. Our friend Mr. Peal of Burton is in some doubts about an Ingin being Smooth enough!' Their fears were partly justified, for until the governor was applied to the engine rotary motion was irregular. For this reason many spinners preferred to waste potential power by using a pumping engine and to ensure regular motion by using a water wheel. Entrepreneurs who adopted rotary power at an early date generally used it for the preparatory textile processes and not for spinning.

A regular motion was also necessary in flour milling. Smeaton was responsible for the Navy Board countermanding an order for a Wasborough/Pickard engine in 1781[88] 'I apprehend that no motion communicated from the reciprocating beam of a fire engine can ever act perfectly steady and equal in producing a circular motion, like the regular efflux of water in turning a water-wheel. . . . In the raising of water . . . the stoppage of the engine for a few strokes is of no other consequence than the loss of so much time, but in the motion of mill-stones grinding corn, such stoppages would have a particular ill effect.' Boulton and Watt and Rennie had many teething troubles to overcome in the Albion Mill before the steam engine worked the machinery efficiently.[89]

The small pumping engines generally worked satisfactorily but the larger ones tended to be less efficient, as E. Aldred found: 'The engine I have erected for returning water upon my wheels does not answer so well as I expected.'[90] Salvin's Savery engine 'was to have been equal to the power of about twenty horses but upon trial it is found to fall short half of that power.'[91]

Wrigley found difficulty over the amount of smoke his engines produced and could only get round the problem by building high chimneys.[92] It was probably to these engines that Peter Atherton[93] was referring when he told Boulton and Watt of a proposal to apply to Parliament for an Act obliging all proprietors of steam engines in Liverpool to erect large high chimneys. He added 'it would be very disagreeable to me to be obliged to Build a chimney as high as some which are erected to some Common steam engines which appear to be as high as spire steeples of churches.' S. D. Chapman[94] has suggested that it was opposition to the quantity of smoke produced by Newcomen engines that led to the large number of Watt engines erected in Nottingham.

The first cotton mill engine by Boulton and Watt was erected for Robinsons at Papplewick, Notts. Richard Dayus[95] the fitter reported that 'we find no ill conveniency at all attend it, any more than the fly wheel which is very liable to go the contrary way, in setting on and stopping the engine which is very hurtful to the machinery in the mill.' Although the fitter was calm enough, the damage done to machinery by the fly wheel going in the reverse direction would have been considerable. Marshall[96] of Leeds was also worried on this point and asked 'whether in your engines the crank is ever liable to turn the wrong way round as we understand it is in engines of the common construction.'

A recurring problem was the fact that few entrepreneurs were able to do anything to remedy the situation if their engine was giving trouble. They lacked the necessary technical knowledge and skilled engine men were extremely hard to come by. Many letters were written to Boulton and Watt by desperate factory owners asking to borrow an engine man until someone at the factory had learned how to manage the engine. Boulton and Watt did not have men to spare; they had enough problems getting skilled fitters for themselves, and could only allow their men to repair engines in cases where there had been faulty workmanship. Boulton and Watt tried to overcome this problem by sending written instructions for repairs but the lack of technical know-how at the factory meant that it was difficult for them to assess what was wrong. Aitcheson and Brown[97] reported sadly 'The engine erected here does not work with that sweetness that others of the same construction does that we have seen . . . she always makes a great clank which puts us in fears that some part of the machine is not properly put together.' In another letter they apologised for being 'under the necessity of troubling you but none of us has understanding the engine so well as we could wish.'[98] Fortunately the problems were often comparatively minor ones and only small adjustments were necessary, a high consumption of fuel, for instance, was often due to poor packing in the piston. When an engine man could be spared he sometimes found that the trouble was caused by overloading. Lawson found Robert Peel of Burton's engine 'very dirty and much loaded' and Strutt's engine was 'much loaded'. Spode's engine man attempted to increase the power of his engine with disastrous consequences.[99]

The sun and planet wheels were troublesome. Gorton and Thompson[100] found that their sun wheel broke within a month of the engine being erected. Aitcheson and Brown's first pair of gear wheels broke and then the second pair broke within ten

Gardner and Manser, King and Queen Foundry, Rotherhithe, 1790

Some important experiments in rolling cast steel were carried out at this works: 'We wrote to our friends in Sheffield to send a ton of cast steel to your works to be tried in rolling and to try the powers of your mill in this respect; but as they will wish to be present at the experiment we shall be very much obliged to you to inform us as soon as the mill is at work.' (James Watt to Robert Gardner, 16 May 1792, Foundry Letter Book.)
Portf: 58

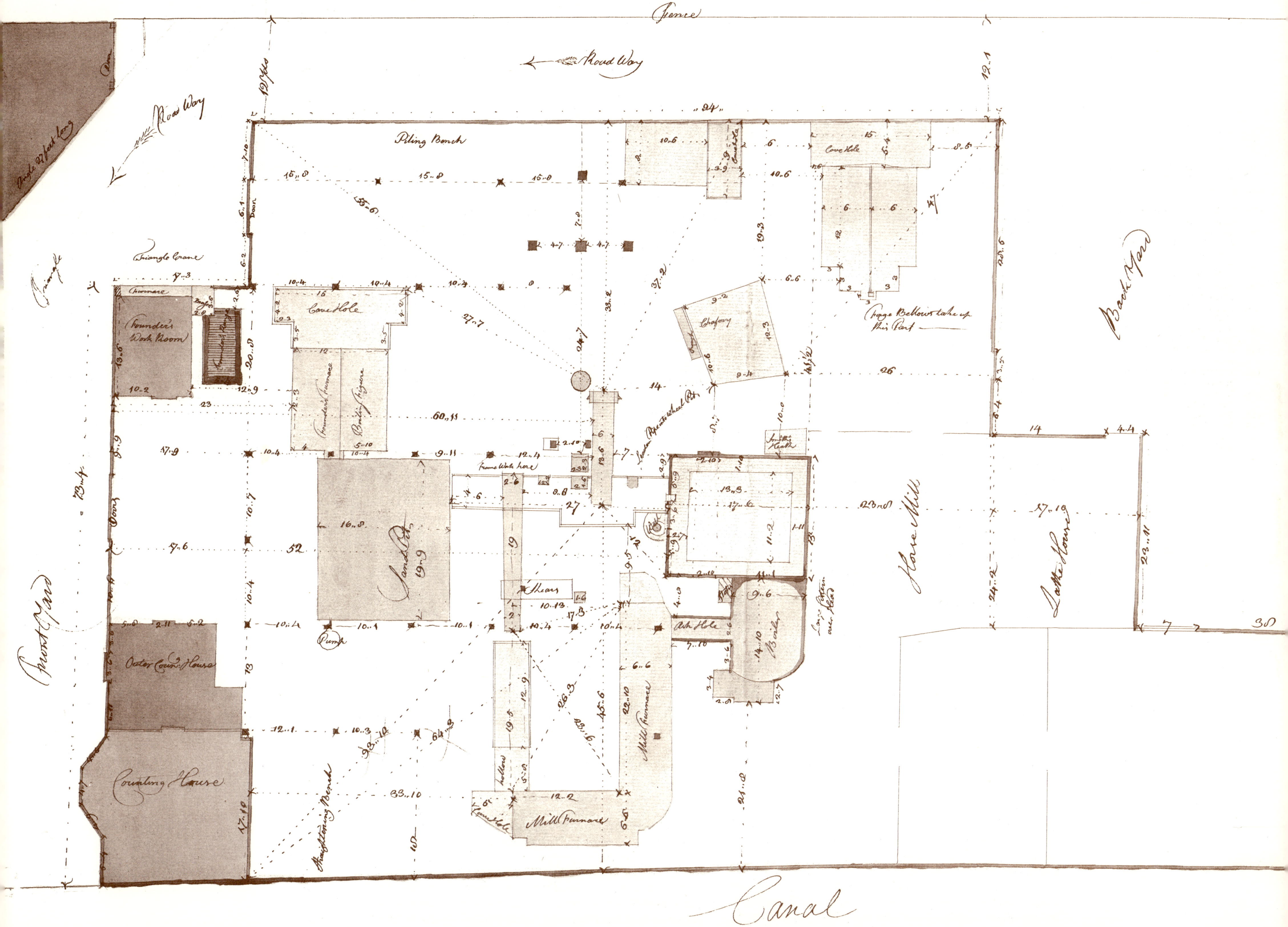
Fence
Road Way
Road Way
Piling Bench
Cove Hole
Triangle Crane
Founder's Work Room
Cove Hole
Chafery
Forge Bellows take up this Part
Back Yard
Frame Work here
Sand Pit
Shears
Pump
Outer Counting House
Counting House
Mill Furnace
Mill Furnace
Ash Hole
Boiler
House Mill
Lathe House
Front Yard
Straightening Bench
Smiths Hearth
Canal

days of being fitted.[101] The bearings on Gott's engine had to be replaced[102] and the bottom of W. Osborne's cylinder blew out twice.[103] Boilers caused problems. P. Atherton of Liverpool and Budd and Goadsby complained of the holes in theirs and Boulton and Watt had to request Wilkinson to test boilers for holes by filling them with water before sending them out to customers.[104] Strutts had problems with their boiler:[105] 'One of our boilers appearing to lose steam at the top we took off the covering which consisted of three layers — horse dung, clay and bricks — and found the outside covered with scale or calcined iron 1/8 of an inch thick, which being beaten off left the boiler with large holes in various places.'

At the root of many of these technical problems was the variable quality of the castings used in steam engines. Few foundries in the country were capable of making engine parts to the required degree of precision. Had Boulton and Watt been in a position to make all their engine parts from the beginning this particular difficulty might have been avoided. 'We have hitherto found every instruction we can give insufficient to make these people accurate and notwithstanding we have given them drawings and dimensions to work by they will make such deviations from them as to give us the justest cause of complaint.'[106] When faults were found the factory master was not always clear who he should complain to and Boulton and Watt sometimes tried to avoid their responsibility in this matter by telling the owner of the engine to correspond direct with the manufacturer of the part in question. If the maker was Wilkinson this could be difficult. Liptrap was under no illusions as to who was ultimately responsible.[107] 'I am exceedingly sorry to send you for the Birmingham coach so bad a sample of your own work . . . I would not have troubled you with these iron barrs if I thought a simple narration of the fact *could* have been believed; but to the eternal disgrace of the manufacturer such is the heart of our gudgeon, and as such I send it to you that you may proceed against him as you think proper, but in my opinion speaking at large pro bono publico such a rascal should not go unpunish'd.'

Management problems

A number of entrepreneurs under estimated the power that would be required to drive their machinery. If Boulton and Watt were provided with enough data on the previous power system such as the number of horses and diameter of the horse wheel or the dimensions of the water wheel and fall of water they could suggest an engine of equivalent power. If the steam engine was the first power system to be installed in the factory there were likely to be complications. The experience of Arkwright, the Robinsons, Atherton and others soon showed the approximate number of spindles that could be turned by a given horse power. Boulton and Watt's own experience at Albion mill enabled them to help others building steam corn mills. But in industries of which Boulton and Watt had no direct knowledge they were not in a position to make an accurate estimate. If the power requirements were underestimated the entrepreneur might be unable to utilise all his means of production. H. Coates and Co,[108] oil millers of Hull, reported 'We found her not capable of turning all our works together which caused us to make several alterations, one in particular of removing the pair of stampers which may serve our purpose for the present but we have not yet made a tryal. . . . The greatest inconvenience to us is that the little mill stones cannot work at the same time with the other by which we are prevented crushing both linseed and rapeseed together which we frequently want to do.'

Once an order for an engine had been placed with Boulton and Watt the entrepreneur could wait up to ten months for delivery. Some people were not aware of this and only ordered their engine when the factory was nearly complete. Charles Lees of Stockport, having ordered an engine in July wrote anxiously in November[109] enquiring when it would be ready as 'our tenants also are very tedious about its not coming being greatly inconvenienced by having their machinery in different places.' The fitter arrived to put the engine up at the end of December. James Meredith[110] of Manchester countermanded his order because his tenant was not prepared to wait several months. Delivery dates shortened once Soho Foundry was in production and four months became usual by the end of the 18th century. If a manufacturer required a steam engine urgently and was not prepared to wait for a new one, Boulton and Watt could occasionally assist him to find a second-hand one. Markland Cooksen and Fawcett[111] of Leeds had completed their factory and millwork and were 'pritty forward with an engine of so inferior a construction we have only to lament we did not see the work at Cupers Bridge sooner. We have built a very large and expensive water wheel with every other requisite to abandon which would be attended with a very heavy loss.' Cooksen added that if Boulton and Watt could supply an engine within three months they would be prepared to bear the loss of abandoning their other engine and wheel. Boulton and Watt could not supply them with a new engine in the time but negotiated with another firm[112] who were selling theirs and Markland, Cooksen and Fawcett bought a second-hand Boulton and Watt engine.

Once the engine arrived there could be a delay of up to six weeks before it was ready to work. During this period the fitter was paid by the purchaser of the engine and this could add up to £60 to the bill. In the brewing industry the installation of the engine had to be fitted into the brief period between the end of one brewing season in June and the beginning of the next in August. Goodwyn's engine parts arrived at the end of July and by 9th August the two fitters from Soho had got the engine going. Barclays were less fortunate and their engine did not begin work until November; amongst their many problems Boulton and Watt took the fitter away for another job before he had completed the erection of Barclay's engine.[113] Aitcheson and Brown[114]

---

101. Aitcheson and Brown to Boulton & Watt 21 Feb, 1789.
102. Southern to B. Gott 6 Aug, 1795 Foundry Letter Book.
103. W. Osborne to Boulton & Watt 6 July, 1784.
104. Boulton & Watt to Bersham Iron Wks. 22 June, 1792 Foundry Letter Book.
105. W.G. and J. Strutt to Boulton & Watt 23 March, 1797.
106. Boulton & Watt to R. Emerson 7 Sept, 1795 Office Letter Book.
107. Liptrap to Boulton & Watt 30 Jan, 1787.
108. H. Coates and Co. to Boulton & Watt 5 May, 1785.
109. C. Lees to Boulton & Watt 27 Nov, 1795.
110. J. Meredith to Boulton & Watt 20 Oct, 1796.
111. J. Cooksen to Boulton & Watt 7 Feb, 1792.
112. Salvin Bros.
113. Peter Mathias op cit pp. 89-91.
114. Aitcheson and Brown to Boulton & Watt 21 Feb, 1789.

Woolwich Smithery, Commissioners of the Navy, 1814, 1833
This plan is identical to a rough drawing dated 1814 and was copied by Boulton and Watt from Charles Dupin's *Voyages dans la Grande Bretagne*. Two steam engines were ordered for the works; one was to operate two forge hammers for anchors, a drilling and boring machine, a lathe and to give occasional blast to fires. The other was for blowing about forty two fires. (Holl to Boulton and Watt 28 April 1814, A.O., Rennie Box.) Portf: 452

84

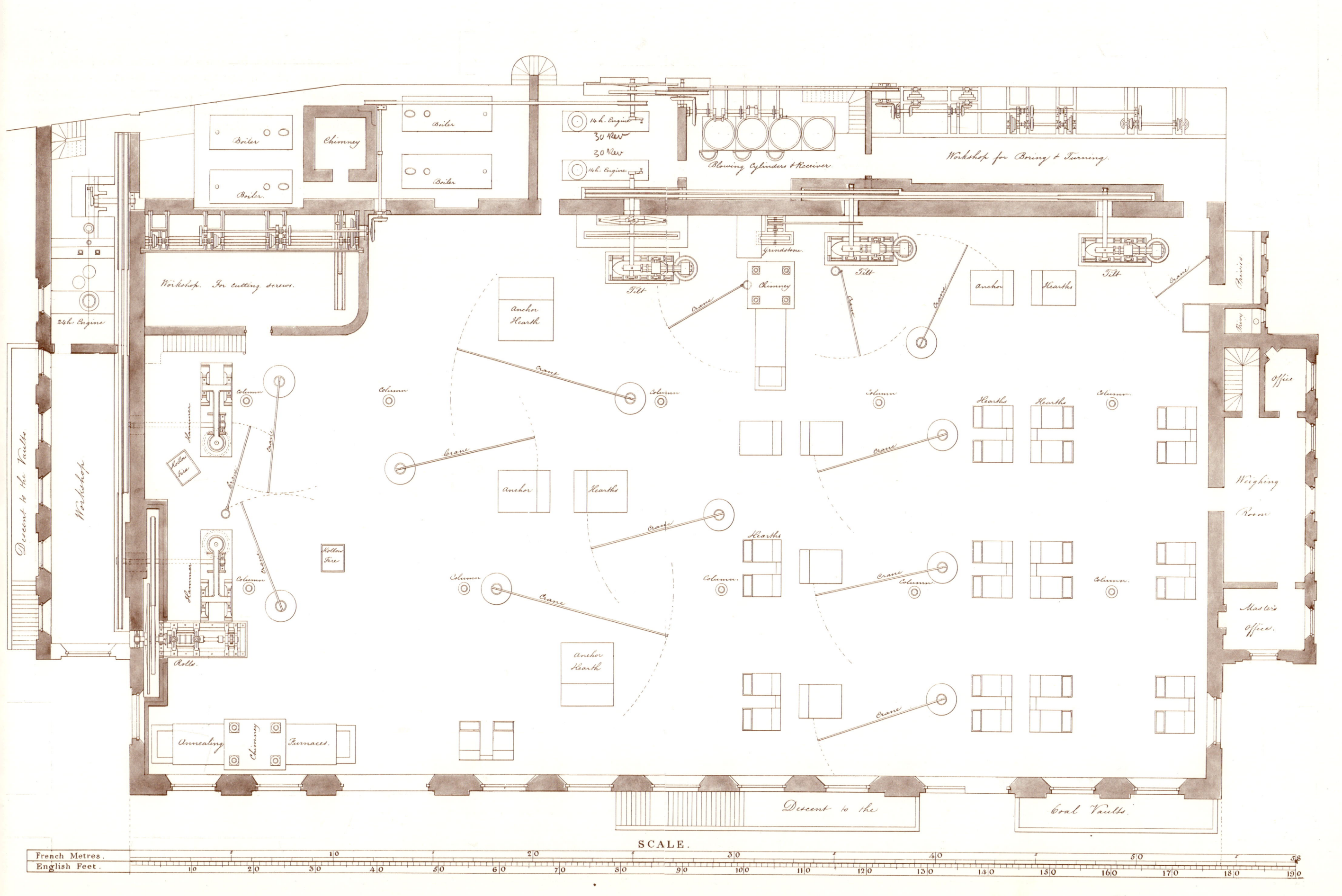

115. Liptrap to Boulton & Watt 30 Jan, 1787.
116. Underwood Spinning Co. to Boulton & Watt 19 Nov, 1794.
117. J. Wilkes and Co. to Boulton & Watt 8 Aug, 1798.
118. A. Raistrick, 'The Steam Engine on Tyneside 1715-1778', *Trans Newcomen Soc* 27, 1936-7.
119. A. E. Musson & E. Robinson op cit p. 405.
120. J. Farey op cit p. 122; *Encyclopaedia Britannica* 1797 article 'Steam'.
121. S. Unwin to Boulton & Watt 30 Jan, 1798.
122. Kent Myers and Co. to Boulton & Watt 9 Dec, 1795.
123. Leeds City Archives DB 233.
124. I am greatly indebted to Dr. S. D. Chapman for help with these.
125. See appendix; Marshall MSS 57 p.1.
126. M. M. Edwards, *The Growth of the British Cotton Trade*, p. 211; A. Young, *Tours in England and Wales* (L. S. E. Reprint, 1932), p.279.
127. Science Museum, Goodrich Papers, Journal 1804.
128. C. Clowes to Boulton & Watt 13 Feb, 1785.

were worried about delays:'our whole works will be at a stand for want of grist and we have the Duty to pay on our stills at work or not.' Liptrap had other worries:[115] 'I assure you for the consequence of stopping a corn distillers with 2000 hogs in the yard is so great that we are not without our daily fears.' The Underwood Spinning Company at Paisley[116] insisted that their Boulton and Watt engine should be at work before their common engine was removed 'as we cannot be six weeks idle without a considerable loss, it will therefore have to be placed at the corner of the backfront of the mill.' Wilkes of Measham[117] made sure that they had a sufficient stock of carded and scribbled cotton to tide them over whilst their engine was altered.

Cost of steam power

Through the work of Raistrick[118] and others more is known about the cost of early 18th century Newcomen pumping engines than about later Savery or Newcomen engines used in factories. A Savery engine for pumping a domestic water supply cost £50 in about 1712. Newcomen engines were far more costly, one for Edmondstone Colliery, Midlothian cost £1007. 11. 4d in 1727. Estimates in the 1730s dropped to around £800 or £900 largely as a result of the introduction of cast iron cylinders and wrought iron boilers and the end of Savery's patent in 1733. A Joseph Young Savery engine of *c.* 4 hp. cost a little over £200 including the water wheel in 1790 and Watt jun.[119] reported 'The gross sum which your engines cost at first startles all the lesser manufacturers here, and it is scarcely possible to make them comprehend the advantage to be derived from a regular motion, from a machine liable to a few repairs, and from an annual saving of fuel when weighed against 2 or £300 more of ready cash.' Apart from the matter of cost, and the Savery engine was much cheaper than other kinds, it was a simpler engine. The engine had 'great durability' on account of its having 'few moving and rubbing parts'. Farey noted that 'From the simplicity of its construction it is not liable to wear out for a very long time.' One maker of these engines, Joshua Wrigley, 'contrived his engines to work without an attendant',[120] and this was no small advantage in an age when experienced engine men were hard to come by. Another advantage of the Savery engine was that as Samuel Unwin[121] said, it was 'less expensive than a Rotative one because we had only to pay for the coals consumed in proportion to the power wanted to make up temporary deficiencies in the Resevoir or stream.' There is little evidence of the cost of a Bateman and Sherratt or a Thompson engine but they were certainly cheaper than Boulton and Watt engines of equivalent power. Kent Myers and Co of Liverpool,[122] for example, decided against a Boulton and Watt engine because of the greater initial cost. Richard Paley of Leeds paid Sturgess of Bowling Ironworks £127. 2. 7d for a steam engine. The total cost of the engine was £457. 2. 4d but this appears to have included the engine house as well.[123]

Insurance valuations[124] of Savery engines range from £50 to about £200, whereas valuations of Newcomen engines are usually £250 upwards. Nash and Abbot of Manchester had an engine valued at £400 and Pooley and Hallam had one valued at £700 but these are unusually high valuations for atmospheric engines. Joseph Thackery had a huge atmospheric engine with a 120 in cylinder but this was only valued at £100.[125] Thackery's valuation epitomises the problems of using insurance valuations to arrive at an estimate of the capital cost of steam power. Steam engines depreciated rapidly, at a faster rate than waterwheels, and depreciation must be taken into account when using insurance valuations. In the absence of enough dating data it must be concluded reluctantly that without supporting evidence an insurance valuation of a steam engine gives little indication of its original capital cost. Boulton and Watt steam engines also depreciated quickly. An eight year old engine costing about £525 when new was being sold at an asking price of £250 in 1806. Wilkes allowed 25% p.a. depreciation on his Boulton and Watt engine.[126] The following table shows the cost of Boulton and Watt engines in about 1795:

Cost of rotative engines c. 1795

HP.	Stroke	Metal Materials £	Iron Boiler £	Premium 5 yrs. £	Total charge £
8	4	295	40	155	490
10	4	320	50	192	562
12	4	342	60	232	634
14	5	434	70	270	774
16	5	460	80	310	850
18	5	480	90	350	920
20	5	500	100	388	988
24	6	620	115	466	1201
28	6	660	130	544	1334
32	7	845	150	622	1617
36	7	885	170	700	1755
40	7	927	195	776	1888
45	8	1060	220	874	2154
50	8	1112	240	970	2322

Source: Roll, p.312

Prices fluctuated according to the price of iron and dropped slightly for larger engines at the end of the century. In the early 19th century they rose again. In 1804 a 30 hp Boulton and Watt engine cost £1,418, a 24 hp engine cost £1,276 and a 20 hp engine cost £1,083.[127]

The relatively high cost of a low horse power Boulton and Watt engine was a deterrent to the smaller manufacturer who found the decision of whether to use steam power at all a difficult one. Charles Clowes,[128] a London brewer wrote 'We have no doubt on ye large scale your Engine would produce an annual saving but we have doubt where it will on our small scale.' By the time the cost of transport from Birmingham had been paid, the engine house built, the framing for the engine (not included in the estimates) constructed, and the fitter paid what might

Williams and Jones, Britannia Nail Works, Birmingham, 1814
This factory was built as a brewery in the late 18th century and as such is depicted on a copper trade token. It was converted to a nail factory in the early 19th century when nail making was still predominantly a domestic industry.
Portf: 446

86

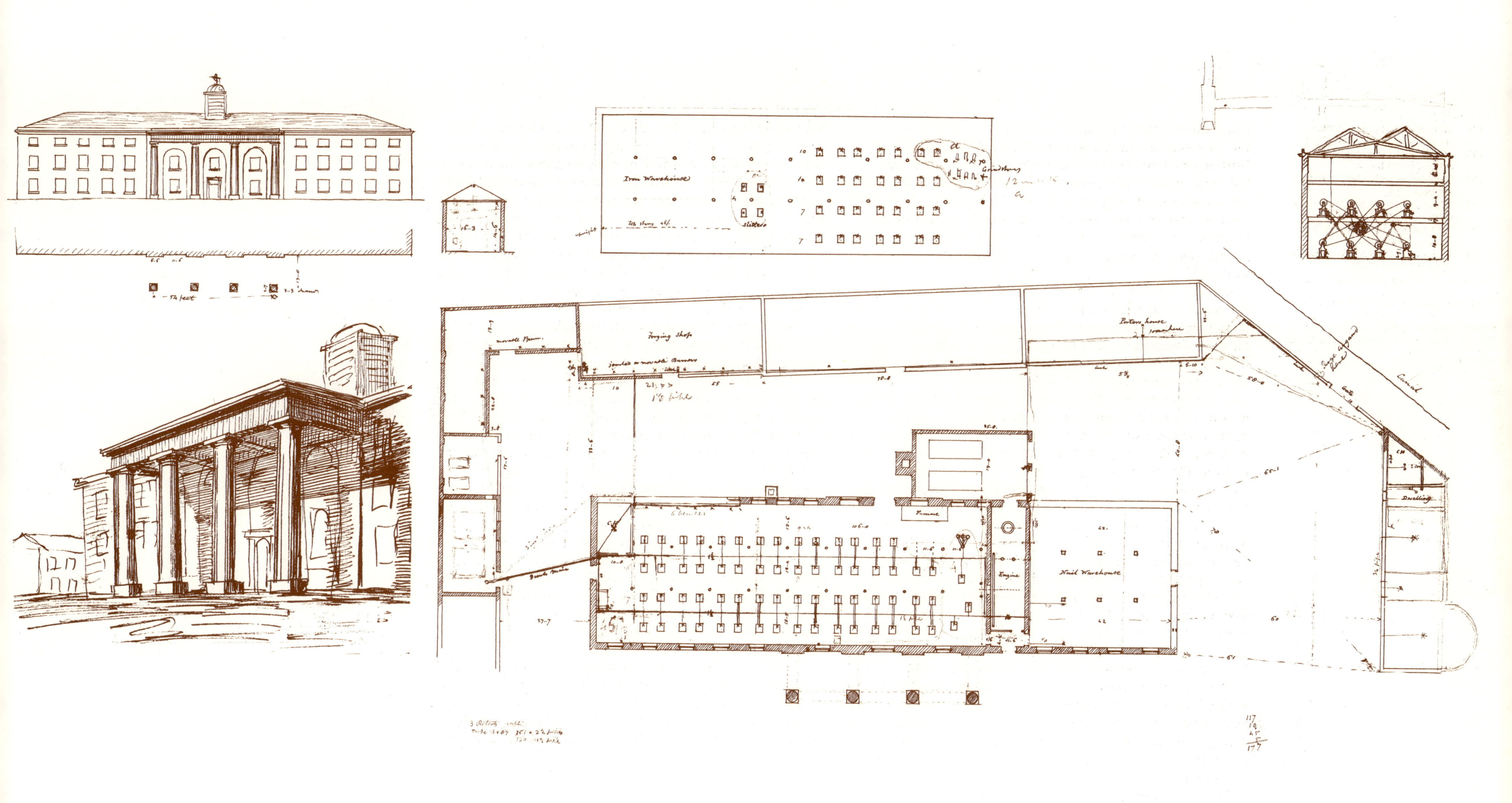

129. H. Pearson to Boulton & Watt 30 Jan, 1789.
130. R. Arkwright to Boulton & Watt 23 Dec, 1790 Box 4.
131. S. Oldknow to Boulton & Watt 22 Dec, 1798.
132. T. Jones to Boulton & Watt 18 Oct, 1796.
133. J. Cartwright to Boulton & Watt 15 Sept, 1788.
134. Claytons and Walshman to Boulton & Watt 7 July, 1785.
135. C. Spearman to Boulton & Watt 25 Feb, 1797.
136. C. and J. Morehouse to Boulton & Watt 28 June, 1787.
137. Boulton & Watt to J. Wilkes 25 April, 1799 Foundry Letter Book.
138. J. Watt Jun. to Gregory Watt 7 Dec, 1802.

once have seemed an advantage could become a millstone round the lesser manufacturer's neck. Pearson and Grimshaw[129] of Nottingham wrote 'The engine works quite to our satisfaction but must confess the expense is far greater than we expected, the whole amount being about £800' – for a 5 hp engine. The ancillary expenses were a shock to Arkwright too and he questioned Lowe's charges for the framing of his Nottingham engine adding 'this large expense will prevent me having one at Wirksworth.'[130]

Boulton and Watt demanded a premium until 1800 which was computed at the rate of £5 per hp. per year in the provinces and £6 in London. Entrepreneurs facing great demands on their capital were often hard pressed to pay it. Samuel Oldknow[131] wrote 'I know the best apology for not having remitted in due time for my steam engine premium wd have been a bill . . . [but] I crave your kind indulgence a little longer.' Wright[132] gave up the use of Watt's patent principles on his Bateman and Sherratt engine because he could not afford the annual premium. Some engines were not used all the year round but no abatement of the premium was allowed. Many large engines were not worked to capacity for a number of years but the entrepreneur had to pay the full premium. From the outset the entrepreneur was faced with the choice of buying an engine of the size he would require when his factory was eventually full of machinery and paying a high annual premium; or of buying a smaller engine with a lower annual premium in the knowledge that it would not provide enough power to drive all his machinery when his factory was filled. John Cartwright[133] faced this problem: 'Before I can judge between erecting an Engine of full or of half power for our intended works it will be right for me to know the comparative expense . . . could we have been relieved from the expense of that part of the power that must lie dormant until the second wing of our mill is set to work it would have been far more agreeable to me to have had one engine than two: nor should I think it very difficult to prove that all the looms and other machinery in our first wing would not require half the power of the engine you offered me.'

The coal consumption of a Boulton and Watt engine was lower than for a Savery or Newcomen engine. This could tip the balance in favour of a Boulton and Watt engine if it was to be erected at some distance from a coal field. But if coal was cheap as it was in some parts of Yorkshire, Lancashire and Scotland the coal consumption of a non-Boulton and Watt engine was not of paramount importance. Claytons and Walshman[134] of Keighley reported 'We find your calculations over rated so far in the consumption of coles by the common engine that the Extra Expenses attending the Erection of yours with the Yearly payments of your premium will not be saved.' But the heavy coal consumption of some atmospheric engines made them uneconomic even when they were placed on a coalfield. C. Spearman[135] described how Barnes 'laughed at me for having erected two [common engines] on a different construction near Leeds . . . which to my cost I have found a constant taxation in repairs and requiring a chaldron of coals for a spoonful of water.'

Boulton and Watt stipulated that their permission should be sought before a steam engine was put to another use or was sold. The latter was a reasonable request since their income was derived chiefly from premiums until 1800. But having to seek permission for a change of use brought forth a sharp comment from C. and J. Morehouse[136] of Gainsborough: 'It is incumbent upon us to lay out our money upon as safe and independent principles as possible . . . with regard to the moral certainty or the stability of the trade for which the engine is erected, we beg to leave to remark trade is no inheritance, time and circumstances make great alterations.'

The most important factor operating in favour of Boulton and Watt engines until 1800 was their greater efficiency and reliability. High horse power Newcomen engines were inefficient on the whole and manufacturers could ill afford lengthy stoppages due to engine failure. Boulton and Watt engines had their problems but it must be born in mind that it was generally the dissatisfied customer who made his feelings known. The advantages of the Boulton and Watt engine increased away from coalfields where the saving in coal was greater.

When Watt's patent expired a number of engineers and founders such as Bateman and Sherratt, Matthew Murray and Bowling Ironworks began to produce low pressure steam engines on Watt's principle. Boulton and Watt were expected to reduce their prices as competition increased but they refused to do so maintaining that they made only a 'fair' profit on their engines and that their engines were far superior to the others on the market.[137] In 1802 they gave an estimate for a 40 hp engine for a Mr. Swann of London and discovered later that Murray Fenton and Wood got the order because their estimate was £400 lower than Boulton and Watt's.[138] Whereas a 20 hp Fenton Murray and Wood engine cost £600 in 1804, a 20 hp Boulton and Watt engine cost £1,083.

Mints
Matthew Boulton's contribution to minting technology has yet to be studied in detail. It is clear that he designed the machinery and he received orders to supply complete mints to a number of foreign countries. The following drawings demonstrate clearly the fact that considerable status was attached to a mint.

British Mint, 1806
Rennie was millwright for the mint but Boulton designed the layout and probably supplied the machinery although he was bedridden at the time. Boulton insisted on designing the functional buildings, leaving an architect to provide an imposing facade.
Portf: 716

reverse B. Mint
15 December 1806
⅙ inch to the foot

138. J. Watt Jun. to Gregory Watt 7 Dec, 1802.

Prices for steam engines 1804

Trevithick (guineas)

hp	cost	hp	cost
1	120	8	470
2	210	9	490
3	270	10	510
4	330	11	530
5	380	12	550
6	420	13	570
7	456	14	590 rising by 20 gns per hp.

Fenton, Murray and Wood

hp	£ cost
20	600
24	—
25	715
30	830
35	938
40	1045

Boulton and Watt

hp.	£ cost
20	1083
24	1276
30	1418

Source: Science Museum, Goodrich papers, Journal 1804.

Before Watt's patent expired the manufacturer had the choice between the higher capital cost and lower running cost of a Boulton and Watt engine and the lower capital cost and (in many situations) the higher running cost of a Savery or atmospheric engine. These distinctions could no longer be made after 1800. There was a much wider choice of engines for the manufacturer, Boulton and Watt were heavily undercut by other firms and their share of the steam engine market fell.

Soho Mint, Birmingham, *c.* 1800

It was here that Matthew Boulton coined large quantities of coins and medals. The building adjoined Soho Manufactory and is shown on the extreme left of the engraving of the manufactory. Boulton wrote jubilantly to Rennie in 1800 'The building is built, the coining and cutting out presses are all finished, the engines will be finished by Lady Day.' (Boulton to J. Rennie 24 Jan. 1800, A.O.).
Portf: 714

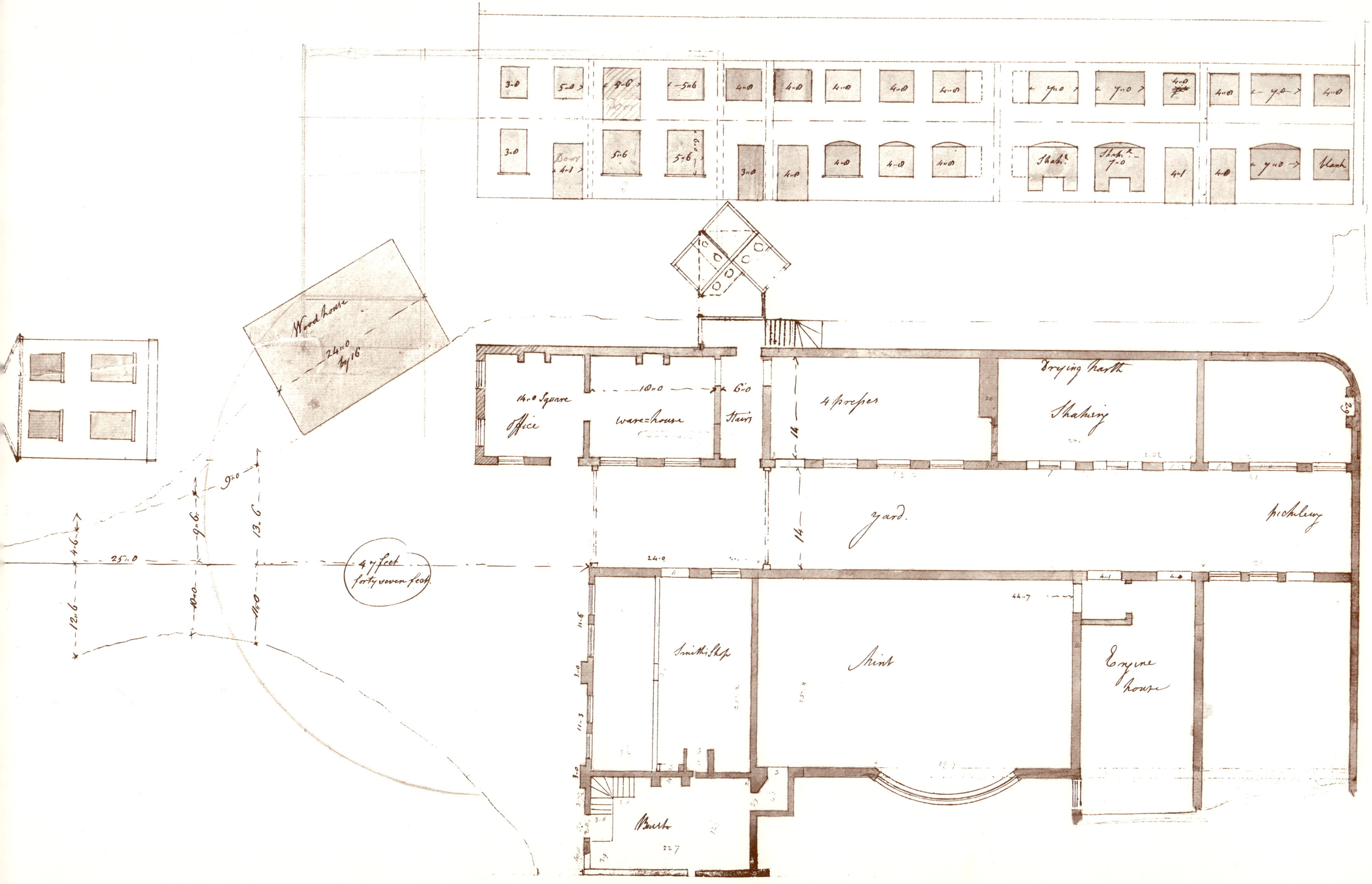

Wood house
24.0 by 16
Office
ware=house
Stairs
4 presses
Shaking
yard
pickling
Smith's Shop
Mint
Engine house
Busts
47 feet
forty seven feet

Some Newcomen and Savery factory engines in Lancashire, Yorkshire and Scotland

Name of firm	Location of factory	Date	Maker/hp
Ainsworth, R.	Bolton	c.1794	
Arkwright, Simpson, Whitenbury	Manchester	1782-3	
Asburner, T.	Bolton	1796	
Ball, T.	—	1795	
Barton, Dumbell & Co.	Stockport	1796	
Bateman & Sherratt	Manchester	1796	Bateman & Sherratt
Burton, D.K.	Manchester	1796	Bateman & Sherratt
Brindle, R.	Leyland	1796	
Campbell, D.	Manchester	1794-7	
Cannon, W. & Hodgkinson, R.	Atherstane	1797	
Carlile, J.	Bolton (2)	1795-7	
Deansgate Mill	Manchester	—	
Dunkerley	Oldham	1796	
Faulkner, S. & Co.	Bolton le Moors	1798	
Goodier, J.	Manchester	1796	Bateman & Sherratt
Grime, G.	Bolton	c.1792	
Haigh Forge	Wigan	1790	
Hardman, G.	Manchester	1795	
Haughton, J.	Stockport	1795-7	
Hill	Pendleton	—	
Holt, D. & Co.	Manchester	1793-7	
Holywell Co.	Holywell (Flint)	1797	
Horridge, J. ?	—	1799	Wrigley
Horrocks, J.	Preston (3)	1794	Bateman & Sherratt
Horrocks	Bolton	—	Thompson
Houldsworth, W. T. & H.	Manchester (2)	1796	Thompson
Jordain, T.	Manchester	c.1786	Wrigley
Joulet, W. & Son	Salford	1797	Bateman & Sherratt
Kennedy, J.	Manchester (2)	1793	Savery type & Thompson
Kirkman, J. & R.	Liverpool	1795	
Lees, Cheetham & Co.	Stalybridge	1800	
Lees, D.	Oldham	1797	
Lees, James	Oldham	1794	
Lees John	Oldham	1795	
Lees, John	Stockport	1799	
Lightoller & Hilton	Chorley	1796	
Lyon, T. et. al.	Warrington	1791-7	
Lythorpe, W.	Ashton-in-Makerfield	1790	
Marshall & Reynolds	Manchester	1790	
Mort, J.	Manchester	1796	
Moss	Manchester	c.1787	
Nash & Abbott	Manchester	1797	
Nightingale, Harris & Co.	Manchester	1796	
Norton	Salford	—	
Parr	Liverpool	c.1786	
Patten	Cornbrook, Manchester	—	Thompson
Peel, Ainsworth & Co.	Bolton	c.1792	
Peel, Yates & Co.	Burnley	1796	
Pemberton, E.	Liverpool	1797	
Pooley & Hallam	Manchester	1797	
Prestnall, Oldham et. al.	Stockport	1795	
Rigby, W.	Wigan	1795	
Runcorn, R.	Manchester	1796-7	
Sandford, B. & W.	Manchester	1784	Wrigley
Smith & Townley	Manchester	1797	
Taylor, Weston & Hill	Pendleton	1797	
Thackery & Whitehead	Chorlton (2)	1784, 1789	Wrigley, Bateman & Sherratt
Thackery, J.	Manchester	1795-8	
Thackery, Stockdale & Co.	Cark (2)	1791	Rowe, Bateman & Sherratt
Thorneley, J.	Warrington	1797-8	
Topping & Harrison	Warrington	1795-7	
Unsworth, E.	Chorley	1797	
Watson, J.	Preston	1794-7	
Watson, Fielding, Myers	Preston	1795	
Worsley, J.	Warrington	1795-7	
Wright	Manchester	1796	Bateman & Sherratt
Wrigley, J. & Co.	Manchester	1799	Wrigley

YORKSHIRE

Name of firm	Location of factory	Date	Maker/hp
Berry, G. & N.	Hanley	1796	
Beverley, Cross & Co.	Leeds		
Blagborough & Holroyds	Leeds		
Capley, W.	Leeds	1790	
Cartwright, J.	Doncaster	1788	
Chadwick, J.	Leeds	1796	
Chambers, J.	Wakefield	1796	
Chester, Wilson, Rhodes	Dewsbury	1796	
Churwell Mill	Leeds	1784	
Clayton & Walshman	Keighley	c.1785	
Crowder, I.	Morley	1797	
Coupland & Wilkinson	Leeds (2)	1792	
Eastwood, J. & Co.	Horbury	1797	
Firth, Blackburn, France	York	1795	
Firton, J.	Gildersome	1795	
Fisher & Nixon	Leeds	1796	
Garnet, J. & R.	Leeds	1796	
Grimshaw, J.	Leeds	1795	
Holroyd	Shipscar	c.1800	32½ in cyl. 24 in pumps
Houldsworth, S. & J.	Wakefield	c.1790	
Ibberson & Co.	Farnley	1794	
Markland, Cooksen, Fawcett	Leeds		
Marshall & Benyon	Leeds	1791	Wrigley
Metcalf	Silcoats, Wakefield	1794	
Roberts & Co.	Farnley	1794	
Rogerson & Co.	Hunslet		
Tarbottom & Carr	Thorner	1795	
Troughton Bros.	Driglington	1796	
Turton, J.	—	1797	
Walker, Ard	Leeds	1795	
Weatherill, J., J., & J.	—	1796	
Wells, Heathfield & Co.	Sheffield	1795	
Wilkinson & Paley	Leeds	1795-7	
Wormald, J.	Gomersall	1797	
Wood & Cook	Wakefield	1796	

SCOTLAND

Name of firm	Location of factory	Date	Maker/hp
Brand, J.	Glasgow	1797	4 hp
Campbell, Spier & Co.	Paisley	1797	10 hp
Henery, A.	Glasgow	1797	4 hp
McCleod, Twigg & Co	Paisley	1797	8 hp
Pattison, J.	Glasgow	1797	8 hp
Twigg, J. & Co	Paisley	1797	24 hp
St. Mirran Co	Paisley	1796	
Scott Stevenson & Co	Glasgow	1797	3 hp
Smith & Currie	Stravether	1797	6 hp

Source: Boulton & Watt MSS; Musson & Robinson; Sun Insurance Registers; Royal Exchange Registers.

Russian Mint, 1800
This mint has the appearance of a fortress and was probably so designed for reasons of security. Unfortunately the key to the drawing is missing. There is a wash drawing of one of the facades to this mint in the Boulton and Watt papers.
Portf: 713

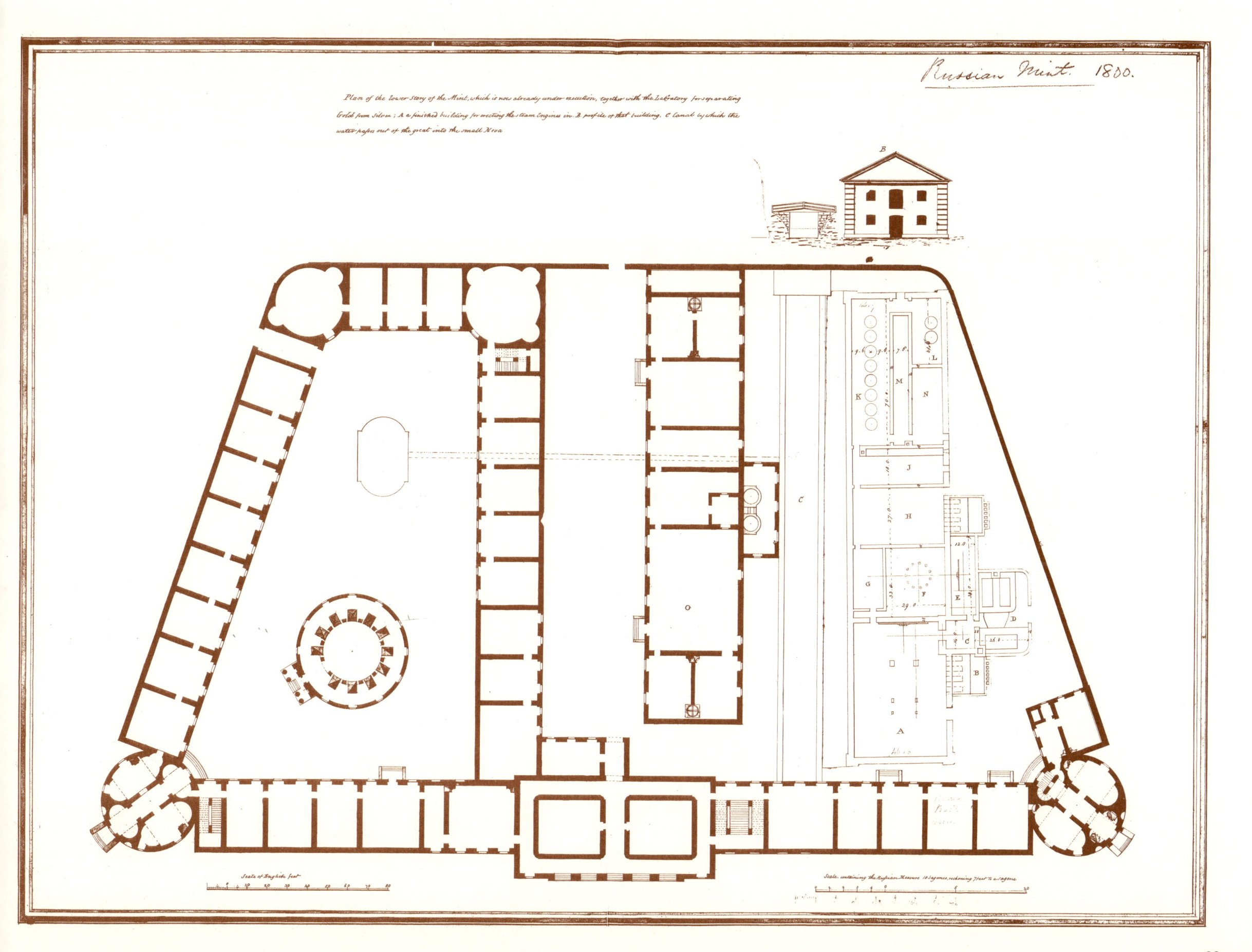

Russian Mint. 1800.

Plan of the Lower Story of the Mint, which is now already under execution, together with the Laboratory for separating Gold from Silver. A a finished building for meeting the steam Engines in B. profile of that building. C canal by which the water passes out of the great into the small Neva

Danish Mint, 1805
This was a comparatively small mint. Each hearth had one or two pairs of bellows.
Portf: 709-710

Brazil Mint, 1811
This was one of the largest mints designed and supplied with machinery by Boulton. Boulton left some rooms vacant 'for any purpose which the peculiar arrangement of the Portuguese establishment may require.' The coining rooms were to be built first but plans had been made for a completely integrated melting and rolling plant to adjoin them.
Portf: 715

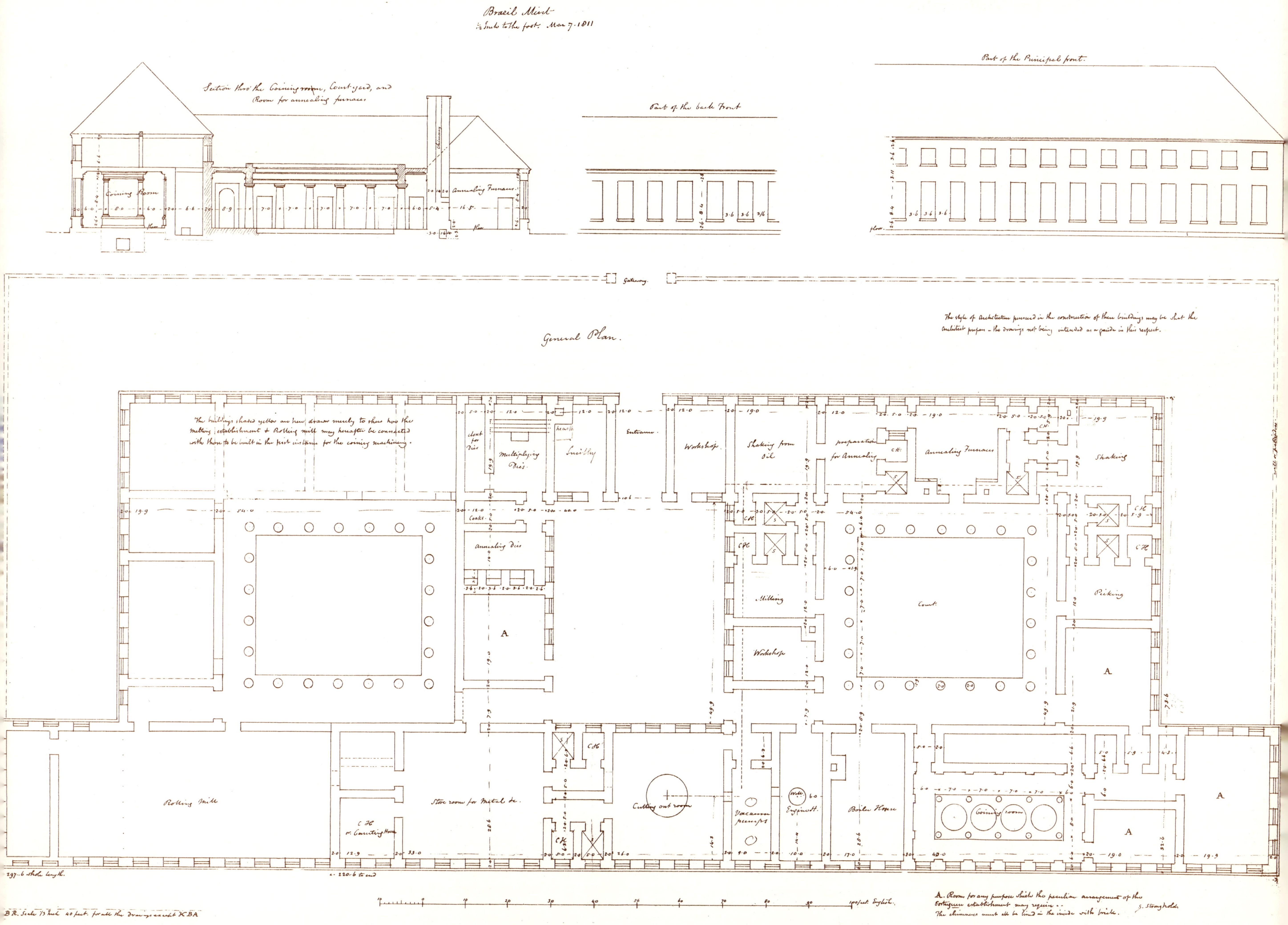
Brazil Mint
Section thro' the Coining room, Court-yard, and Room for annealing furnaces
Coining Room
Annealing Furnaces
Chimney
Part of the back Front
Part of the Principal front
Gateway
General Plan.
The style of Architecture pursued in the construction of these buildings may be what the architect prefers – the drawings not being intended as a guide in this respect.
The buildings shaded yellow are here drawn merely to shew how the melting establishment & Rolling mill may hereafter be connected with those to be built in the first instance for the coining machinery.
Multiplying Dies
Smithy
Entrance
Workshop
Shaking from Oil
preparation for Annealing
Annealing Furnaces
Shaking
Annealing Dies
Milling
Court
Pickling
Workshop
A
Rolling Mill
Store room for Metal &c.
C.H. or Counting House
Cutting out room
Vacuum pumps
Engine
Boiler House
Coining room
A. Room for any purpose which the peculiar arrangement of the Portuguese establishment may require. –
The chimneys must all be lined in the inside with brick.
S. Strongholds

6. The Millwrights

[1]. For an essay on a millwright see Eric Robinson's 'The Profession of a Civil Engineer in the Eighteenth Century' in A.E. Musson and Eric Robinson, *Science and Technology in the Industrial Revolution*; and Musson and Robinson's work on Thomas Hewes *ibid*.

[2]. *A General Description of All trades*, London, 1747.

[3]. Prof. J. Robison to James Watt 10 Feb. 1784, A.O.

[4]. Salop Record Office, Attingham Papers, T. Bell to T. Hill 4 Jan. 1757. I am indebted to Mr. R. Chaplin for this reference.

[5]. Information from Mr. R. Sherlock to whom I am indebted.

[6]. Cyril T. Roucher, *James Brindley, Engineer 1716-1772*.

[7]. F. Williamson, 'George Sorocold of Derby, a Pioneer of Water Supply,' *Derbyshire Arch. Soc.* 10, 1937.

[8]. Cyril T. Boucher, op cit pp31-5; Birmingham City Library, Diary of James Brindley, 1755-58

[9]. *Reports of the Late John Smeaton; A Catalogue of the Civil and Mechanical Engineering Designs of John Smeaton*

The person upon whose judgement rested the technical success or failure of a factory was the millwright. He discussed with the manufacturer the machinery to be used, designed the shafting and gearing, recommended the power system and designed it if a horse or water wheel was to be used, and probably decided the basic dimensions of the factory. In view of this surprisingly little known about the millwright. Insufficient records hamper a discussion of the jobs he performed and the place he held in society, and so far little has been written of the part played by the millwright in the factory movement. The infrequency with which a millwright appears in the directories from the 1770s had led two recent authors[1] to assume that 'the general skill of the millwright was being replaced by more specialised crafts.' But this is to confuse the role of the millwright.

The 18th century millwright was a worker in wood and metal. Sometimes he was itinerant, sometimes he was attached to a factory. There is an excellent description of the millwright's trade in the 1740's:[2]

'The Trade is a Branch of Carpentry (with some assistance from the Smith) but rather heavier work, yet very ingenious, to understand and perform which well a person ought to have a good Turn of Mind for Mechanics, at least to have some knowledge in Arithmetic, in which a lad ought to be instructed before he goes to learn this Art; for there is a great variety in Mills, as well as in the Structure and Workmanship of them, some being worked by Horses, some by Wind, others by Water shooting over and others by its running under. And why not in Time by Fire too, as well as Engines? They take an Apprentice £5 to £10, work from six to six, and pay a journeyman 12/- to 15/- a week; but £50 or £100 worth of timber and £50 to spare will make a Master of him.'

Although by tradition the millwright was associated with corn milling work a great variety of other industries claimed his attention. He might be approached by the steward of an estate to design a corn mill capable of producing a certain number of bushels of flour per week. He might be asked to design a saw mill to cut the timber produced on the estate. Or he might be approached by a merchant clothier to design a fulling mill, or by a potter to design a flint grinding mill. He would decide on the machinery and the speed at which it should rotate; he would choose the materials and, with his assistants or with local labour make the shafting and gearing. By the mid 18th century he was calling in specialists to perform tasks. The coppersmith or backmaker would be brought to the brewery, the spindle-maker to the spinning mill. As the mechanisation of industry progressed during the 18th century this trend continued and joiners, filers and clockmakers were employed to make certain kinds of machinery whilst the master millwright was concerned with the whole mill or factory. His responsibility and authority made the master millwright exclusive and he could command high prices for his work. An increasing number of machine makers was required and the demand for men skilled in filing and turning and making gear wheels seemed insatiable. Many craftsmen were recruited from clock and watch manufacture and it is no accident that the machines in a factory were described as 'clockmakers work' in insurance valuations. 'Clockmakers work' was distinguished from 'Millwrights work' in the valuations of factories and the distinction is important.

Apprenticeship was usual in the millwrighting trade; Brindley, Ewart, Rennie and Fairbairn all served their term with a master millwright. But it was a trade where a journeyman or master could become well known at an early age; Rennie, for instance, had been a master millwright for four years after his apprenticeship with Andrew Meikle when he became known as the builder of the best corn mill in Scotland.[3]

The good master millwright generally became a specialist in one or more branches in his field; he might concentrate on corn milling or brewing or flint grinding or cotton spinning. Builders of mills and factories brought millwrights from considerable distances. A millwright involved in the building of Tern Mills at Attingham, Shropshire was brought from Tewkesbury, Gloucestershire[4] and one Staffordshire landowner sent to Scotland for his millwright.[5] James Brindley operated in Staffordshire, Lancashire, Cheshire and Worcestershire[6] whilst George Sorocold of Derby,[7] best known for his millwork in Lombe's silk mill, installed public water works driven by water wheels at Derby, London, Leeds, Macclesfield, Wirksworth, Yarmouth, Bristol and several other towns.

Even in the mid 18th century there were exceptional millwrights who were also engineers. Brindley, Smeaton and Yeoman belong to this class. Leaving aside their canal and other engineering work they performed a wider range of millwrighting than the typical master millwright. Brindley, working between the Potteries and the silk area of Macclesfield, Congleton and Leek was commissioned to build and repair flint mills and silk throwing mills. In 1735 while still an apprentice he repaired the silk mill belonging to Michael Daintry of Macclesfield. In 1758 he built a large flint grinding wind mill in Burslem. In 1755 he was employed to make the water wheel and 'the commoner sorts of millwork' in a new silk mill being built at Congleton, and at about the same time he erected a paper mill on the River Dane at Wildboarclough. His most frequent millwrighting work was with corn mills. In 1752 he built Leek corn mill, part of which still survives, and between 1755 and 1757 he built Ashbourne Mill and either built or carried out repairs on corn mills at Tatton, Abbey Hilton, Wheelock and Congleton. He was also involved in the erection of several steam engines.[8]

The range of Smeaton's millwrighting activities was even wider. He made reports on corn, dyewood, oil, paper, powder, flint, copper, boring, rolling, slitting and hammer mills.[9] To-

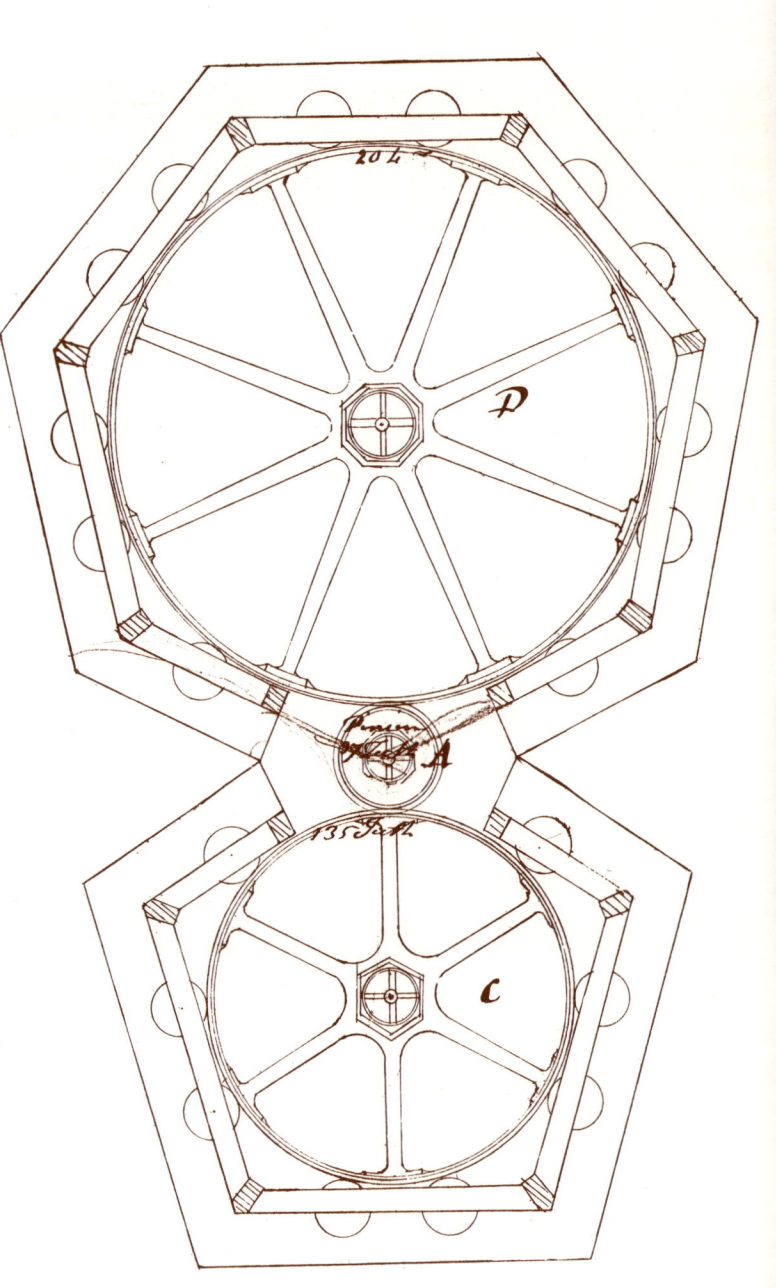

Snuff and Tobacco
Fish & Yates, 1786
Snuff and tobacco grinding were often carried out in a corn mill by taking a drive from the water wheel. But this drawing shows a purpose-built factory.
Portf: 12. Scale approx 1in : 4ft 6in.

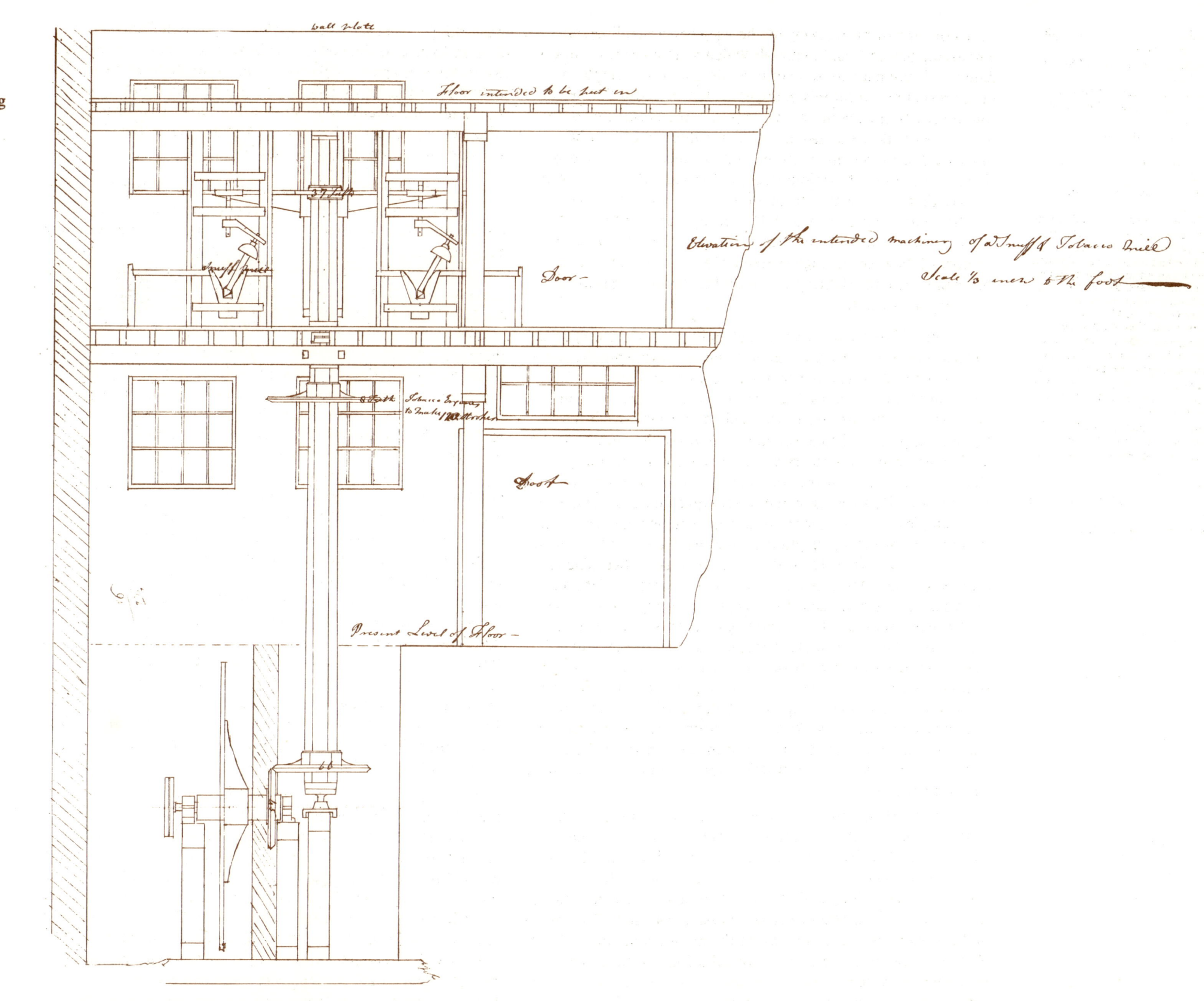

10. J. Rennie to Matthew Boulton, 28 Jan. 1785, A.O.
11. J. Rennie to Matthew Boulton, 19 June 1784 A.O.
12. Bill to Matthew Boulton from J. Rennie August 1784 – Feb. 1786. A.O.
13. J. Rennie to M. Boulton, Jan. 8 1788 A.O.
14. M. Boulton to J. Watt 18 April 1786 A.O.
15. J. Watt to M. Boulton 20 April 1786 A.O.
16. The Robinsons came from Scotland and would have known of Rennie. The water at Robinsons' mill was said to be 'set or laid on by a watch' which would mean Rennie's sliding hatch; on the other hand the 'watch' might have been a workman. Notts. County Record Office DD4P 79/63.
17. 'In a new mill they are just now building where there is an overshot wheel of 42 ft. diameter they have taken care, agreeable to my advice . . . to provide a proper place for a steam engine.' J. Rennie to M. Boulton, 5 Sept. 1785, A.O.
18. Brotherton Library, Gott Mss. 52, J. Rennie to B. Gott, 27 May 1793
19. J. Rennie to M. Boulton 3 Nov. 1804, A.O.
20. Portf 55
21. Boulton & Watt to J. Rennie 1 Aug. 1814, A.O.
22. J. Rennie to J. Watt 7 Jan. 1792

wards the end of his working life he was recommending and employing methods and materials which were considered out of date by his younger contemporaries. Rennie, for example, was unimpressed by Smeaton's millwork at Hilberts Mill,[10] Carshalton, Surrey 'I must confess I was not very much satisfied with her; indeed I allow she had an excellent water wheel but her inner work is extremely confused and ill executed.' He added 'I examined some other flour mills on the same River but I cannot say that I got much information from them, it appears to me they excell more in their water wheels than any thing else.' Smeaton also improved and was involved in the erection of a number of Newcomen engines (see p71); his influence on the profession of millwrighting and upon the design of millwork was very important.

By the 1770s the demand for able millwrights was very great indeed. Boulton and Watt wanted to employ a millwright and such was the shortage that they had to write to Professor J. Robison at Edinburgh asking if he could recommend anyone. As is well known, Robison suggested John Rennie who, having been apprenticed to a well-known Scottish millwright Andrew Meikle, had been in business on his own account for four years. Rennie, being aware of the contacts that could be made through Boulton and Watt was prepared to relinquish his independence to be employed by so important a firm. Rennie when accepting Boulton and Watt's offer admitted that he had carefully planned his career.[11] The first step had been to practise in Scotland and gain experience. After this he hoped to be employed for a year or two as foreman or agent to an engineer in England in order to see the best methods of constructing machines there. The third stage was to be a tour of Holland, Germany and France to see how millwrights work was executed there. Rennie's terms with Boulton and Watt allowed him scope to practise as a millwright while at the same time engaging in the erection of steam engines and advising Boulton and Watt on millwork. Rennie for his part entered into a bond not to erect any engine constructed on Watt's principle or to apply Watt's rotative motion to any engine without permission.

He established himself in London and one of his first jobs was to advise Matthew Boulton on his rolling mill. Rennie charged £2. 9. 10½d[12] for twelve and a half days work in preparing an estimate. He later sent a written estimate for the millwork of a new rolling mill at Soho Manufactory which came to £1327. 5. 9d.[13] Another early commission was that of designing and superintending the erection of the millwork at Albion Mills. Rennie made several important contributions to the design of corn mills of which the most important was the use of cast iron for gear wheels. He was a difficult character to work with and succeeded in upsetting the other craftsmen at Albion Mills. Matthew Boulton remarked[14] 'If Rennie had been as attentive to his own department as he has seemed to be to other peoples

many things might have been better . . . you will find it most prudent to give to Engineers such instructions as you find relative to Engines and to Millwrights such as relate to Millwork.' In reply Watt wrote 'If we have wasps and hornets about us we should not provoke them to sting us but rather give them a little treacle, I hope he is not of a bad disposition though he has too much forwardness and conceit.'[15]

The specialisation in millwrighting that was discernible in the early 18th century became more pronounced towards the end of the century. Rennie having based himself in London unwittingly excluded himself from textile millwrighting. The fact that he was involved in the erection of several steam engines has led to a mistaken view that he was the millwright at these factories. It has often been thought that Rennie was the millwright at Robinson's cotton mills at Papplewick,[16] Nottinghamshire, but there is no direct evidence for this although he was certainly involved in the erection of the engine and was responsible for suggesting[17] that Robinson should make room for an engine in the new mill they were building. Rennie was employed by Gott to design the machinery for a dyewood mill[18] but Gott called in other specialists for the rest of his millwork. Rennie probably achieved greatest renown as a corn mill millwright. Boulton and Watt referred all enquiries concerning corn mills to Rennie and through them he acquired a number of foreign orders. He also executed a considerable amount of work in the metals industry. Boulton employed Rennie to design the millwork of his new rolling mill at Soho Mint and when Boulton designed the machinery for the new London Mint Rennie was employed as millwright and engineer. Rennie wrote patronisingly on one occasion.[19] 'Your plan came safe to hand – it would be presumption in me to say anything in it is wrong but I trust you will give me credit when I say it meets my entire approbation.' He was employed as millwright by Gardner and Manser[20] a large firm of iron founders and in 1814 he was appointed to design the Woolwich Smithery for the Commissioners of the Navy.[21] (pages 82, 84)

But neither corn milling nor iron founding was expanding at the rate that the textile industry was growing and there is the strong possibility that Rennie was not obtaining the number of commissions that he had hoped for. Brewing was a major London industry but there was already a well established millwrighting family, James and John Cooper, specialising in this work. In 1792 Rennie was told of the death of John Cooper and wrote immediately to Watt[22] 'as part of his business will no doubt be at liberty if you could conveniently assist me in procuring a part of it you would confer a very particular favour on me if I could succeed to a sufficiency of mill business – it would make me to stay at home and avoid the inconvenience of the canal and other wandering employments.' Rennie probably made a tactical error in deciding to live in London. Cotton was the

Brewing and Distilling
Goodwyn & Co, London, 1784
This is a section and plan of only part of the brewery but the plan is useful in showing the degree of machanisation reached in one of the largest London breweries.
Portf: 2; Misc. Mills

Section No. 1 — Messrs Goodwyn & Co.

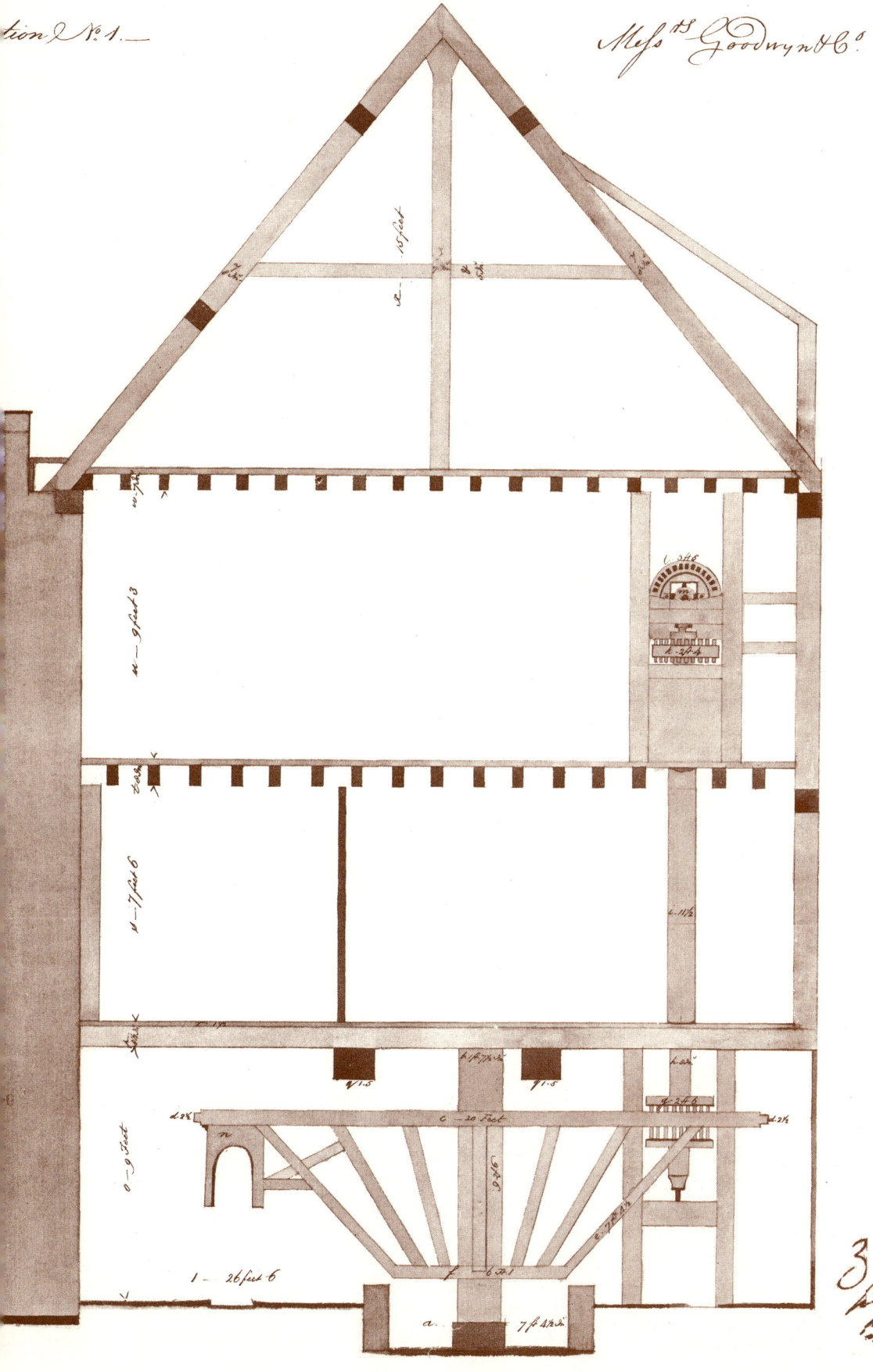

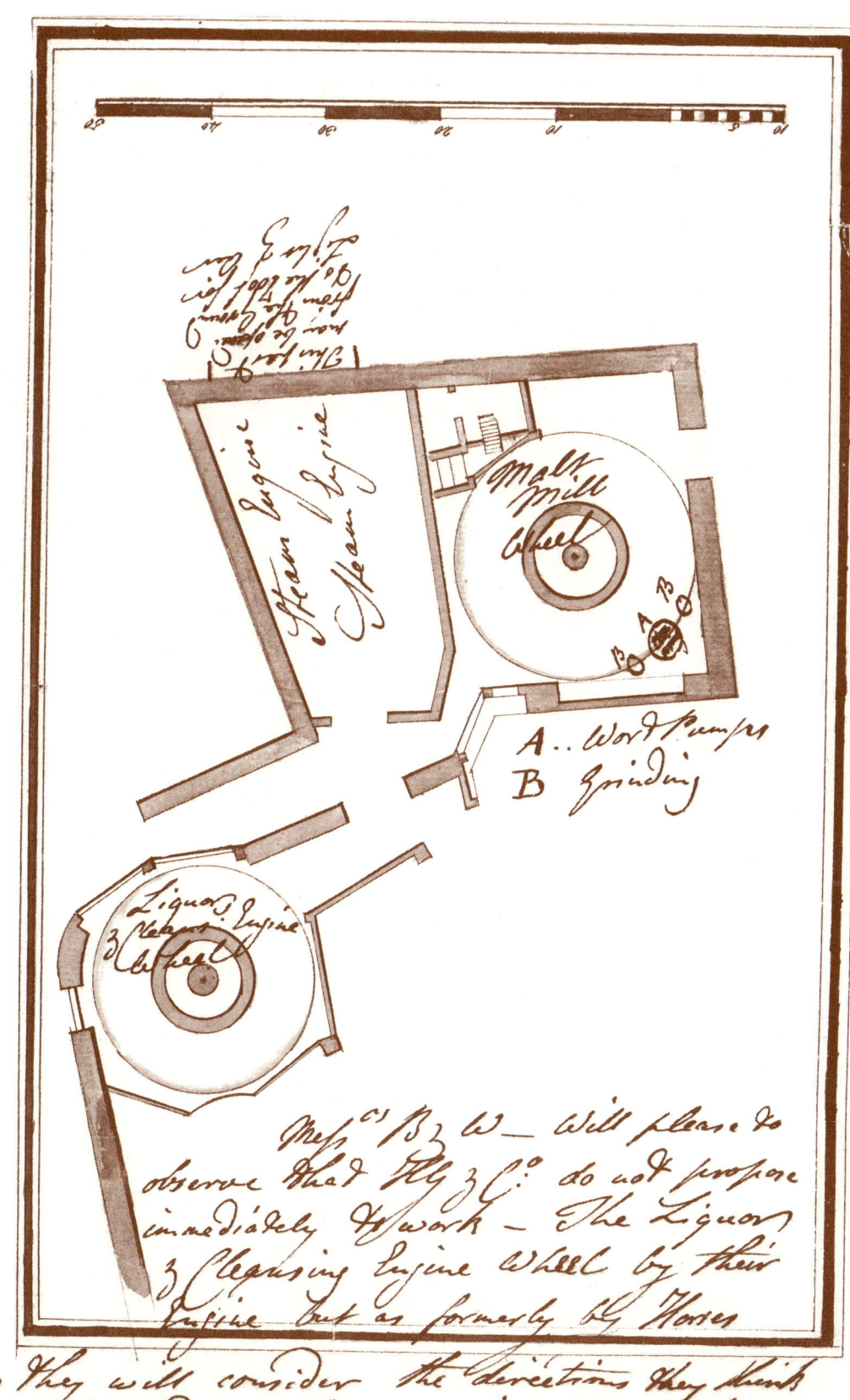

A .. Wort Pumps
B Grinding

Messrs B & W — will please to observe that Hy & Co. do not propose immediately to work — The Liquor & Cleansing Engine Wheel by their Engine but as formerly by Horses

& beg they will consider the directions they think proper & send Mr Cooper — Millwright at Poplar for strengthening or altering the Malt Mill Wheel

23. J. Rennie to M. Boulton 19 Nov. 1791 A.O. John Sutcliffe moved to Lancashire where he was involved in canal work but presumably executed work for cotton spinners too. William Jessop, better known as a canal engineer, became a partner in a cotton spinning firm at Newark.
24. Peter Mathias, *The brewing Industry in England 1700-1830.* p.86
25. J. Southern to J. Thompson, 5 Dec. 1793, Foundry Letter Book
26. Portf 22, portf 3
27. W. Pole, ed. *Life of Sir William Fairbairn*, p.121
28. Derby Mercury 13 Dec. 1771, quoted in RS Fitton & AP Wadsworth, *The Strutts and the Arkwrights*, p.65. Boulton and Watt refer to Lowe as 'carpenter' and they bought timber from him on several occasions.
29. Portf 277.
30. J. James to Boulton & Watt 24 Sept. 1787
31. P. Drinkwater to Boulton & Watt 13 Dec. 1789
32. C. Lees to Boulton & Watt 27 Nov. 1795
33. R. Owen & Co. to Boulton & Watt Oct. 1795
34. Boulton & Watt to Lowe 10 Feb. 1796, Foundry Letter Book
35. S. Marsland to Boulton & Watt 16 Mar. 1798
36. W. Pole, op cit p.121
37. P. Ewart to Boulton & Watt 25 Sept. 1792
38. Boulton & Watt to Lowe 13 Nov. 1801; 1 Dec. 1801, Foundry Letter Book
39. J. Rennie to M. Boulton 26 Jul. 1788 A.O.
40. P. Ewart to Boulton & Watt 18 Aug. 1792
41. W. Douglas to Boulton & Watt 21 Feb. 1792
42. Boulton & Watt to D. Holt 13 April 1792 Foundry Letter Book
43. P. Ewart to Boulton & Watt 18 Aug. 1792
44. R. L. Hills, *Power in the Industrial Revolution* p.113
45. Parkes, Brookhouse & Crompton to M. Boulton 25 Nov. 1796, A.O.
46. M. Boulton to Parkes, Brookhouse & Crompton 28 Nov. 1796, A.O.
47. Sir Eric Roll, *An Early Experiment in Industrial Organisation*, p.162
48. A. E. Musson and E. Robinson, pp 441-2
49. Portf 140
50. Portf 165
51. Portf 39
52. Leeds Intelligencer 14 April 1789, Leeds Mercury 23 Nov. 1784, 6 Jan. 1789.
53. Brotherton Library, Marshall MSS 57, p.4
54. Collected under the title 'Treatise on Canals and Reservoirs...', 1816
55. Portf 91.

most rapidly expanding industry and one in which there was a great demand for the services of a millwright. Had he based himself in Lancashire as Peter Ewart did he might have found no necessity to pursue 'canal and other wandering employments.' He obviously wished to execute cotton millwork for when he heard that a Mr. Bowen was building a cotton factory in London he asked Matthew Boulton[23] 'I doubt not with a little of your assistance I might be employed to do the millwork — will you have the goodness to lend me a lift in this business[?]'

The Coopers had a near monopoly of London brewery millwrighting. Goodwyn noted that James was 'much employed in the Brewery and Distillery and with general approbation,' and requested that he should be concerned in the adaptation of the horse wheel at Goodwyn's brewery.[24] Cooper was well aware of the growth in ancillary millwrighting created by the introduction of the steam engine and was anxious to obtain commissions in this field. In 1793 Boulton and Watt[25] recommended James and Thom Cooper to a prospective customer, James Thompson, as they 'have been most concerned where our engines have been and know the work'. James Cooper was millwright to Calvert's brewery and he was also commissioned to do the millwork for several flour mills in and near London.[26]

Fairbairn attributes all the important early cotton factory millwork to one man, Thomas Lowe of Nottingham.[27] Writing in the latter part of the 19th century he said 'His work, heavy and clumsy as it was had in a certain way answered the purpose, and as cotton mills were then in their infancy he was the only person qualified from experience to undertake the construction of the gearing. Mr. Lowe was therefore in demand in every point of the kingdom where a cotton factory had to be built.' This is difficult to prove or disprove but Fairbairn's statement should be examined for if it is true Lowe's influence on the evolution of the cotton factory must have been very great indeed. Several problems complicate the issue. Fairbairn exaggerated a number of important points in his books and he naturally wished to give the best impression of his own millwork by contrasting it with the cruder millwork performed in the early factories. There are few references to the millwrights of the early spinning mills so that it is difficult to associate Lowe with more than a few factories with any certainty. On the other hand the dimensions of many of the early spinning mills were similar and the 30 ft width was almost universal. This could point to the early spinning mills being the work of one man. The explanation of the similar dimensions lies in the fairly standard dimensions of the machines that the factories contained and the crucial question is whether the millwright of the 1770s and 1780s was responsible for machine making. Lack of evidence hampers the discussion but Arkwright and Strutt, for example, were advertising in 1771 not for millwrights but for 'two journeymen Clock-Makers or others that understands Tooth and Pinion well:

Also a Smith that can forge and file — Likewise two Wood Turners that have been accustomed to Wheel-making Spole [spool]-turning etc.' It could be argued that these craftsmen would have been working under the direction of a millwright — they probably would have done in many cases — but Lowe's interests lay in the timber trade, in carpentry and in general millwrighting; he was employed to make and erect the engine framing for a number of Boulton and Watt's Midland customers.[28] Lowe is known to have been millwright to John and William Elliott,[29] cotton spinners of Nottingham and to John James,[30] son of Hargreaves' partner, of Nottingham. James had driven his machinery by horse wheel initially and when he ordered a Boulton and Watt steam engine this posed a problem for Lowe. The engine could either have been connected to the horse wheel as in some London breweries, entailing little alteration to the millwork, or the horse wheel could have been by-passed by new shafting. Lowe decided on the latter scheme in which less power was lost by friction. Lowe may have been millwright to the Robinsons of Papplewick and no doubt he received commissions from other local Nottinghamshire and Derbyshire spinners. He was employed by several important Lancashire cotton spinners including Peter Drinkwater,[31] the first spinner to use a Boulton and Watt rotary engine in Manchester. In 1795 he worked for Charles Lees[32] of Stockport and for Robert Owen[33] at Chorlton. In 1796 he undertook the millwork for Salvins[34] of Durham and in 1798 he was commissioned by Samuel Marsland.[35] Fairbairn records that Lowe executed the water wheels and, one may guess, the millwork at Catrine Mills in Scotland.[36] As he was known in Scotland he was probably employed by other spinners too. Lowe was occasionally employed in other kinds of millwork. He was commissioned by Gott[37] of Leeds in 1792, for example, and in 1801 he executed some millwork for Wedgwood and Byerley.[38] These examples show that Lowe was widely known to cotton spinners but they do not prove that he was the only person qualified from experience to undertake the construction of the gearing.

There was another millwright specialising in cotton factory millwork in the 1790s who was of equal importance with Lowe. This was Peter Ewart. Ewart was sent, sounding almost like second choice, to Boulton and Watt in 1788 'instead of Robert McArthur who I am informed . . . is in a very bad way'.[39] After a spell as an engine erector Ewart went to Lancashire where he practised as a millwright remaining, like Rennie, an agent of Boulton and Watt's. In 1792 Ewart was commissioned by one of the most important Lancashire cotton spinners:[40] 'I undertook to execute all the millwork and drums of Messrs Atherton and Cos mill in Salford for £600', but he lacked business experience 'I am now sorry to find that it will amount to £200 more than I expected it would.' Ewart assisted William Douglas[41] of Pendleton 'to lay out our new mill' in 1792 and in the same year he advised David Holt and Co[42] on their millwork. Benjamin Gott chose Lowe as his millwright after a disagreement with John Sutcliffe and Ewart told Boulton and Watt 'they requested that I should assist them in planning it'.[43] He installed water wheels at Tutbury cotton mill, Staffs and was responsible for building a new dam and wheel at Styal.[44] Ewart 'rendered essential service'[45] to Parkes Brookhouse and Crompton, worsted spinners of Warwick and Matthew Boulton, when asked what gratuity should be given, suggested ten guineas and commented that Smeaton, whose judgement and knowledge of spinning mill was not equal to Ewart's, would have charged more.[46] Ewart entered the cotton trade in 1792 (see below) but returned to millwrighting after a year and between 1795 and 1797 assisted Boulton and Watt in the planning of Soho Foundry.[47] He went into partnership with another cotton spinner in 1798 continuing with this partner and finally alone until 1835 when he left Lancashire to become Chief Engineer and Inspector of Machinery at H. M. Dockyards.[48]

Henry Gardner of Liverpool executed the millwork in several cotton factories including Hodgson and Capstick's Caton Mill[49] and Thomas Ridgeway's[50] factory near Bolton. He was also millwright to the British Plate Glass Co. at Ravenhead.[51]

The picture is less clear in the woollen and worsted industries. Lowe was probably commissioned to execute work for some of the larger manufacturers but many of the woollen factories were small scribbling and fulling mills in which the millwork was comparatively crude and simple. A minor local millwright was probably engaged to make this. John Jubb[52] of Leeds was a millwright and machinery maker who specialised in the manufacture of scribbling and carding machines and fulling stocks but nothing is known of his millwrighting activities. The major Yorkshire millwright was John Sutcliffe of Halifax who was engaged by two of the largest Yorkshire manufacturers of the 1790s, Wormald, Fountaine and Gott, woollen manufacturers, and Markland Cooksen and Fawcett, worsted and cotton spinners. He was consulted by John Marshall[53] but it is not known whether he was commissioned to do any work for him. Sutcliffe is author of several essays of which the most significant ones from the millwrighting point of view are *Instructions for Designing and Building a Corn Mill* and *Observations on the Carding, Roving, Drawing, Stretching and Spinning of Cotton.*[54] He was a man of strong opinions which he could express forcibly on occasion. His reputation was obviously high, yet in retrospect his judgement was faulty on some major points of millwrighting. He prepared plans for Markland Cooksen and Fawcett's spinning factory in Leeds[55] and as an example of late 18th century draughtsmanship they have few equals amongst factory drawings (page 18). Sutcliffe was presumably responsible for recommending the motive power and chose a water wheel and common pumping engine. Markland Cooksen and Fawcett decided to abandon the

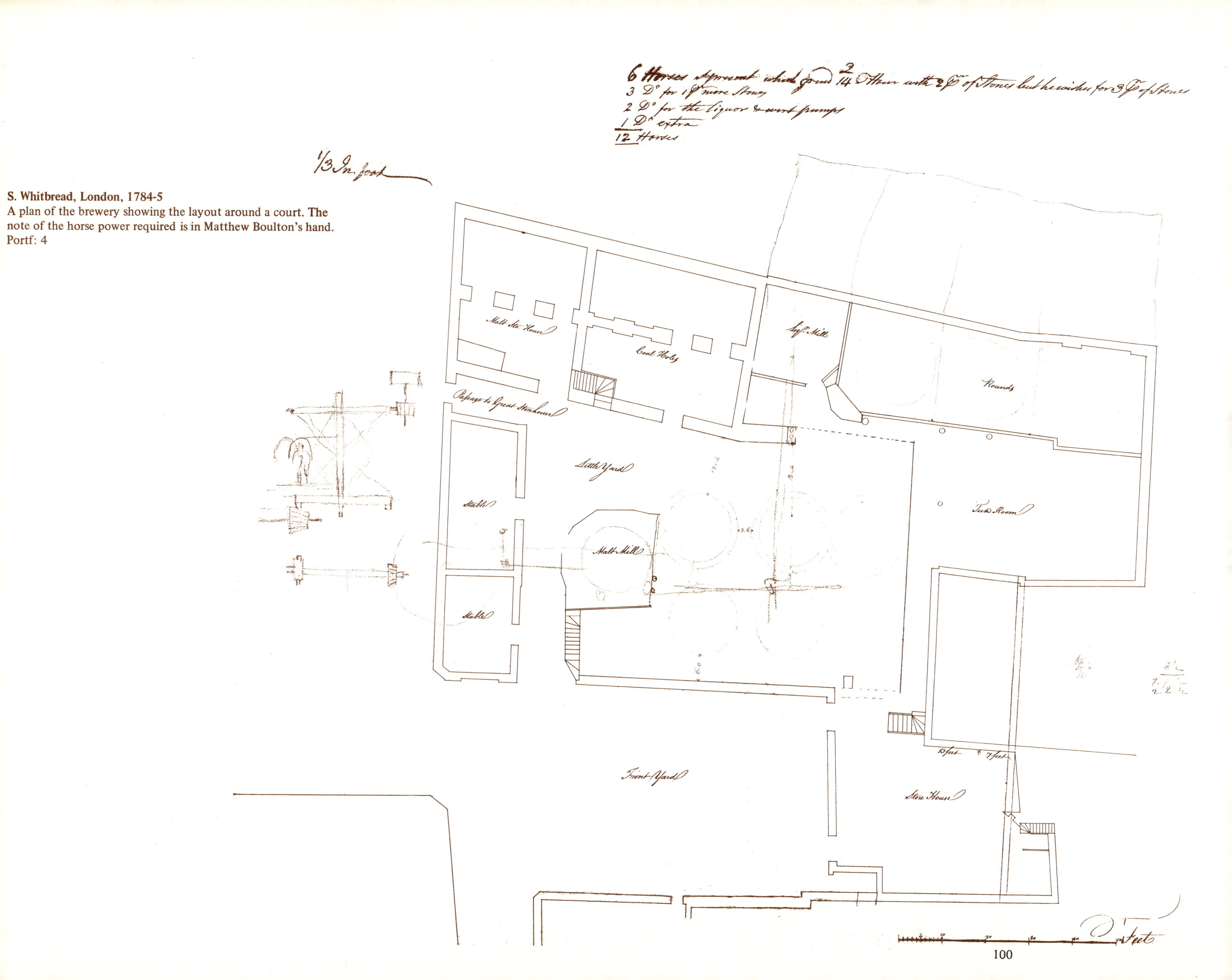

S. Whitbread, London, 1784-5
A plan of the brewery showing the layout around a court. The note of the horse power required is in Matthew Boulton's hand.
Portf: 4

water wheel and engine in favour of a Boulton and Watt rotative engine which Sutcliffe does not seem to have considered.[56]

It is possible to trace Gott's reaction to Sutcliffe's work in detail. In December 1791 Sutcliffe was asked to prepare a plan and estimate for the whole of Gott's factory at Leeds. The estimate being accepted, Sutcliffe engaged masons and bricklayers to begin work on the foundations and had detailed correspondence with Gott on the size of windows and other aspects of factory design.[57] He meanwhile prepared two sets of plans 'The reason why so much time was spent upon them was because Mr. Gott desired they might be drawn two or three times over that no improvements may be omitted. I tould Mr. Gott that I had spent near 7 weeks in drawing the plans'. Gott sent Sutcliffe to London to investigate the different kinds of patent steam engine then in use and also to several textile manufacturers to report on their machinery. He visited Mr. Bryon at Netherton Hall to examine the shears and then travelled to Mr. Beard in Derbyshire 'to examine his shears and consult with him about shearing cloth by water'. In April he visited Mr. Hollings' mill 'to get dimensions of the dyehouse vessels'. When Sutcliffe sent his list of charges amounting to £162. 13s. to Gott towards the end of 1792 he felt that it was necessary to justify his high charge for planning and drawing, almost as if he anticipated trouble: 'I believe there be few Engineers perhaps none but myself in the same situation but what would have charged from £30 to £40 for the design of the plans, over and above the time spent in drawing them. This Mr. Wyate [Wyatt] of London and Mr. Car [Carr] of York have always done. If I did myself justice in this charge the sum ought to be three times as much as I have set down for it'. The sum in question was £10. 10s.[58] Gott reacted as Sutcliffe must have expected: 'The sum far exceeds what we supposed' and he proposed to put the matter to arbitration, suggesting Carr of York; adding 'if Mr. Jesop the Engineer of Newark or Mr. Gardner a considerable millwright of Liverpool, or Mr. Rennie of London a millwright and Engineer be preferred by you we have no objection'.[59] The list of millwrights and engineers drawn up by Gott may indicate the few millwrights whom he considered to be superior to Sutcliffe; at all events it illustrates the exclusiveness of the millwrighting profession. Rennie was chosen and reported some months later that it was impossible to say whether the charges for visits to Bean Ing were excessive or not but 'As to the drawings I can with truth say they are much beyond what I should have charged on what I conceive to be their real value'.[60] He added that a young clerk in his office had copied them in fourteen days. The disagreement over Sutcliffe's charges was not the only point of dispute. Sutcliffe, as at Markland Cooksen and Fawcett's factory, recommended a common engine, in this case a rotary one and immediately ordered it from Bowling Ironworks. Gott asked for the order to be cancelled as he had decided in favour of a Boulton and Watt engine. This prompted Sutcliffe to resign 'as you have got another person to examine and make such alterations in your works as he thought proper my future attendance would be quite unnecessary this being the case you will Excuse me for declining to have anything more to do with them.'[61] He added in a later letter what must be a classic piece of invective: 'You have acted the part of Dark Desyning Hypocrite and have shown yourself to be a man in whose narrow shrivl'd contracted Bosom there is no Room for the admirable virtues of truth Charity Humanity Honour and integrity to dwell in.'[62] Gott thereupon employed Lowe and Ewart.

All the more important millwrights had workshops and employed a number of assistants — Brindley employed at least eight men, Rennie brought men from Scotland with him. By the early 19th century several millwright/engineers employed 150 to 200 men.[63] But it was difficult to keep journeymen millwrights or workers in a millwright's shop. A seemingly endless demand in Lancashire and the major iron and engineering works in the country made the millwright's workman one of the most sought after and best paid craftsmen in the 18th and early 19th centuries.

In June 1786 the London millwrights struck work for higher wages.[64] Rennie's works came to a standstill and he requested Boulton to help him by sending two men to London. It was usually Matthew Boulton who was short of millwrights and workmen, partly because his own men were being constantly tempted away by higher wages from manufacturers who hoped to learn industrial secrets. In July 1788 Ewart who had been trained by Rennie was sent to Soho to work for Boulton and Watt. By October Boulton was needing more assistance and Rennie sent three workmen, presumably from his own establishment, to do a job at Soho for 16/- per week 'they also want their lodgings to be paid but this you must not do'.[65] In September 1791 Boulton was again writing to Rennie[66] 'I must therefore beg the favr of you to lend a man for 9 days and send him down pr coach or if you could part with a good hand I should be glad of one, as James means to return to Scotland when this job is done: however I will not serve you as we are constantly served by persons we lend our men to. I will not bribe or keep your Man without your consent. . . . If you should hear of any good workmen send them to me I want such men as Ramsden wants and employs I also want a good millwright for a constancy — adieu.' Rennie could not help:[67] 'I have not a man in my service besides the ostensible Thore man fit to send from here on any trifling business far less the management of your mill — the fact is some Danish and American pimps that have been for sometime strolling about London have deprived me of several of my best workmen — and I am reduced to the necessity of making foremen of men scarcely fit to be hindsmen.' He continued 'In respect to workmen the cotton trade has deprived this place of many of the best cloth makers and mathematical

---

56. See p
57. Cusworth Hall MSS, J. Sutcliffe to B. Gott 27 May 1792
58. J. Sutcliffe charges 22 Dec. 1792
59. B. Gott to J. Sutcliffe 24 Dec. 1792
60. J. Rennie to Wormald, Fountaine & Gott 2 March 1793
61. J. Sutcliffe to B. Gott 4 Aug. 1792
62. J. Sutcliffe to B. Gott 13 Aug. 1792
63. A. E. Musson and E. Robinson op cit pp 477-480
64. J. Rennie to M. Boulton 28 June 1786, A.O.
65. J. Rennie to M. Boulton 26 Oct. 1788 A.O.
66. M. Boulton to J. Rennie 26 Sept. 1791 A.O.
67. J. Rennie to M. Boulton 19 Nov. 1791 A.O.

Gyfford, London, 1787
This is the only good section of a brewery in the Boulton and Watt Collection. It clearly shows the large amount of room required for cooling.
Portf: 27

Gifford

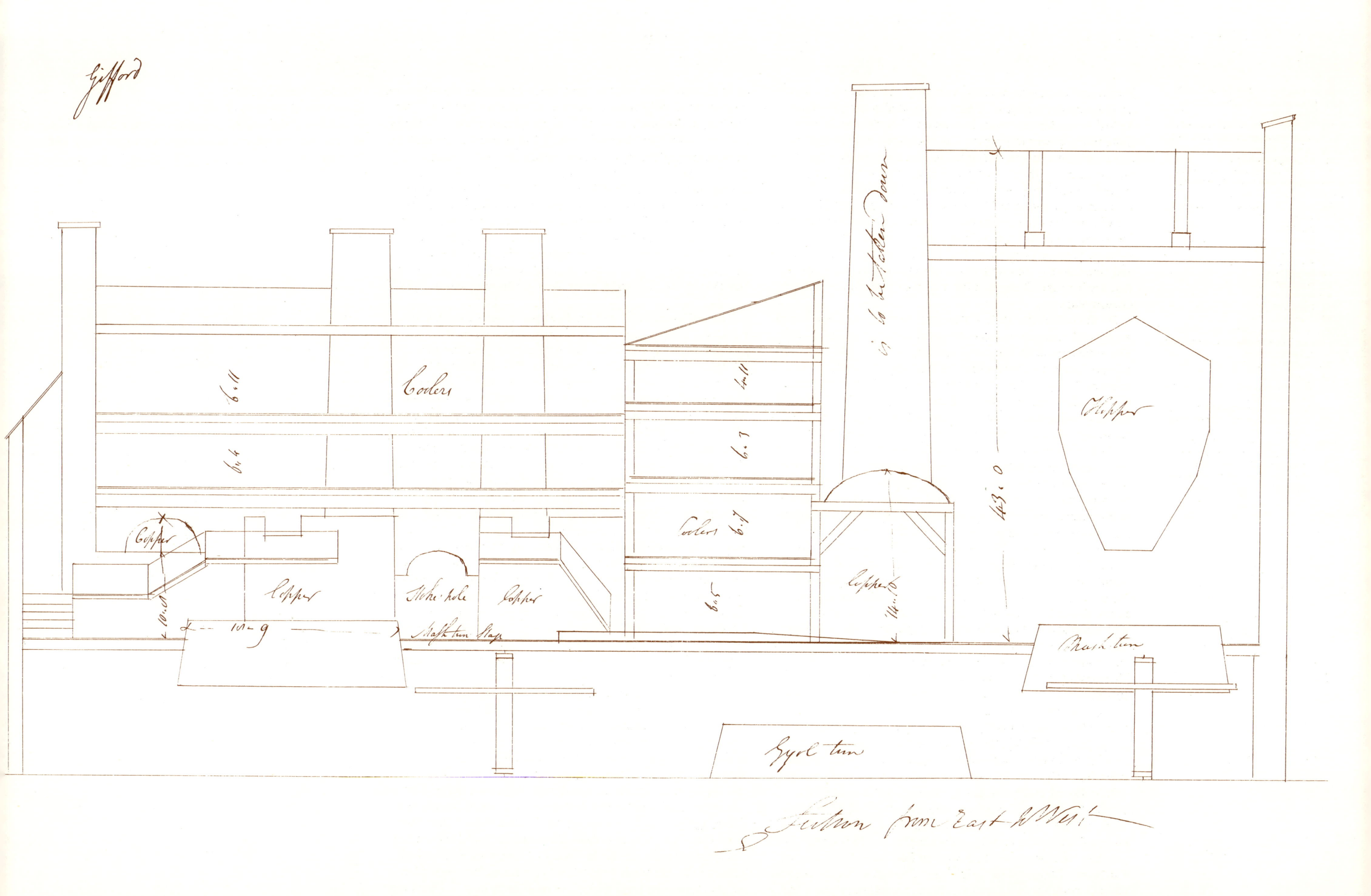

102

68. P. Ewart to J. Watt 28 Sept. 1791
69. P. Ewart to M. Boulton 12 Dec. 1791 A.O.
70. The two Hewes wheels replaced an earlier wheel in the mill, R. L. Hills, op cit. p.112
71. R. L. Hills op cit p.109
72. W. H. Chaloner and A. E. Musson, *Industry and Technology* fig. 12
73. P. P. 1824, V, 340-50
74. A Ure, *The Philosophy of Manufactures*
75. Portf 488
76. Portf 303
77. Portf 525
78. Portf 531
79. Portf 1050
80. Portf 357
81. Portf 510
82. Portf 516
83. Portf 508
84. W. Fairbairn, *On the Application of Cast and Wrought Iron to Building Purposes*, pp.5-6
85. Portf 515
86. Portf 522
87. A. E. Musson and E. Robinson, op cit pp 480-81
88. W. Pole op cit pp III-115
89. A. E. Musson and E. Robinson op cit pp 446, 462
90. S. D. Chapman, 'Fixed Capital Formation in the British Cotton Industry' pt IV. I am indebted to Dr. Chapman for showing me the typescript before publication.

instrument makers so much so that they can scarcely be had to do the ordinary business.' Ewart who had also been contacted replied in similar vein from Lancashire:[68] 'I have done everything in my power to procure him [Boulton] a millwright but, as yet, without success. I applied to Mr. Lowe of Nottingham but he has none that he can recommend that will go — I have engaged some for myself but I am sorry to say they are such as I cannot recommend to Mr. Boulton. I shall apply to Mr. Gardner of Liverpool but doubt he can spare none.' In another letter Ewart reported that it was almost impossible to get good millwrights, filers or turners and that he was having to take 'joiners and carpenters — The very few general good filers and turners that are here are all engaged for a term of years in the different Cotton Mills.'[69]

By the end of the 18th century the distinction between millwright and engineer or machine maker was becoming less obvious. The millwright with his wide knowledge of mechanics was admirably suited to pursue either of the other occupations and many millwrights played a dual role. Earlier in the 18th century Brindley and Smeaton had shown how closely allied mill building and millwork was with civil engineering and had branched into canal, dock and harbour schemes as an extension of their millwrighting occupations. Fairbairn and Rennie began their careers as millwrights but became mechanical engineers while continuing to function as millwrights. Thomas Hewes on the other hand was a millwright and machine maker. Hewes is best remembered for his improvements in water power technology. Belper West Mill was powered by three wheels two of which were Hewes'.[70] It was in Hewes office that Fairbairn worked as a draughtsman when he moved north and it was no doubt Hewes' inspiration that prompted Fairbairn to develop the water wheel further when he went into partnership with Lillie. Hewes built a water wheel for Greg at Styal in 1820[71] and sometime in the 1820s went into partnership with Wren. The partnership of Hewes and Wren and later of Wren and Bennett marked the zenith of the millwrighting trade. Hewes and Wren installed at 25 ft diameter 18 ft wide wheel at Bakewell cotton mill in 1827[72] and when he gave evidence to the Select Committee on Artisans and Machinery in 1824[73] Hewes stated that he had executed millwork at factories in Lancashire, Yorkshire, Gloucestershire, Devon, Scotland and Ireland. He had erected entire fire-proof factories, water wheels, gearing and spinning machinery. The proportion of his business concerned with millwork can be gauged by the fact that he employed about 150 men of whom forty were employed on heavy millwork and the rest on machine making. Wren and Bennett had no equals as cotton factory millwrights in the 1830s and 1840s yet they have been overshadowed by Fairbairn. Ure eulogises Fairbairn's fire-proof factories[74] and Fairbairn's own claims to having revolutionised factory building and millwork have led historians to neglect Wren and Bennett. They were commissioned to execute millwork for some major Lancashire manufacturers such as the Oxford Road Twist Co, Manchester in 1824,[75] McConnel and Kennedy in 1833,[76] G. Cheetham in 1840,[77] Birley and Co 1840[78] and Samuel Greg.[79] They executed the millwork for major textile factories outside Lancashire such as the Mold Twist Co in 1836,[80] The Norwich Yarn Co,[81] and Clarke Maze and Co,[82] of Bristol in 1840. They were employed to design and make the millwork for John Wood's factory at Bradford in 1833.[83] This is particularly interesting because the factory was designed by Fairbairn[84] yet he apparently did not obtain the commission for the millwork. Wren and Bennett also executed work in at least two major glassworks, the Union Plate Glass Co in 1838[85] and Chance Bros of Birmingham in 1839.[86]

The claims made by and for Fairbairn require examination. He is said to have been the man who put factory architecture on a sound scientific basis, the man who played a leading role in the revolution in factory construction. He claimed that his firm were 'the leading millwrights of the district'.[87] Fairbairn's first important commission for a cotton factory was to renew with horizontal cross shafts the millwork that turned the mule spinning in A. and G. Murray's factory. Fairbairn claims that he recognised the great defects in the millwork of cotton mills which were 'all' driven with large square shafts and wood drums; the speeds being increased by a series of straps and counter drums, and he proposed to lighten the shafting and treble the velocity. From Murray's factory Fairbairn and his partner Lillie 'an old shop mate' were engaged to execute some of the millwork at McConnel and Kennedy's factory. Fairbairn says that he believed the new system of double speeds used in the factory was John Kennedy's invention.[88] But iron shafting and gearing had been introduced before Fairbairn 'revolutionised' millwork. Rennie introduced cast iron gearing into Albion Mill in the 1780s and at about the same time Peel used cast iron gearing in one of his mills. Musson and Robinson have shown that Hewes used wrought iron shafting in place of the heavy timber shafting in factories and as early as 1806 Peel, Williams and Co, engineers of Manchester, were producing wrought iron shafting and cast iron gearing. As Musson and Robinson say 'Fairbairn's later claim to have inaugurated the revolution in millwork by substituting slender wrought iron constructions, operating at much higher speeds, in place of the cumbrous old wooden shafts and drums is open to question.'[89] It is certain that transmission systems had been considerably improved before Fairbairn commenced as a millwright on his own account in Manchester and to this extent his claims are unfounded. But on the other hand there is evidence to show that until the mid 1830s, when the spinning mule was rapidly adopted following Roberts' improvements, the small factory unit containing improved (Arkwright) frames or hand mules was more economic.[90] Firms were tending to hold back on machinery replacement and, on bring-

Madder, Dublin, 1809
Constitution Brewery, London, 1807
Examples of the smaller breweries which managed to survive the competition of the large breweries, in the early 19th century.
Portf: 767; 398

Constitution Brewery – 1809.

A PLAN of part of Mr MADDER'S
Brewery
By A. Nevill, 50 Camden St. 1809

References

A — Mill House for Grinding Malt, with three Pair of Stones
B — Steam Engine House — Sky light in Roof
C — Boiler House with room for two Boilers — The Liquor Back over the same.

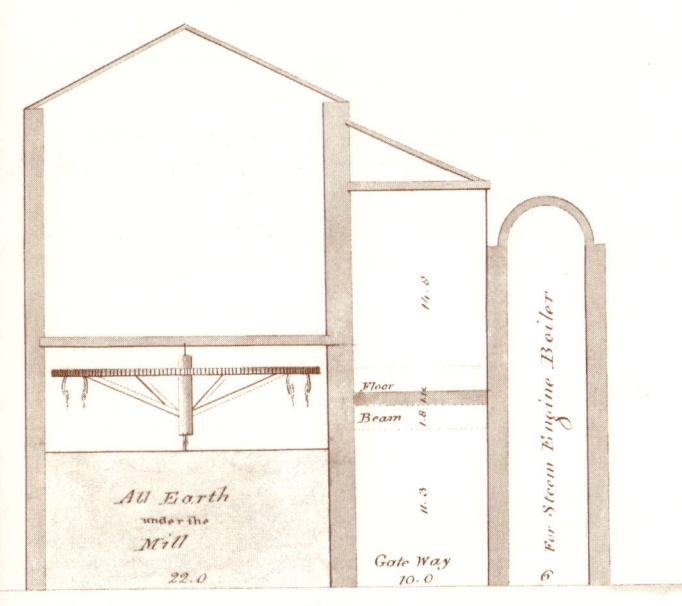

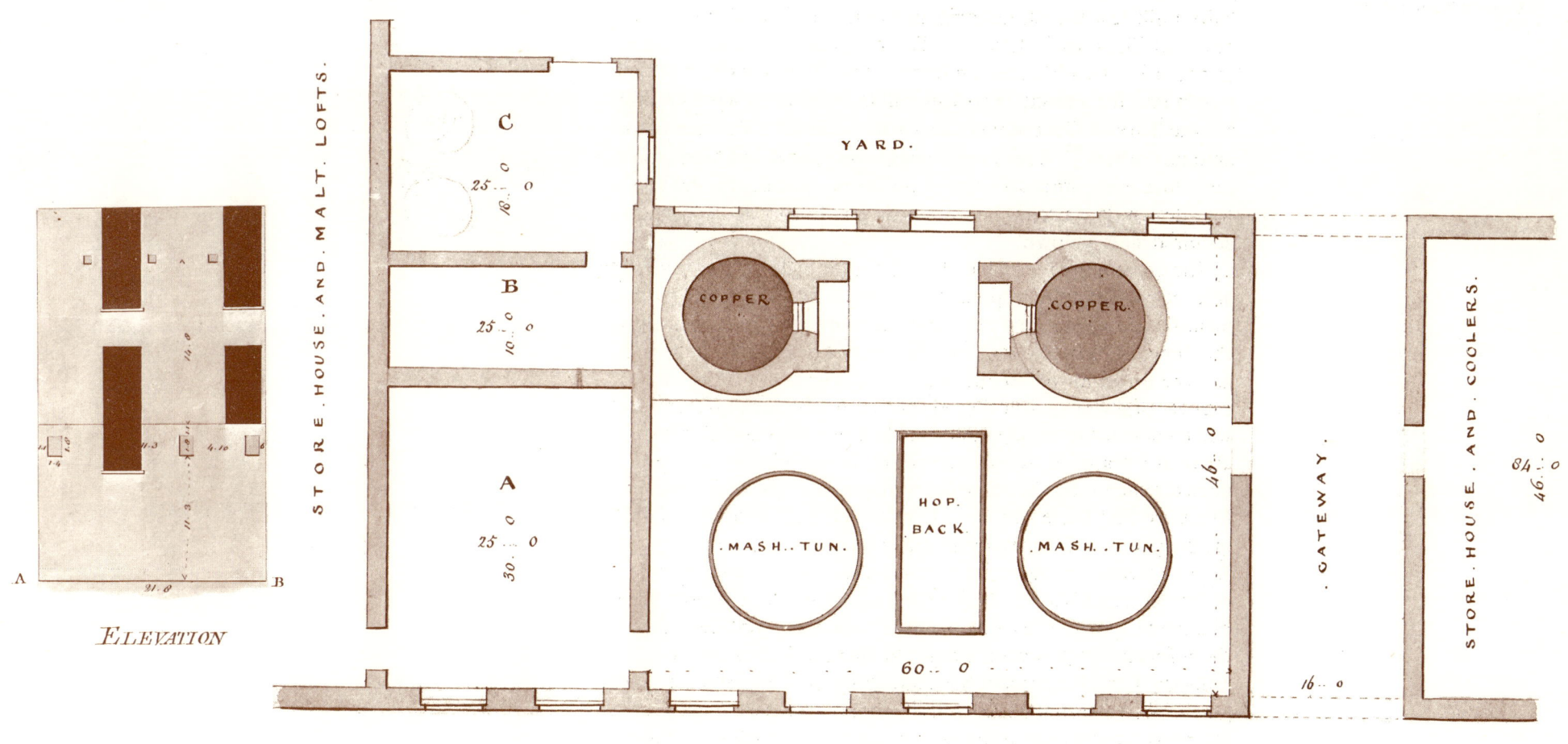

Watling Street
SECTION

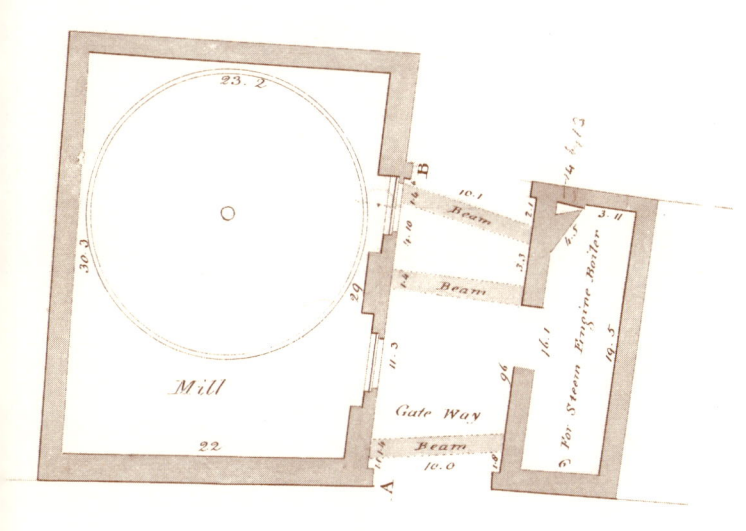

Watling Street
GROUND PLAN

ELEVATION

STORE HOUSE AND MALT LOFTS

YARD

COPPER   COPPER

MASH TUN   HOP BACK   MASH TUN

GATEWAY

STORE HOUSE AND COOLERS

BELL LANE

91. W. Fairbairn, Mills and Millwork pt 2.
92. C. Clowes to J. Watt 10 Feb. 1785
93. The factory was a commercial failure however. A. E. Musson and E. Robinson, op cit p. 373
94. M.H. Mackenzie, 'Cressbrook and Litton Mills, 1779-1835,' *Derbyshire Arch. Journ.*, 1968, pp.4-5; Sun Insurance Register CS 644067, I am indebted to Dr. S.D. Chapman for this reference.
P. Ewart to Boulton & Watt 19 Jan. 1792.

ing their millwork up to date. When Fairbairn installed his first iron shafts and gear wheels the multi-storey fire-proof factory was a comparative rarity and the adoption of improved transmission systems was therefore slow. Fairbairn's important millwork coincided with the building of larger factories to house the automatic mule in the 1830s and 40s. To this extent it was perhaps natural, although misleading, that he should be given the credit for the innovation of improved millwork. Fairbairn built few British factories but both Orrell's mill in Stockport and Titus Salt's Saltaire mills were greatly praised by contemporaries. He obtained a number of prestigious export orders: a corn mill for a pasha, a cotton mill in Bombay, a woollen mill for the Turkish Government and a flax mill for a Russian baron amongst others.[91] Where no expense was spared Fairbairn could introduce every improvement in millwork, machinery and factory design. It is probably not a little due to this that Fairbairn acquired such renown.

The entrepreneur depended on the millwright to a considerable extent. Many entrepreneurs had little or no knowledge of manufacturing and they relied on the millwright to recommend the dimensions of the factory, the kind of machinery to be used and the prime mover to operate it. Millwrights, as others, were capable of giving conflicting advice and unless the entrepreneur was mechanically minded like Marshall of Leeds and able to arrive at a reasoned decision he could involve himself in unexpected expense by choosing the wrong millwright or by failing to ensure that there was co-ordination within the factory building programme. Even Marshall made mistakes; he chose a Wrigley steam engine which within a year was replaced by one of Boulton and Watt's. Charles Clowes[92] wrote 'our people have blundered greatly I say this much because I am confident that the information you want cannot be had until our surveyor, millwright and backmaker have altogether had a conference with you.' Some manufacturers were fortunate in succeeding in tempting a millwright to become a manager or partner in their concern. T. Yeoman was manager of Cave's spinning mill at Northampton in the 1740s[93]; in *c* 1783 William Newton, millwright and formerly a head carpenter on the Duke of Devonshire's building project at Buxton, built Cressbrook Mill for Richard Arkwright and subsequently managed it. William Jessop, millwright and engineer, was a partner in the cotton spinning firm of Sketchley, Handley, Jessop and Marshall of Newark in *c* 1795 and Peter Ewart went into partnership first with Oldknow and later with Greg. Oldknow was well aware of the advantages of having a millwright as partner 'he [Oldknow] had already done a good deal in printing by rollers and being confident that there is a great deal more to be done which requires good machinery he expects me to be of considerable use in that way.'[94]

The master millwright was a man of considerable status in the 18th century; in great demand, well paid and with good prospects of partnership with men of means. But he left few personal records and until more detailed work has been done on, for instance, country estate papers and 18th century business records, he will remain something of a mystery.

Barclay and Perkins, Anchor Brewery, Southwark, 1786.
A detailed plan of part of one of the largest London breweries. The drawing is misleading for a plan reproduced in Peter Mathias' *The Brewing Industry in England 1700-1830* (plate 2) shows the site to have been much larger than it appears to be from this plan.
Portf: 10

Gosport Brewery, Hampshire,
A plan of part of the naval victualling yard. Smeaton installed a horse wheel at the brewery in the 18th century.
Portf: Misc. Mills

106

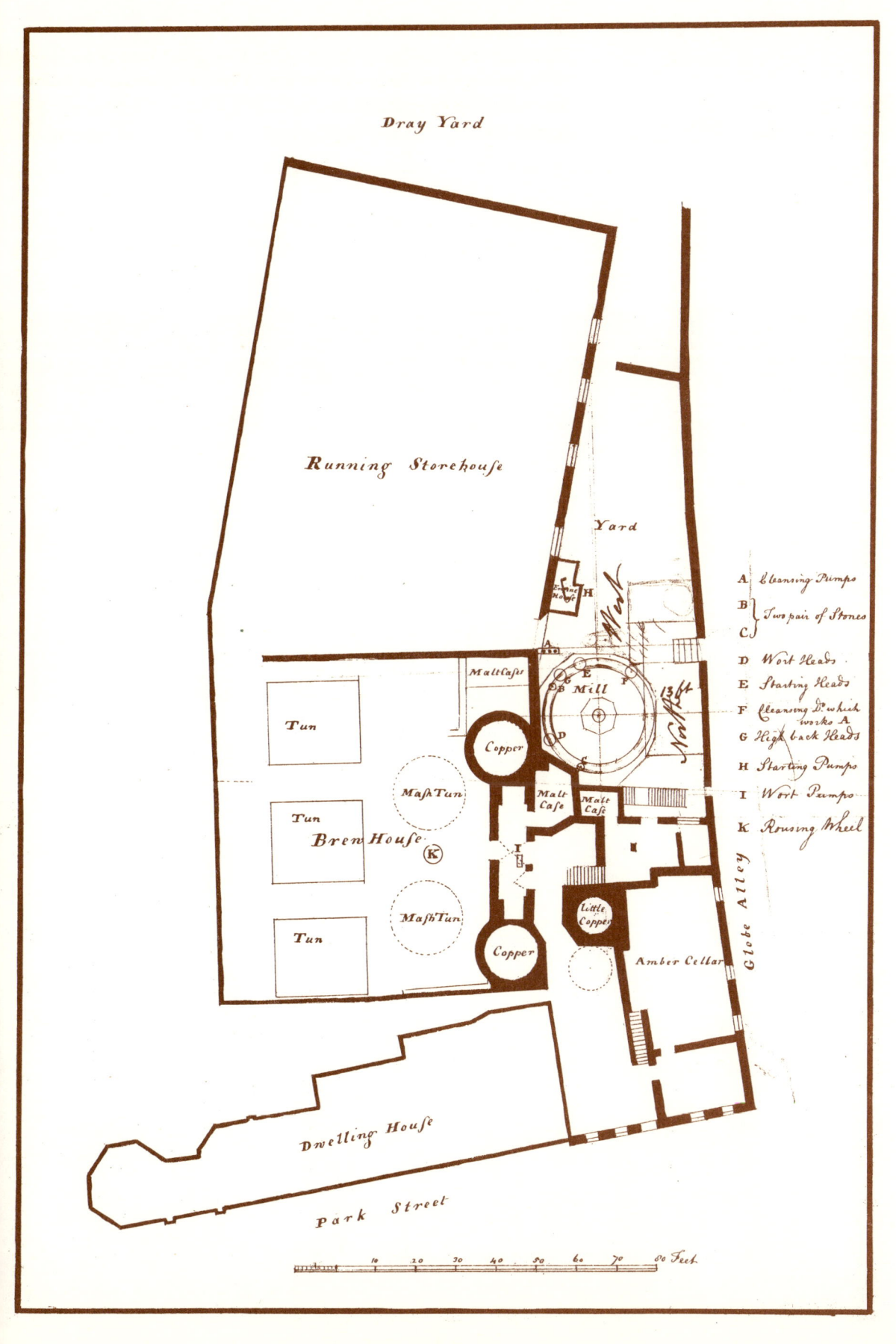

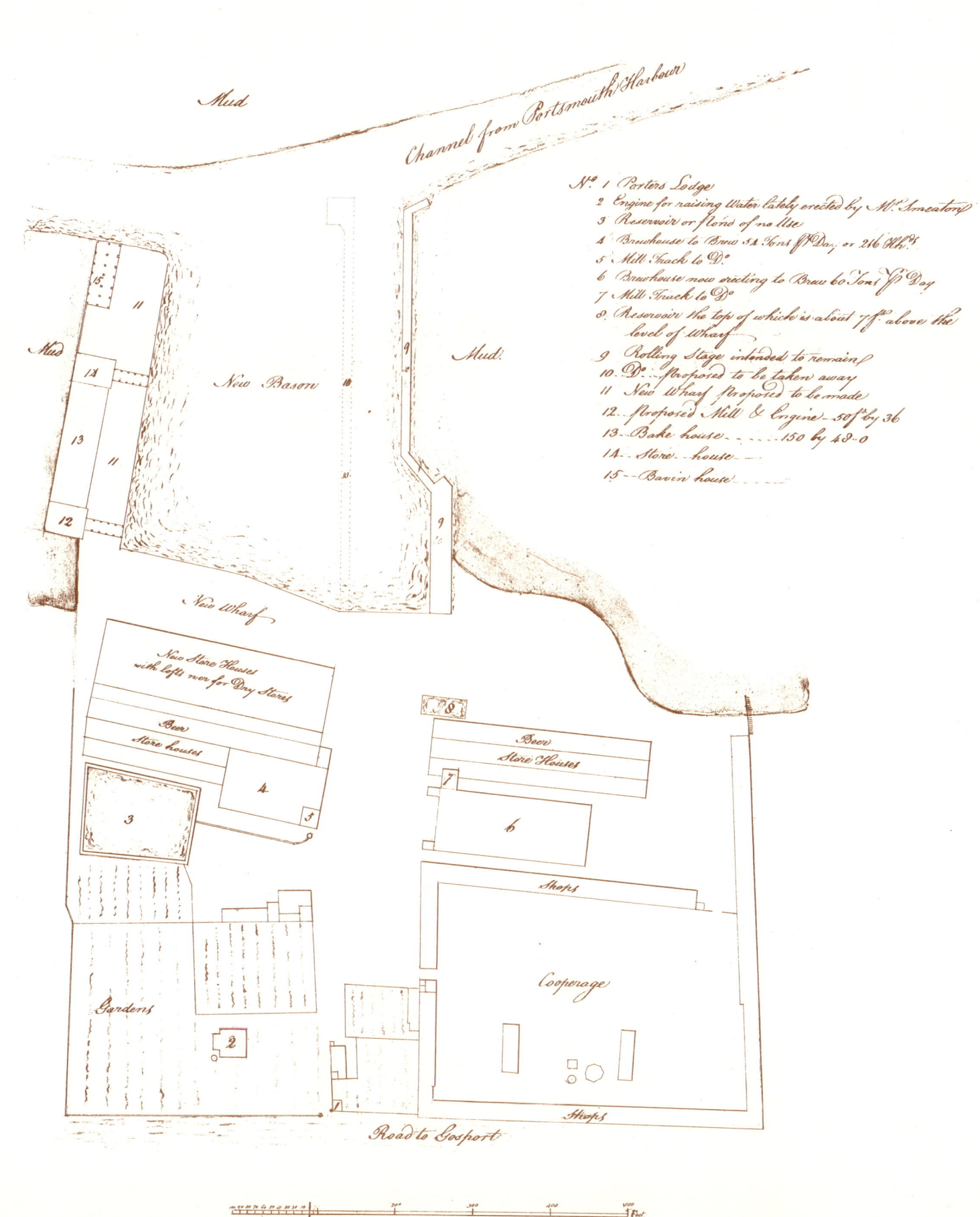

Aitcheson & Brown, 1787
A grinding mill for a malt distillery.
Portf: 18. Scale approx 1in : 9ft 8in.

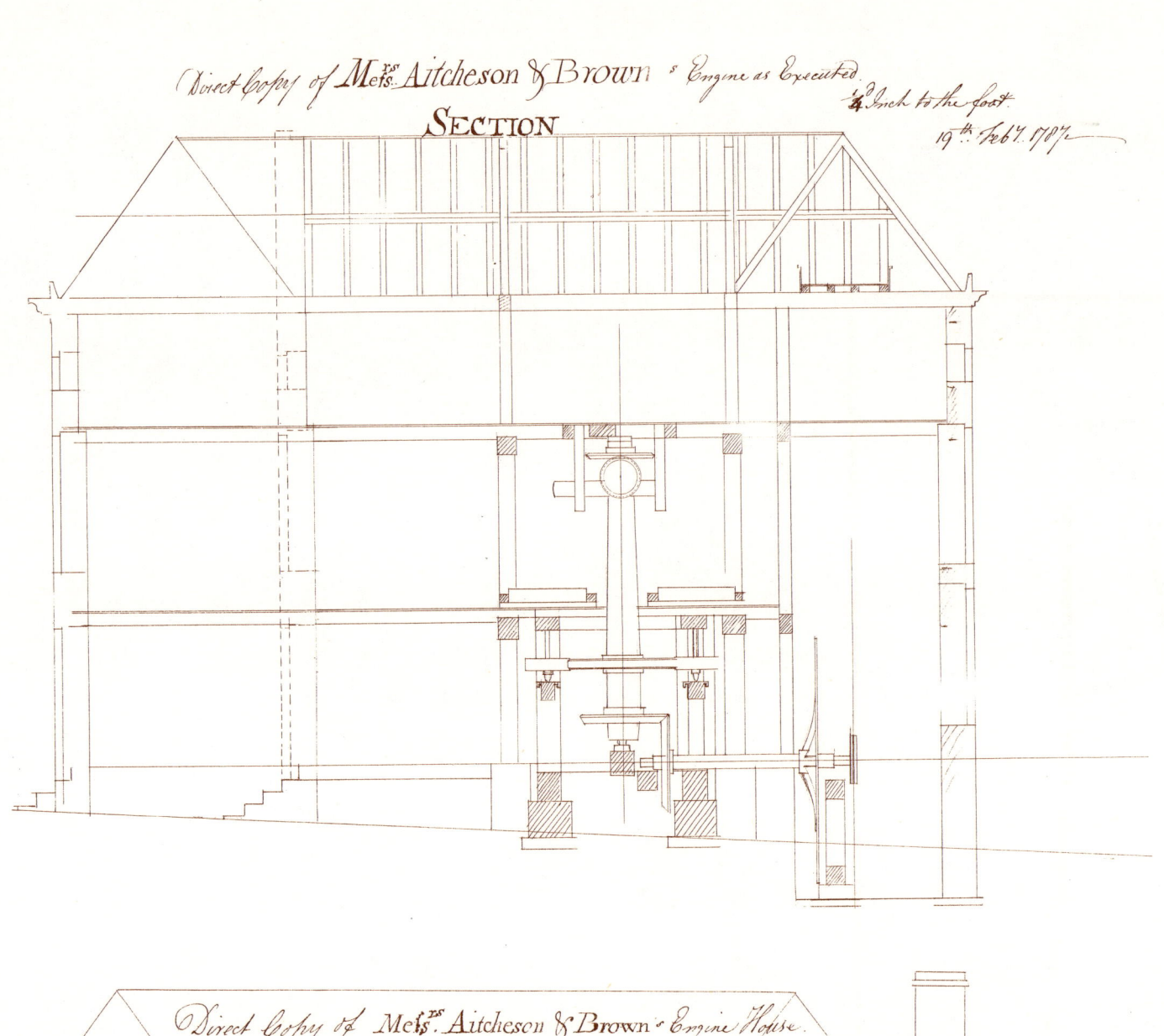

John Busby, jun., 1837
This drawing provides a contrast to Aitcheson and Brown's small 18th century distillery. By the mid 19th century the distillery covered a considerable area.
Portf: 519

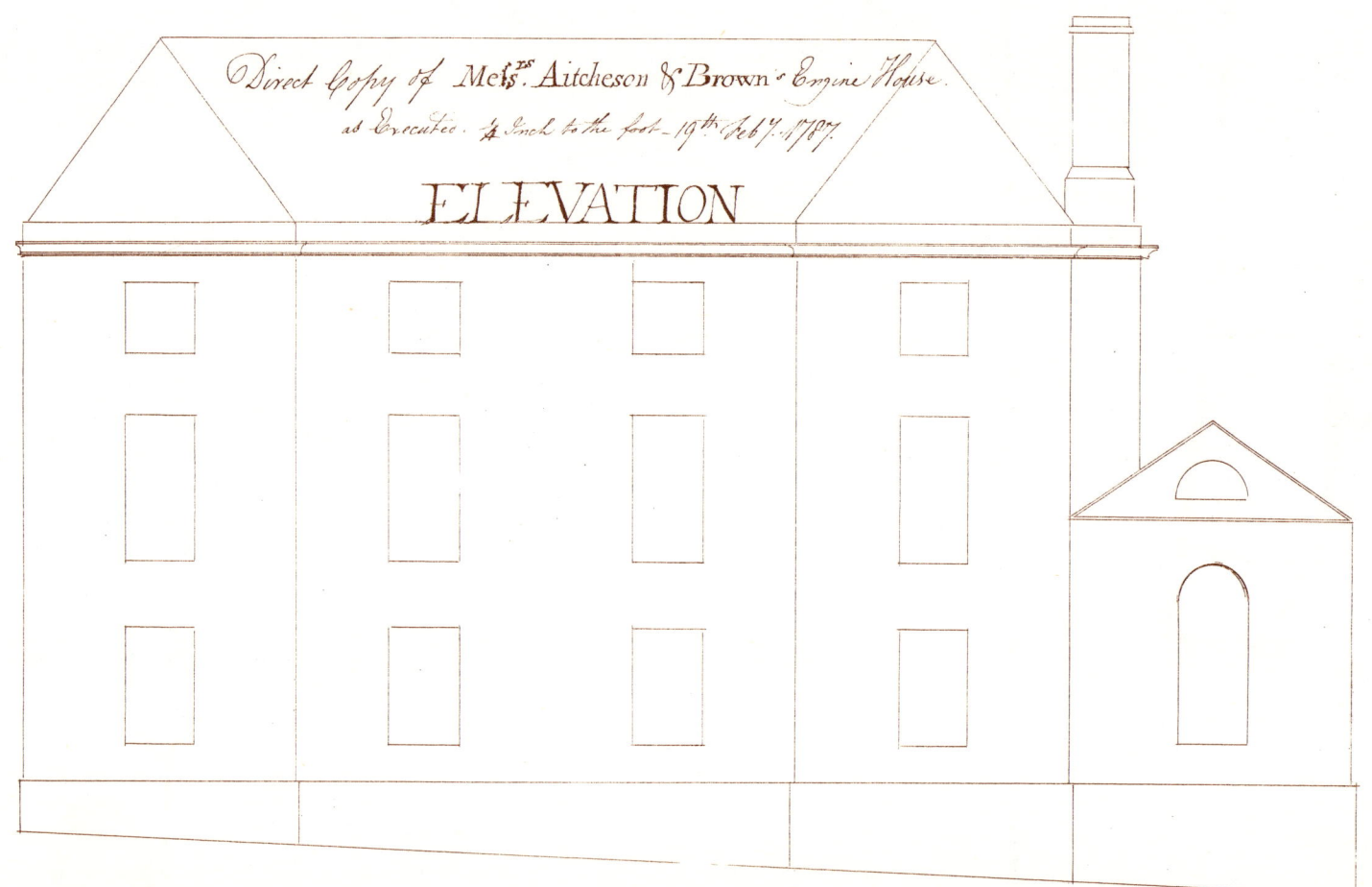

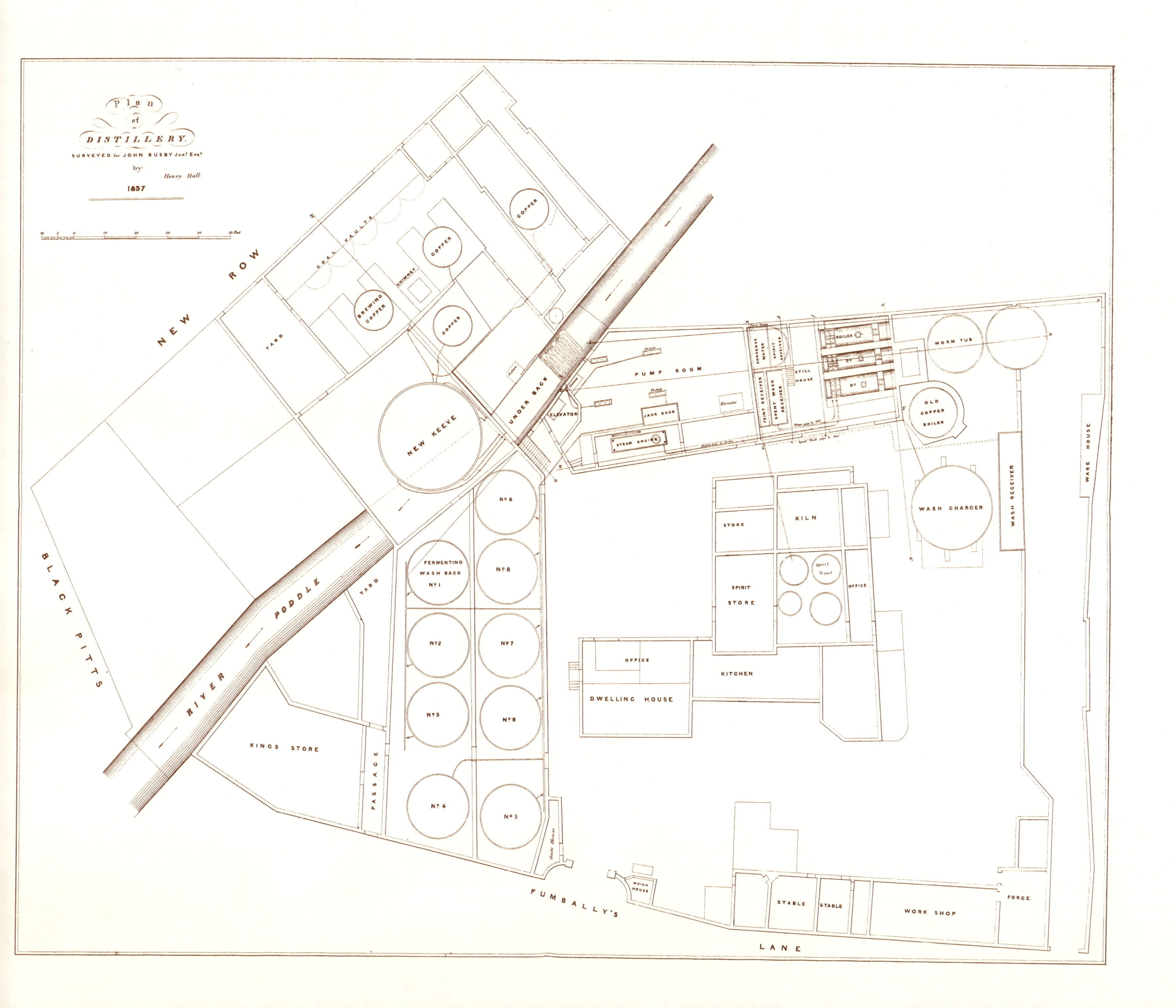

7. Factory Heating

1. Box 1334. For a brief introduction to the history of heating see A. F. Dufton, 'Early Application of Engineering to the Warming of Buildings,' *Trans Newcomen Soc.* 21. 1940-1.
2. Thomas Tregold, *Principles of Warming and Ventilating Public Buildings, Dwelling Houses, Manufactories, Hospitals, Hot Houses, Conservatories, etc.* 1824, p. 167.
3. ibid p. 289.
4. *Philosophical Transactions* 1745.
5. *Journal Royal Institution*, vol. 3.
6. Patents nos: 1816, 1968.
7. Box 30. un-named pamphlet p. 173.
8. *Rees Cyclopaedia*, art 'Manufacture of Cotton; M.C. Egerton, 'William Strutt and the application of convection to the heating of buildings,' *Annals of Science*, 24, 1968.
9. C. L. Hacker, 'William Strutt of Derby,' *Journ. Derbys Arch. and Nat. Hist. Soc.* 80, 1960, pp. 57-9.
10. G. Salvin to Boulton and Watt 26 March 1796.
11. Shrewsbury Library, Bage Papers, Bage to Strutt, Oct. 15, 1802.
12. Ibid, Nov. 11, 1802.

There were obvious humanitarian motives for the provision of factory heating by the entrepreneur. At Samuel Oldknow's Mellor Mill it was noted that 'the place is merely wanted to be a comfortable temperature for the workpeople'. Hick's woollen mill at Stonehouse was to be heated to about 60°, a reasonable temperature to work in.[1] The pessimist might consider the provision of heating from another angle for no doubt there was an optimum temperature at which child apprentices and the rest of the labour force could achieve their highest output.

But leaving aside the human dimension, the heating of some part of the factory was necessary in a number of industrial processes. In the pottery industry, for example, slip had to be dried in buildings known as stoves. Stove buildings were required in the bleaching, dyeing and printing of cotton and also in the woollen and worsted industries. A temperature in the 70s was necessary in sugar refining and for drying paper in paper mills. A certain amount of moist heat was required for cotton and flax spinning. 'Until the machinery acquires a certain degree of warmth, the spinners find it nearly impossible to keep their work in order; and this is most felt on Monday mornings, when every thing has become cold and adhesive, through being a longer time at rest. It is an evil because in addition to producing bad work, it too frequently occasions the children employed to be treated with unmerited severity.'[2]

Until the latter part of the 18th century the open fire was the only source of heat available and this was unlikely to raise the temperature greatly in a large room. It was costly to have a separate fire on each floor and inconvenient too. Shute's Silk Mill at Watford was heated by thirteen separate stoves in 1816.[3] Nevertheless, the smaller firms continued to use this method of heating throughout the 19th century. But during the 18th century experiments were conducted on other forms of heating, notably on warm air and steam. As early as 1745[4] steam was suggested as a medium for warming rooms although it does not seem to have been applied to the purpose until the end of the 18th century. Count Rumford asserted that 'this scheme has frequently been put to practise with success in this country, as well as on the continent',[5] but there is little evidence to support his view. Patents were granted to John Hoyle in 1791 and to Joseph Green in 1793[6] for methods of warming buildings by steam and Matthew Boulton became interested in the possibilities of steam heating. In 1794 he assisted the Marquis of Lansdowne in improving the steam heating apparatus erected by Green for warming the Marquis's Library. But the scheme was abandoned because of the defective jointing in the pipes. In 1795 Boulton superintended the erection of a similar system in Dr. Withering's library. But this system was also dispensed with because the pipes being made of copper and softly soldered together emitted 'a most disagreeable effluvium which rendered it unpleasant to Dr. Withering who was then in an infirm state of health and suffering from a disease on his lungs.'[7] These setbacks probably account for the time lapse between Boulton's domestic experiments and Boulton and Watt's first steam heating installation in a factory in 1802.

During the 1790s experiments were conducted on warm air heating in factories, and by 1792 William Strutt had designed a system for Belper Mills. Air was conducted into the building and brought in a current to the external surface of an iron 'cockle', a cubical vessel inverted over a fire. The warmed air rose in ducts to each floor of the building and was admitted to the rooms through hatches. Ventilators ensured that there was an adequate circulation of air.[8] This system was installed in the Derby Infirmary designed by Strutt, and Sylvester used it at the Wakefield Lunatic Assylum.[9] In 1796 George Salvin wanted to heat his Durham factory with warm air but wrote to Boulton and Watt saying, rather pitifully that he was at a loss to know how to set about it.[10] Charles Bage of Shrewsbury was obviously impressed by this method of heating although, characteristically, he subjected it to tests and altered Strutt's plans to suit his own requirements. In the first instance Bage was interested only in warming his drying rooms for 'bleached yarn, dyed thread and yarn brought wet from the reels'.[11] 'But', he added 'I fear I may not fall on the best and most economical plan of constructing the warming parts.' He asked Strutt about the best means of conveying smoke away as he was concerned about the possibility of smoke percolating into the drying room and spoiling newly bleached yarn. The other problem which concerned Bage was the best means of diffusing warmed air to all parts of the room. 'Is any contrivance necessary for dashing and diffusing the heated air to every part of the room?' He enclosed a plan of the proposed bleach house and stove 'on your plan about 8 by 9 feet'. It is interesting to speculate why Bage, considering his knowledge of heat transfer – he was using steam to heat the dyeing and bleaching vessels – did not apparently consider using steam to heat the stove building. Three boilers to be used for supplying a small steam engine and for heating the dyeing and bleaching vessels are shown on his diagram. It is hardly likely that he was unaware of the possibilities of steam heating. In the absence of other evidence one must suppose that he had heard unfavourable reports of it. In November[12] he wrote again to Strutt 'I have annexed a scheme for warming a drying stove, the best that has occurred to me, and will thank you to fulfill your promise of finding all the faults you are able, and also to send me a sketch of your latest improvement which I am fully prepared to expect is much superior.'

The iron cockle in Bage's stove was six feet in diameter, surrounded by a dome of brickwork which terminated at an opening to allow the warm air to ascend. The cockle rested on firebricks and was inverted over a grate. Cold air was admitted through the walls at the four corners of the room and was conveyed in cast iron pipes to the edge of the cockle. Warm air was diffused from the opening in the brickwork which surrounded

Factory Heating
Gardom Pares and Co, Calver Mill, Derbyshire, 1807
This drawing shows one of the earliest methods of heating factories by steam. As this was a water powered mill only a small 6 hp boiler was required. It is interesting that Gardom and Pares chose steam heating rather than hot air heating for it was generally considered more economical to install the latter in a mill that was not driven by a steam engine.
Portf: 1334

Reverse)

Sketch of the heating pipes. 10 Inch to the foot   May 1st 1807.   Messrs Gardon Pares & Co.

Section

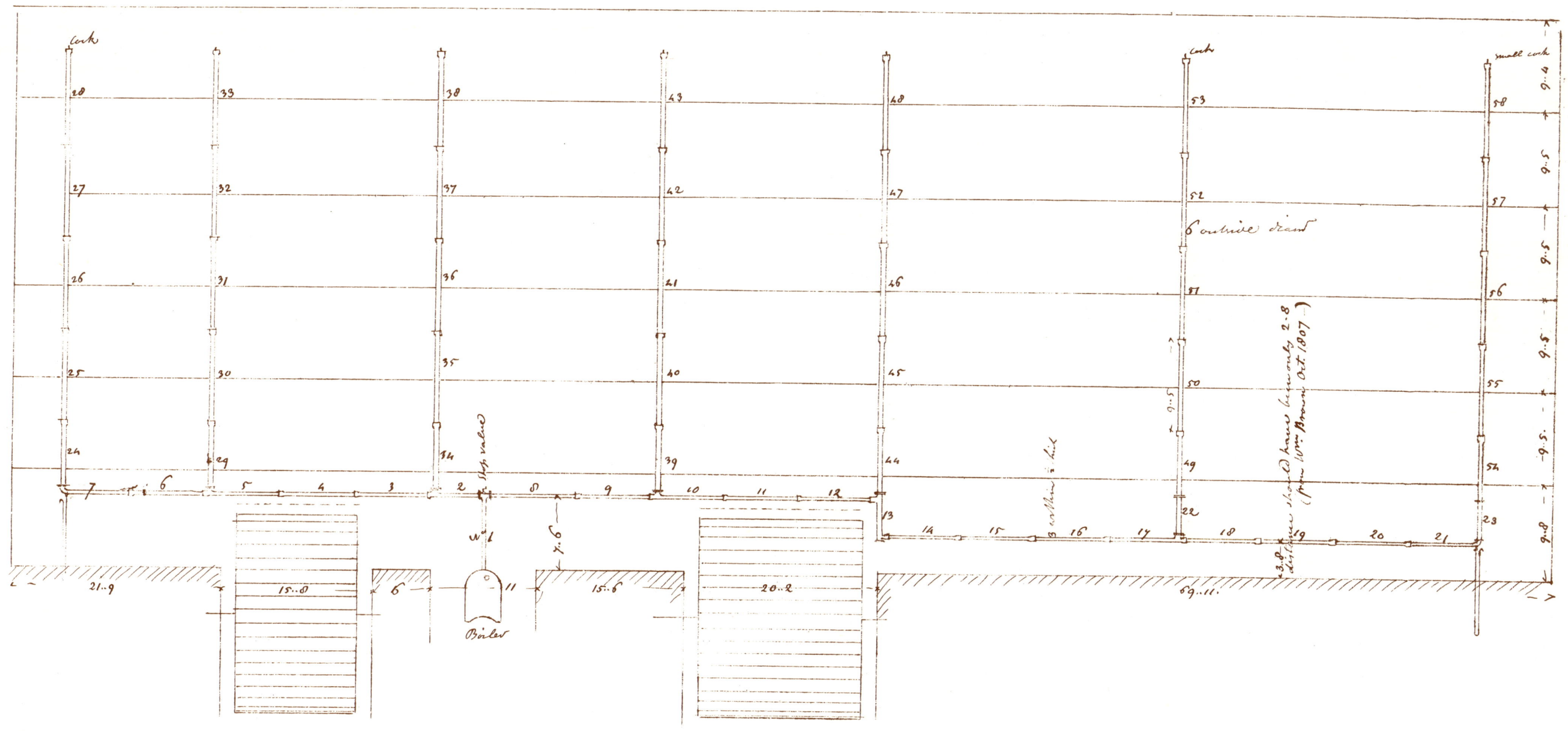

Plan

Mill 160 feet by 30

6 horse Boiler

see drawing of Boiler & setting to a larger scale

160 × 30 × 57 = 275000 cubic feet heated
600 feet of Surface

NB In the rooms through which the pipes pass entirely the space heated by 1 foot of surface of pipe is **436**

110

13. Ibid, no date.
14. Ibid, March 5, 1805.
15. Brotherton Library, Marshall MSS 37 p. 10.
16. James Watt Jun. to J. Mylne, May 11, 1808, Misc. Mills Box.
17. Southern to Lawson 10 March, 1796, Foundry Letter Book. Southern remarked that as Lawson was acquainted with Dale's engineers B. and W. would like him to find out what method Dale employed to warm his mills. The following paragraphs are based on W. Kelly's reply. 24 March, 1796, pf. 1334.
18. Brotherton Library Gott MSS. G. Rennie to B. Gott 22 July 1819.
19. R. S. Fitton and A. P. Wadsworth *The Strutts and the Arkwrights* p. 215.
20. T. Tredgold pp. 166-7.
21. Robertson Buchanan, *An Essay on the Warming of Mills and other Buildings by Steam*, Glasgow 1807, p. 8.
22. G. A. Lee to J. Watt Jun., 17 Jan 1802, M IV L. This is the only evidence that B. and W. were at all involved in the heating of the Salford Twist mill. There are no drawings of heating systems for this firm.
23. Buchanan op. cit. p.8.

the cockle and from four earthenware ducts which led up into the drying room. Shortly after sending his plan Bage wrote again[13] 'you said you would find as many faults with my stove as you could . . . I hope you will not only do this but send me a better. The final arrangement of our plans is delayed till we have your approbation or censure.'

Strutt apparently approved the plan but it was not altogether successful for Bage wrote in March 1805[14] 'For warming your inverted pots [cockles] were adopted with tolerable success. Though I think with most effect when the internal brickwork is carried up some height. You have experienced the same inefficiency of this mode of warming for the room on the same level as the pot', and Bage adopted some kind of fan system to diffuse the warm air to lower levels. When Marshall visited the mill he noted 'I have long thought that our present stove was very improper for drying thread in and too small — steam heat would be far better.'[15]

Strutt's warm air system received the approbation of several noted industrialists of the day. After consultation with Strutt in 1808[16] Boulton and Watt installed the apparatus in the Hunterian Museum, Glasgow in c. 1810; and Boulton's house was heated in a similar way. David Dale with his manager, William Kelly of New Lanark Mills, had tried out three warm air systems by 1796.[17] The first mill built at New Lanark had a simple stove in the cellar at each end of the building and from each stove warm air was conducted up an iron pipe which passed through each storey at the end of the mill. At each storey there was an aperture which could be opened and closed at will. But it was quickly found that 'owing to the propensity of the rarified or warm air to rush to the top; and the smallness of the body of heated air in the pipe; we found it impracticable to warm more than one room at a time.'

This system was modified in the next mill. In this case the warm air was conducted in a brick passage enclosed in the end gable of the mill. Smoke, contained in an iron pipe, was also carried up this passage. A damper across the passage at each floor controlled the ascent of the warm air and another vent controlled the admission of warm air to each room. This system made 'a very good stove' according to Kelly but it had the disadvantage of the vents and dampers having to be regulated manually 'which the hands in the room are always disposed to alter as suits their inclinations'.

In the third and most satisfactory heating system at New Lanark a much larger column of heated air was carried up through the end of the building, the column diminishing in size towards the upper storeys. Only one stove was found necessary to heat a building 150 ft by 30 ft. The apertures to each room remained constant but diminished in size storey by storey. By this means 'the heated air is so regulated as to warm all the rooms at once'. This method, Kelly noted, 'we find answers most completely'. He added that the systems that he had described 'are the only kinds [which] have been seen fit for warming large houses upon a safe principle'. George Rennie had no doubts about which method of heating he considered the best: ventilation by warm air is now preferred to steam in most of our public Establishments here. Its superiority consists in a constancy & equability of Temperature which can be regulated to any proper intensity & in its salubrity & economy & from what information I have obtained I should be inclined to recommend the mode adopted by the London Institution as best adapted to your purposes. But then it is necessary to make such arrangements in your Building as shall answer the end required. 1st by placing the Apparatus on the basement storey or below it which is preferable . . . 2nd That the current of Air over the surface of the Stove Cockle should be so copious that the Cockle should not exceed 280 or 300 degrees Fah^t. lest the Air become burnt & vitiated.'[18]

Not only was the stove system safe but it was also cheap to run. It was certainly cheaper than having open fires in each room and George Lee told William Edgeworth in 1811 that Strutt's method of heating was 'the best and cheapest'.[19] For the manufacturer who used water or animals for the motive power in his factory there seems little doubt that a hot air system was the most suitable. But if steam power was used in the factory it was cheaper and effective to use steam for the heating system also, and for cotton mills 'it is found greatly superior to other methods'.[20]

Buchanan claims[21] that a Mr. Snodgrass was the first person to heat a factory by steam when he installed a heating system in Dale and McIntosh's Speyside cotton works in 1799. George Lee followed shortly afterwards by installing a heating system whilst building the Salford Twist Mill. The structural cast iron pillars of the mill also served as heating pipes and steam was conveyed from storey to storey through them. In January 1802 Lee told James Watt jun. 'Please to acquaint your father and Mr. Southern that we have admitted steam into the building and the elongation was as expected 1/10th of an inch per 10 feet nearly.'[22] T. Houldsworth of Glasgow followed suit when building his mill. Richard Gillespie, a calico printer, introduced steam heating to his warehouse and then gradually to other parts of the works. Messrs Orr installed heating in their Stratford upon Slainy (Ireland) works and were shortly followed by other Irish cotton spinners and manufacturers.[23] After the successful heating of Philips and Lee's mill in 1802 Boulton and Watt received a slow but steady flow of orders for steam heating apparatus. Some but by no means all of the orders came from owners of Boulton and Watt steam engines. Pooley, owner of a Boulton and Watt engine, installed a steam heating system in 1804. He was followed in 1805 by Radcliffe and Ross who needed it for drying ground flint. Boyes and Carlill, sugar refiners of Hull, also installed steam heating in 1805. Stevenson and Houston, cotton spinners, installed their heating system in 1806 and Wedgwood

Robert Arkwright, Bakewell Mill, Derbyshire, 1813
Two methods of steam heating are shown; the main block was heated by steam rising in vertical columns whilst in the wing the steam travelled in horizontal pipes at ceiling level.
Portf: 1334

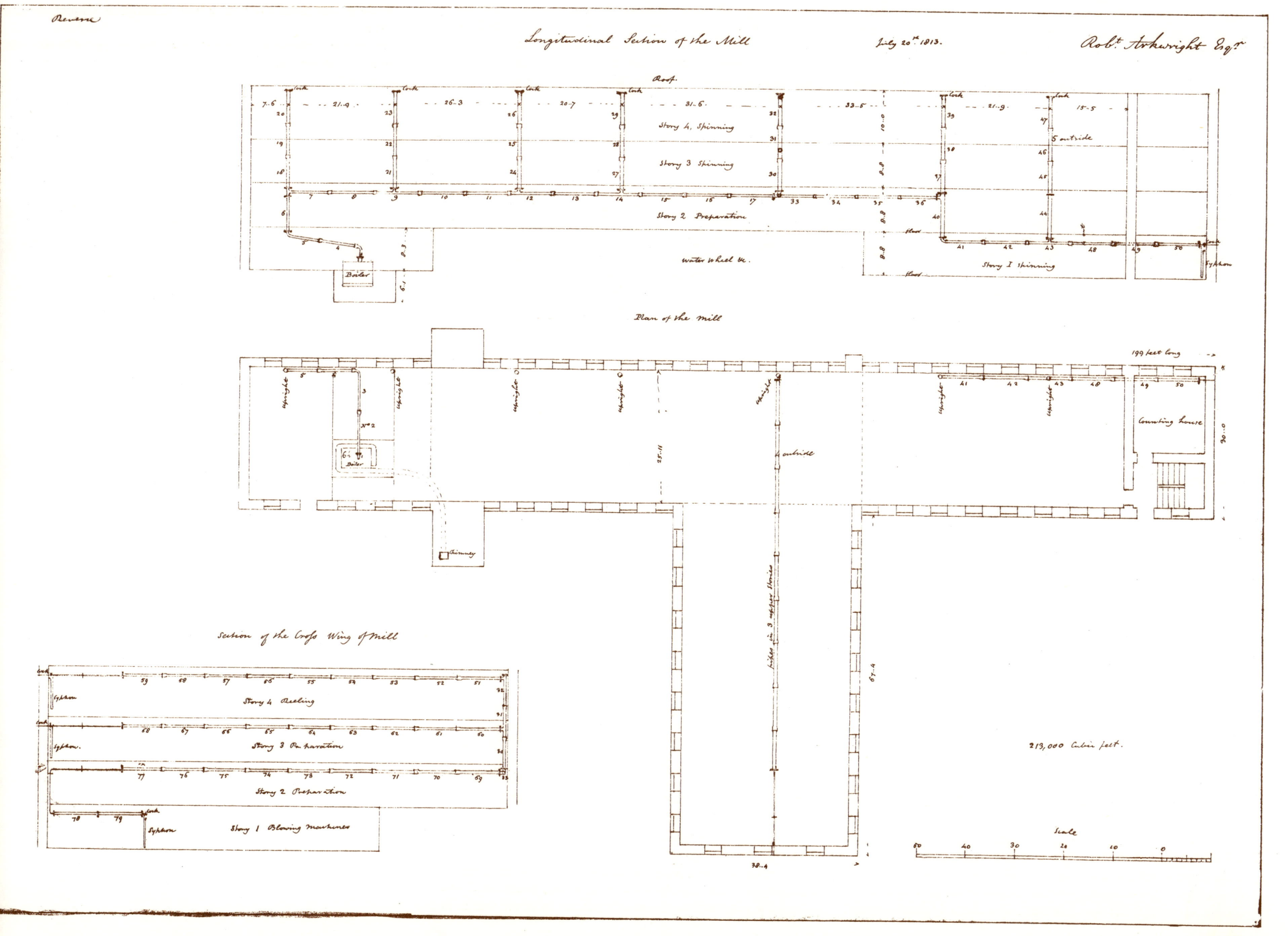

Longitudinal Section of the Mill   July 20th 1813.   Robt Arkwright Esqr

Plan of the Mill

Section of the Cross Wing of Mill

Story 4 Reeling
Story 3 Preparation
Story 2 Preparation
Story 1 Blowing Machines

213,000 Cubic feet.

24. Notes and drawings pf. 1334.
25. Buchanan op cit. pp. 9-12.

and Byerley, owners of a steam engine, employed steam heating in their stone drying house in 1807. Gardom and Pares, owners of a large water driven cotton mill at Calver, installed steam heating in 1807 and they were followed by Samuel Oldknow at Mellor in 1809. In 1810 Boulton and Watt undertook the heating of two important non-factory buildings — Covent Garden Theatre and Lowther Castle.[24]

Steam Heating Apparatus ordered from Boulton and Watt

Firm	Building	Date
Philips and Lee	Cotton Mill	1802
John Pooley	Cotton Mill	1804
Radcliffe and Ross	Stone Drying	1805
Boyes and Carlill	Sugar House	1805
John McCracken	Cotton Mill	1805
Stevenson and Houston	Cotton Mill	1806
Wedgwood and Byerley	Stone Drying	1807
Gardom and Pares	Cotton Mill	1807
Williams Cooper and Boyle	Paper Mill	1808
S. Oldknow	Cotton Mill	1809
Birley and Hornby	Warehouses	1810
Cardwell	Warehouse	1810
Duffy Byrne and Hamill	Print Works	1810
Covent Garden Theatre	Theatre	1810
Lord Lonsdale	Lowther Castle	1810
J. Spode	Stone Drying	1812
Davis	Hat House	1813
R. Arkwright	Cotton Mill	1813
Budgein	Paper Mill	1813
Gardom and Pares	Cotton Mill	1814
H. Hicks and Son	Woollen Mill	1816
La Compagnie d'Ours camp		1823

Source: B. & W. MSS Pf. 1334.

Technical problems were solved in a characteristically empirical fashion. Boiler size and the length of steam piping required to heat a factory were arrived at by a comparison of the efficiency of the heating systems in the first factories to be heated by steam. If a boiler was to be installed solely for the purpose of supplying steam to heat the factory the rule of thumb used was that one cubic foot of boiler would heat approximately 2000 cu ft of space in a factory, assuming that the temperature required would be between $70°$ and $80°$ F, 25 cu ft of boiler required per 1 hp was allowed for a steam engine boiler in the early 19th century so that only a small boiler was required to heat the average sized factory. For cotton factories it was assumed that in most cases 'one superficial foot of exterior surface of steam pipe will warm two hundred cubic feet of space'. Nevertheless a larger allowance was usually made. H. Houldsworth allowed 1 ft of piping to about 179 cu ft, while Kennedy and Watt of Johnston allowed 1 ft to 168 cu ft. At Catrine 1 ft of piping was allowed to about 200 cu ft and a small chapel at Glasgow was adequately heated by allowing 1 ft of piping to 400 cu ft of space.[25]

Data of some steam heating systems

Factory	Steam Pipes	Cu ft Building	Cu ft Boiler	Cu ft Space Warmed By 1 Cu ft Boiler	Cu ft Space Warmed By 1 ft. Steam Piping	Temperature F
H. Houldsworth & Co, Anderston Old Mill	Cast iron	250,000	—	2,000	178	$85°$
Linwood	Cast iron	300,000	120	2,500	168	$70°$
Kennedy & Watts Johnston	Cast iron	289,000	160	1,180	160	$75°$
Catrine	Tin Plate Not Painted	—	—	—	200	—
T. Houldsworth Manchester	Cast iron	—	—	—	195	—
Chapel of Port Glasgow	Cast iron	60,000	10	6,000	400	—
Pt Adelphi Cotton Works	Cast iron	49,140	—	—	182	$65°$
Tambouring Mill, Anderston	Cast iron	—	—	—	240	$60°$
W. King & Sons, Johnston	Cast iron	244,583	180	1,303	200	$70°$

Source: Buchanan P.24.

The main item of expense was the cast iron piping. Several manufacturers experimented with tin plate and copper pipes but cast iron was found to give out more heat than either of the others. H. Houldsworth made some experiments to ascertain the difference between tin plate and cast iron as radiating materials and found cast iron superior in a ratio of 2½ : 1. The arrangement of the piping was an important consideration. Metal expands on heating and this expansion was particularly difficult to manage when pipes had to be joined at different angles. A copper steam pipe 160 ft long became 2 in longer when filled with steam and the expansion of cast iron was found to be about 1/10 in per 10 ft. It was this expansion factor which led to the abandonment of the early plan whereby hollow cast iron structural pillars served as heating pipes too. When a vertical pipe was filled with steam it expanded equally and continued to be straight but there were problems with horizontal pipes for they tended to bend as the upper side became hotter and joints were endangered.

There were other important points. A convenient means of expelling air from the pipes while they were being filled with steam was required. And a small quantity of steam had to be

H. Hicks and Sons, Churching Mill, Eastington, Gloucestershire. 1816

This drawing illustrates a variation of the system of horizontal pipes for steam heating.
Portf: 1334

114

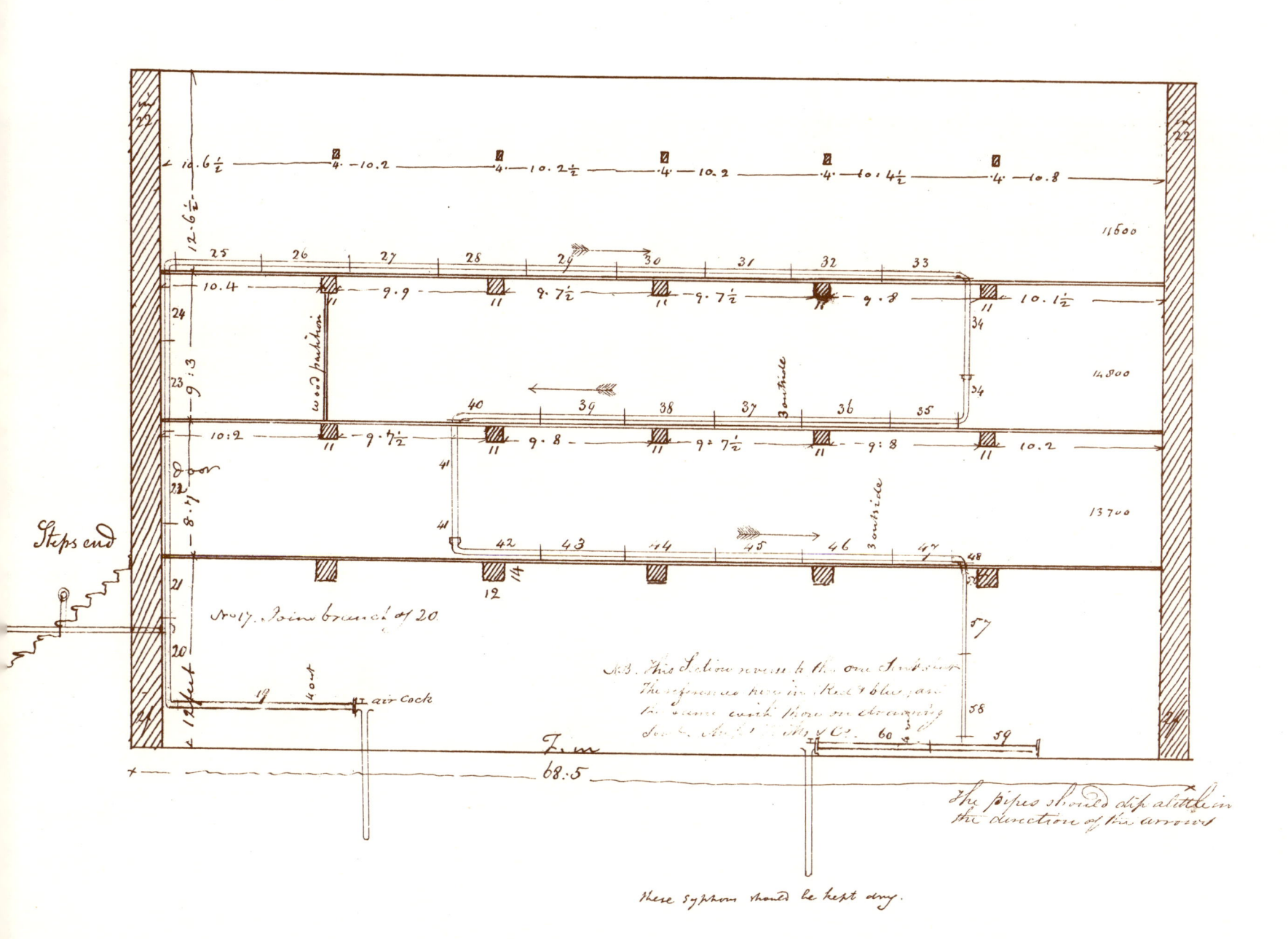

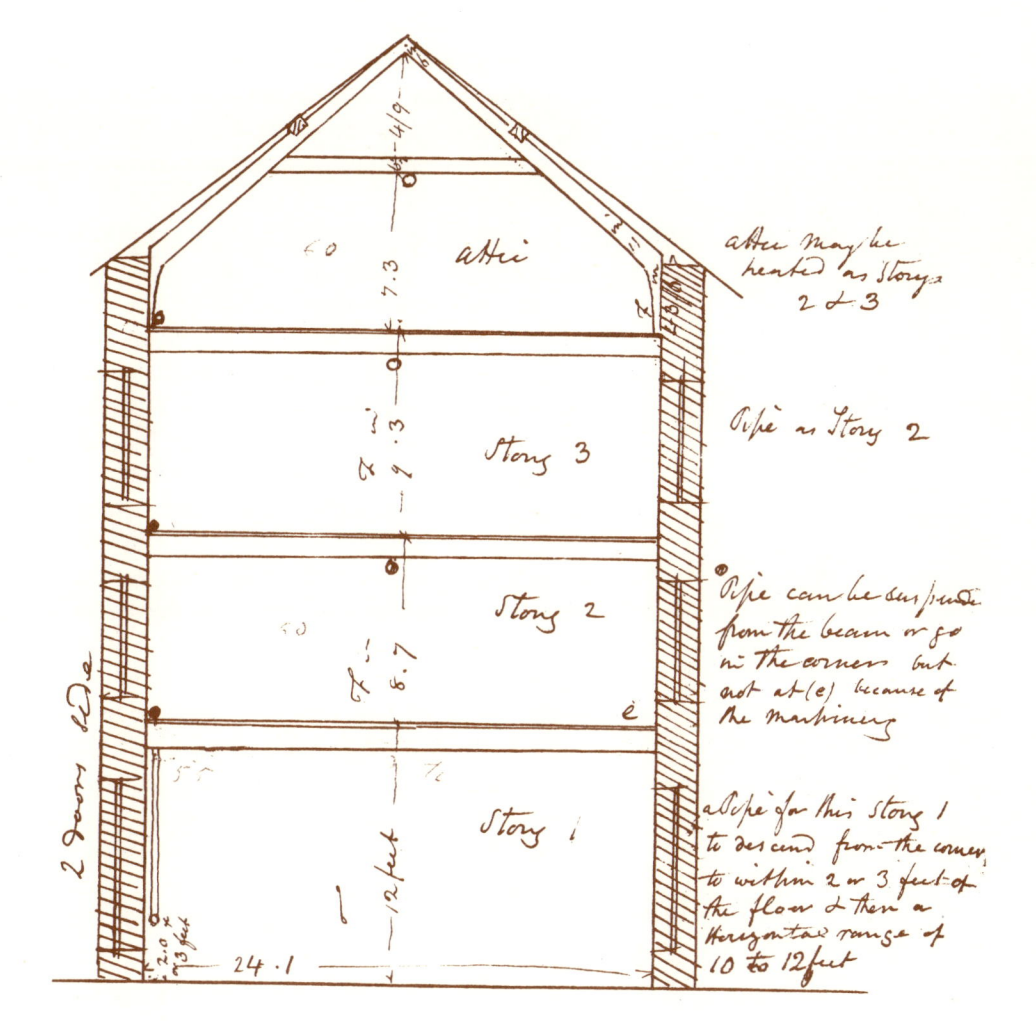

26. Ibid pp.. 14-17; pf. 1334.
27. T. Tredgold op. cit. p. 18.
28. Brotherton Lib, Marshall MSS, 37. p.l.

allowed constantly to escape so as to keep up the heat in the pipes. In a system using vertical pipes a cock was placed at the top of each pipe to allow for this (page 12). When horizontal pipes were used fewer cocks were necessary (pp 114,116). Water produced by condensation had to be removed from the pipes. Buchanan noted that 'when it can be done, it is better to make the water . . . to run in the same direction with the steam'. In a horizontal pipe the steam could drive the water before it. Care had to be taken to prevent water from remaining in any of the pipes after they became cool otherwise when steam was next let into the pipes 'some part of the pipe cracks, and a violent explosion takes place', an accident that had 'not unfrequently happened'. For this reason horizontal pipes were made to slope slightly 'by which means nearly all the condensed steam will collect again in the boiler'.[26]

In both Philips and Lee's and H. Houldsworth's mills the vertical pillars for supporting the floors also served as steam pipes for heating. Because of the expense of iron piping and the ease of installation while the factory was being built, this dual purpose system was often adopted by entrepreneurs building new factories in the early 19th century. Vertical pipes were favoured by Buchanan who noted that they had 'the advantage of diffusing the heat more equally;' and, notwithstanding the fact that they were more expensive to install in an existing building than horizontal pipes, Boulton and Watt used them in nearly all the early heating systems that they installed, whether or not the pipes also served a structural function. The practise of using structural pillars as heating pipes was discontinued after the early 19th century because of the expansion problem. Hot pipes in the centre of a factory could have been extremely dangerous and Boulton and Watt used a system of horizontal pipes in their later installations. The commonest and most simple arrangement (page 12) consisted of a main vertical pipe running up one end of the building with a horizontal pipe running off at each floor terminating at the further end of the building in a cock. A more effective system was that adopted by J. Houldsworth at Manchester and S. Oldknow at Mellor. At Oldknow's mill a boiler fed two central vertical pipes which conducted steam to the attic where from each pipe a branch led along the attic floor to the end of the building, down to the storey below, along that floor, down to the storey below and so on. (page 116) In the early systems, horizontal pipes ran just below the ceiling on each storey which must have greatly reduced the circulation of warm air. In later systems the pipes were laid at floor level.

It is difficult to assess which was considered the better system of factory heating, steam or warm air. George Lee wavered and eventually decided that neither system had a distinct advantage over the other. Tredgold remarked wryly: 'I do not know how the comparison has been made by others, but he must be a novice in the science of heat, that cannot produce nearly the same effect by the one as by the other, all other circumstances being the same. I know that, in either method, it is easy to mismanage things in such a manner, that not more than half the heat will be effective in warming the intended space; and that, by selecting cases for comparison, you may make either appear to be the best method, as far as regards economy of heat.' Those manufacturers who did express a strong preference gave empirical reasons: 'If building a new fireproof mill would have only cockles at first and steam pipes in a year or two — Thought it best to have various means of regulating the heat and moisture: cockles: steam pipes: and steam', John Marshall of Leeds noted after a visit to John Kennedy of Manchester in 1829. He added 'I did not find that any of the cotton spinners had used the steam itself as we do.'[28]

Samuel Oldknow, Mellor Mill, Derbyshire, 1809
Buchanan considered that the system of steam heating illustrated here was the most effective method.
Portf: 1334

Sam.ᵉˡ Oldknow Esq.ʳ Mellor Mills
July 8. 1809. —

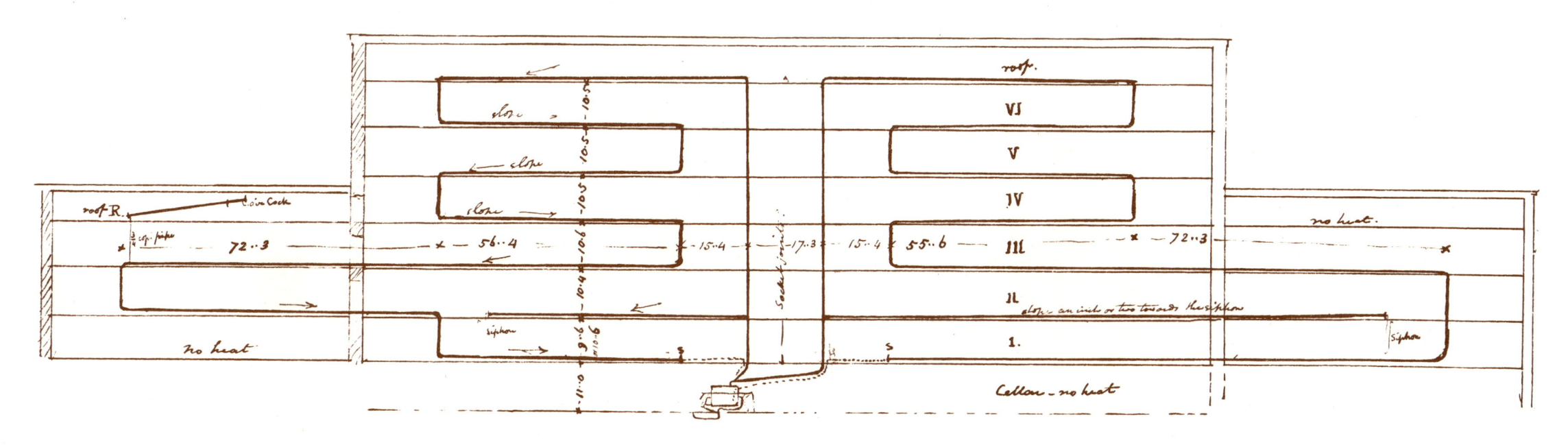

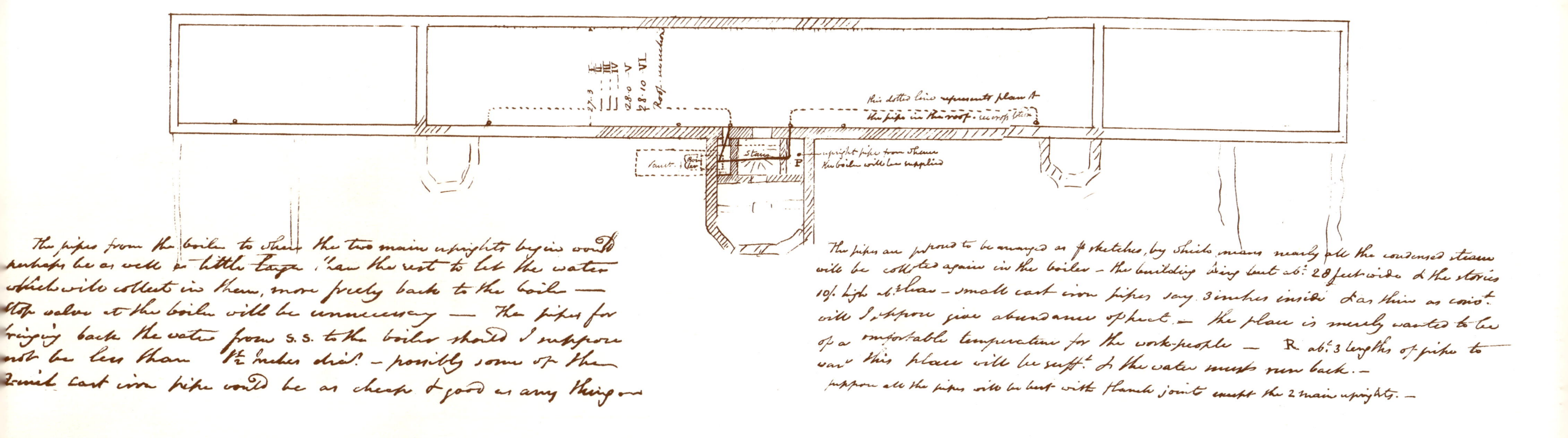

The pipes from the boiler to where the two main uprights begin would perhaps be as well a little larger than the rest to let the water which will collect in them, more freely back to the boiler — A stop valve at the boiler will be unnecessary — The pipes for bringing back the water from S.S. to the boiler should I suppose not be less than 1½ inches diam.ʳ — possibly some of the 2 inch cast iron pipe would be as cheap & good as anything —

The pipes are proposed to be arranged as p. sketches, by which means nearly all the condensed steam will be collected again in the boiler — the building being about 20 feet wide & the stories 10 f. high about — small cast iron pipes say 3 inches inside & as thin as consist.ᵗ will I suppose give abundance of heat. — the place is merely wanted to be of a comfortable temperature for the work-people — R ab.ᵗ 3 lengths of pipe to each place will be suff.ᵗ & the water must run back — I suppose all the pipes will be but with flanch joints except the 2 main uprights. —

117

David Dale, New Lanark Mills, Scotland, 1796
William Kelly, Dale's manager, sent this drawing and the 'Description of Cotton Mill Stoves' to Boulton and Watt in answer to their request in 1796.
'Fig. 1 is an edge view of the Gable of the mill, and represents the situation of the stove inclosed in the building b b. The smoke from the stove a, is introduced at d into the wall and carried up a vent and escapes at the top of the chimney. It is represented in the drawing as going directly from the stove into the vent, but it first takes a turn round the inside of the building b b.

The warm air is introduced into the mill at E, and conveyed into the rooms by the iron pipe ff. In each storey there is a hole or aperture g g, and a valve that shuts quite across upon the top of it at Z Z. The hole in the pipe at g is opened and shut similar to the manner of the Dark Lanterns; by turning round the socket or ferrule g g -.Our first built mill has a stove of this kind at each end of it; But subject to the following objections — Owing to the propensity of the rarified or warm air, to rush to the top; and to the smallness of the body of heated air in the pipe; we found it impracticable to warm more than one room at a time, and had to shut all the rest — When the warm air was admitted into the rooms above, we had to shut the socket g in those below, otherwise the bad air rushed up along with the heated air — It is also obvious the loss of fewel [sic] etc.

The next best way known to me, is that represented by Fig 2d, which is a fau [full] view of the stove as seen from the rooms. In the middle of the Gables of the Mill, there is an opening left in Building about 3 feet by 2 continued from bottom to top and terminates in the chimney heads; which our stone walls in this country easily admit of — At the bottom of said opening is placed the stove A inclosed in brick walls with an aperture on each side with dampers to regulate the admission of the external air to be heated. On the top of the stove are pipes of iron raised to carry off the smoke — these are placed in the middle of the above opening and terminates also in the top of the chimneys. The space round the pipes, serve for the ascent of the heated air which is to be distributed into the rooms. The space cc, in each room are originally left open towards the inside, for the convenience of putting in the pipes; and are faced or built up with brick. In each of these there is an aperture for the admission of warm air at c, which are fitted with an iron valve that can be opened or shut at pleasure. For the more effectual regulation of the ascent of the heated air; just above the aperture c, in each storey, there is an iron plate goes write [sic] across the opening in the gables, which is fitted also with a valve that can be moved up and down by a Quadrant or toothed-circular rack, similar to the moveable plates that are placed in Chimneys above some of the new fashioned fire grates. By opening the above valve at the top storey, the heat is allowed to escape at pleasure to prevent any excessive degree of heat.

The above makes a very good stove; but is attended with some inconveniences, namely the necessity always regulating the admission of the warm air by the valves; which the hands in the room are always disposed to alter as suits their inclinations. There is also a Danger of the foul air below, escaping to the rooms above when the regulating valves are not properly set etc. —

In our 3rd Mill we have a stove different from any of the two preceding ones, as shown by Fig. 3d, aa is a large building on the outside of the gable, taken in at every storey upwards. b is the Stove situated in the bottom, from which the pipe cc. goes up through the building to the top of the chimney; which carried off the smoke and increases the quantity of heated air. At each storey the pipes are supported by a strong cast-iron bar with a hole in the centre that receives the end of the pipe and which rests upon the flange, the ends of the bars being built into the wall. Besides these, there are small iron bars laid across like scantling; upon which a floor of thin rolled iron can be laid; And in which floors, a hole or aperture is made in each, of such dimensions as may be found necessary to check or regulate the ascension of the heated air.

Fig. 4 is an inside view of the Gable, dd are iron doors for admission into the stove when necessary, but always kep lockt. ee are round holes through the wall for the admission of the warm air — these are concentrated in each room upwards, in proportion to the influx of heated air; by which means, and, the apertures in the iron floors at each storey, the heated air is so regulated as to warm all the rooms at once.

Another advantage of this stove, is, that owing to the largeness of the Column of warm air in the building aa, the air is in no degree burnt which is Noxious; but agreeably warm sweet air. The influx or springs of the air into the rooms, owing also I presume to the same cause I find to be more equal than the others. The last described stove we find answers most completely and one only at the end of the Building is sufficient to warm a mill of 150 feet by 30 or upwards. It is proper to add that there should be an arch thrown over ff, in order that the side of the wall may be taken down without injuring the rest of the building, in case of any accident happening to the Stove —

I should also notice that in place of opening a little of the tops of the windows, to let out the foul air in each flat; (which operates against the influx of the warm air) there should be at the farthest distant part of the room, a ventilator to carry off the bad air all the way to the top of the House. It is almost unnecessary to say, that the external air to be heated is admitted in all the stoves thro holes at the bottom of the Building; which must also be regulated by valves.'
Portf: 1334

Wedgwood and Byerley, drying house, Etruria, Staffordshire, 1807
Steam heating was employed in the pottery industry for drying slip.
Portf: 1334

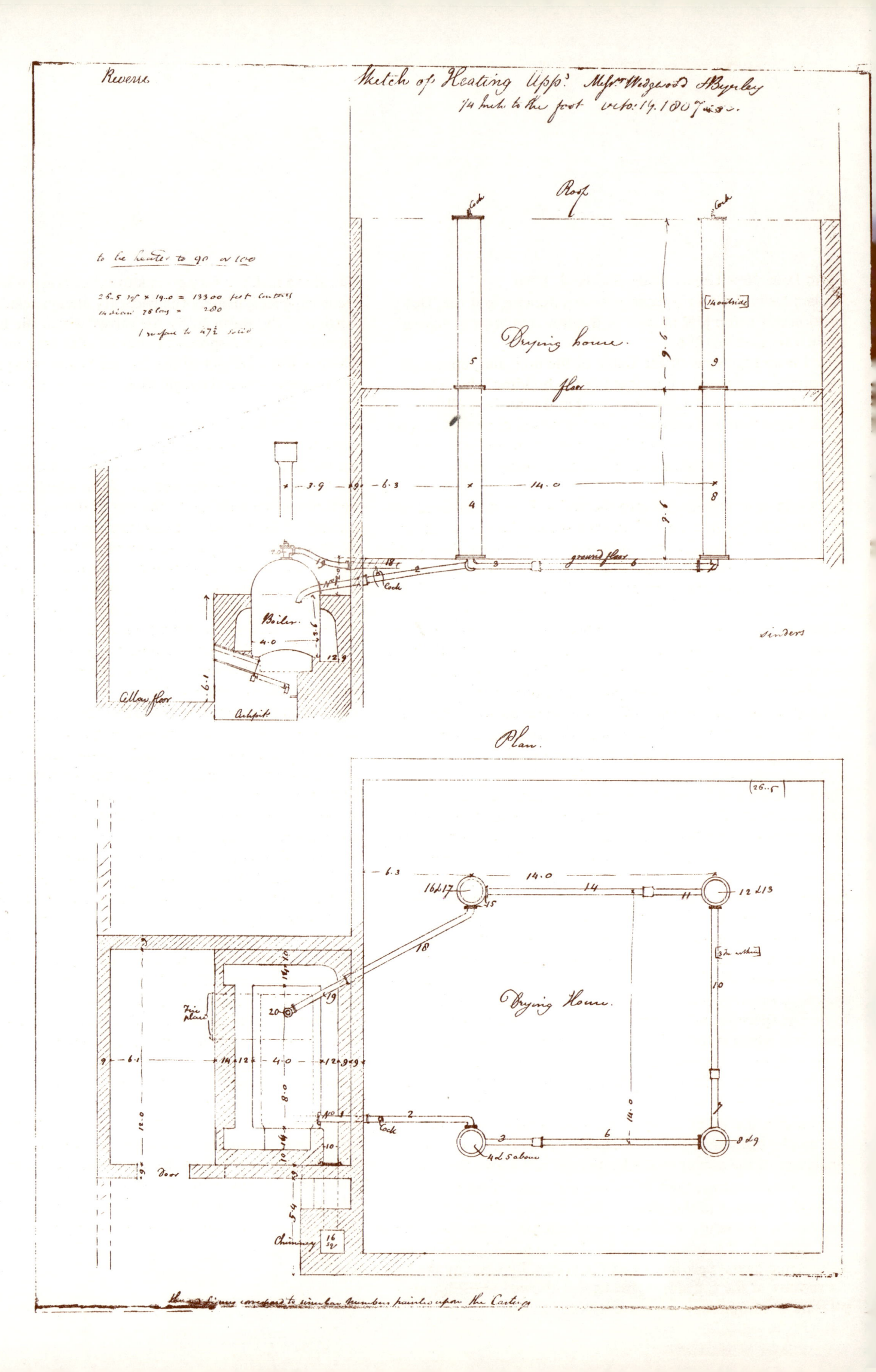

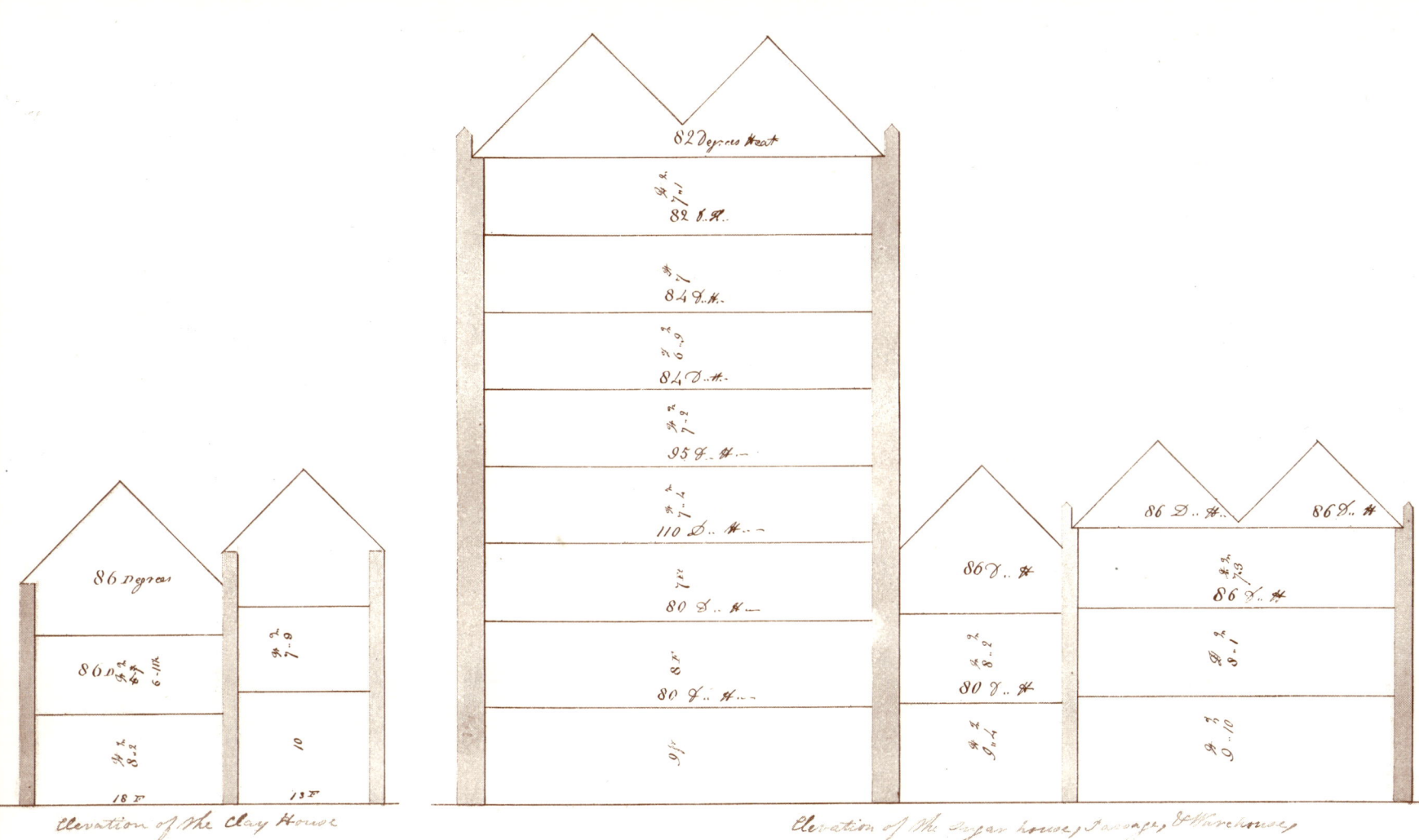

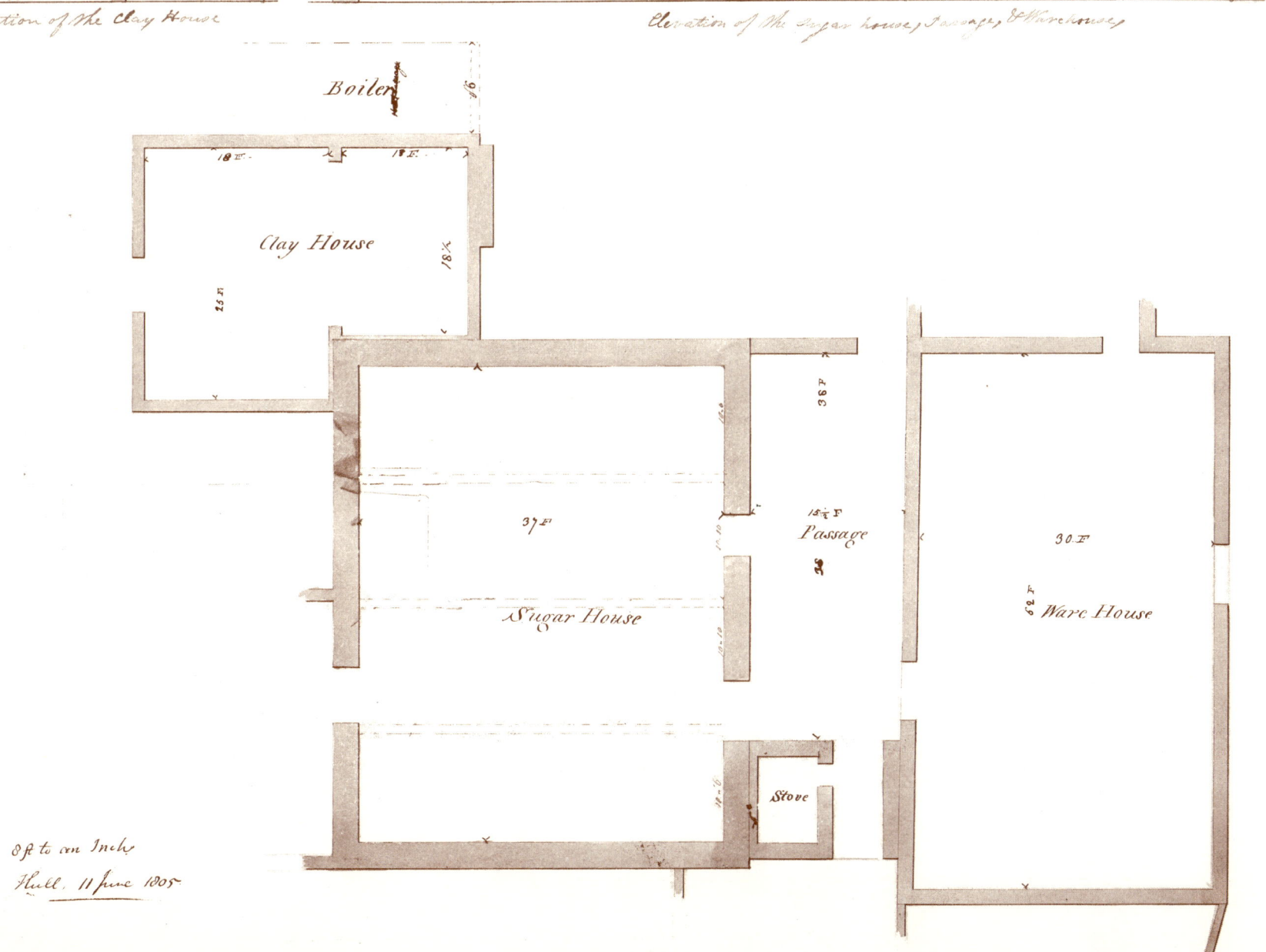

Sugar Refining
Sugar crushing and refining was a developing industry in the late 18th century. A number of London and Liverpool merchants invested in West Indies sugar plantations, and erected crushing mills in the West Indies and refineries in England.
Boyes and Carlill, Hull, 1805
Great heat was required in sugar refining. This drawing of an 8½ storey sugar house is an indication of the large scale of operations.
Portf: 1334

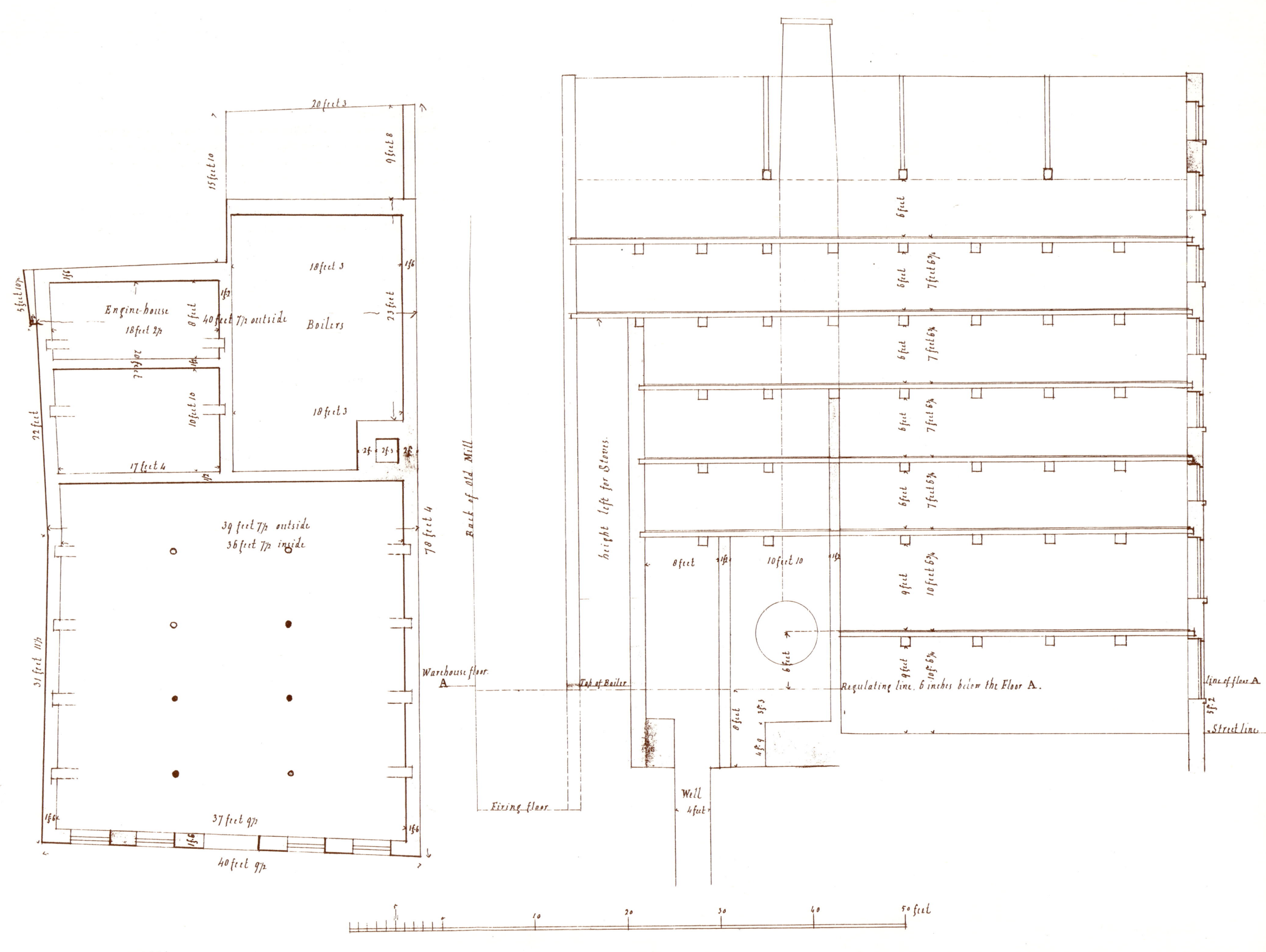

Peter Whitfield Brancker & Co.'s Sugar-house. 8 Aug.t 1815.

Peter Whitfield Brancker & Co, Liverpool, 1815
This refinery was on a similar scale to the one at Hull.
Portf: 824

8. Factory Lighting

1. S.C. on Children Employed in Manufactures, PP 1816, III, pp 280, 277, 134,
2. Notebook of James Watt jun. no pagination, Box 30.
3. Jennifer Tann, 'Some Problems of Water Power', Trans Bristol and Glos. Arch. Soc. 84, 1965, p. 70.
4. Stanley D. Chapman, The Early Factory Masters, p.175.
5. W. Murdock, An account of the Application of the Gas from Coal to Oeconomical Purposes, 22 Feb. 1808.
6. Watt Notebook op.cit.
7. Ibid.
8. Ibid.

For twenty years or so after Arkwright built his first mill at Cromford (1771) night work was common in the cotton factories — Arkwright himself used boy apprentices to work night shifts at Cromford.[1] In Robert ('Parsley') Peel's Burton-on-Trent mills 'the two spinning rooms work all night',[2] and even Sir Robert Peel's factories worked twenty-four hours a day. Mills and factories dependent on water power had to make up for time lost due to water shortages by working at night, for as one Gloucestershire manufacturer commented 'It sometimes happens that we do not get three hours work in the day.'[3] Although perhaps the majority of factories had ceased night work by the early 19th century, a number of mills, if not working all night, at least worked quite late into the night.[4]

The length of time during which artificial light was required each day varied considerably from factory to factory. Murdock estimated that it 'may upon an average of the whole year be stated at least at two hours per day of twenty-four hours. In some mills where there is overwork it will be three hours, and in the few where night work is still continued nearly twelve hours.'[5] Philips and Lee needed light from 6 am to 8 am and from 3 pm to 9 pm in the winter; Peter Marsland required light from 6 am to 8 am and from 4 pm to 9 pm and Radcliffe and Ross needed artificial light in the afternoon and evening only from 3 pm to 8 pm. Gott estimated that seven hours lighting was needed each day in the winter with lighting equivalent to 200 candles for nine hours or more. Wood, Daintry and Co. 'work sometimes until ten in the winter and have no morning work' whilst at W. Douglas's mill at Pendleton eighty-one lamps 'are used all night through the year'.[6]

In the textile factories less light was required in the preparation rooms, weaving shops, engine house and the staircase, than in the spinning rooms. Pooley had light equivalent to sixty-four candles in the carding rooms whilst in the mule spinning rooms between seventy-two and ninety-six candles were required. James Watt jun. reported that in the Salford Twist Mill 'in the upper rooms where they card it is only proposed to have about half the quantity of light'. Horrocks had nineteen candles in the preparation rooms of his new factory and twenty-eight in the mule rooms. Pollard used four candles per mule. Robert Peel of Burton had twenty burners in each weaving shop, twenty-eight in the preparation room and sixty-four in the spinning shop.

Lighting Required in Strutt's West Mill[7]

	Candle Equivalent
4 Mule rooms. 12 mules of 2 lights. Each = 3 candles.	288
1 carding room. 26 lights. Each = 2 candles.	52
Warping Room. 5 warping mills. 6 candles each.	30
Warping Room. 3 reels. 2 candles each.	6
1 cotton picker's lamp	12
3 winding machines. 4 lights. Each = 2 candles	24
1 light for opening machine, 1 for overlooker. Each = 3 candles	6
Sorting Room. 3 lights. Each = 3 candles	9
Staircase	6
Counting House	4
Engine and Boiler House	9
Joiners' Shop	6
	452

Factory lighting was an expensive problem for the entrepreneur. Between 1770 and 1798 there were two alternatives, either the factory could be lit by hundreds of candles or it could be lit by oil lamps. In either case the naked flames and oil, or wax dripping on to wooden floors constituted a severe fire risk. McConnel and Kennedy had approximately 1,500 candles which were lit for four hours per day during the winter half of the year. 125 pounds of tallow were burned each day and, not allowing for wastage, since tallow cost 10½d per pound, the annual cost in tallow was in the region of £681. 'But', James Watt jun. noted, 'it is probable that a considerable addition would require to be made to this for waste etc.' And he added 1/10th to the above mentioned cost. T. Houldsworth's lighting cost him between £310 and £350 a year. James Lees of Oldham had about 300 candles, James Kennedy about 1,000 of six to the pound. Pooley burned, in the depths of winter, two sixty pound boxes of candles a week which over the year came to about £131. Birley and Marsland spent about £187 per year on candles, while Peels at Burton spent about £168; Strutt's lighting at Milford cost approximately £360 per year.[8]

Those manufacturers who used oil lamps spent rather more on lighting their factories. William Douglas of Pendleton spent £255 on lighting 188 lamps in Pendleton Old Mill and his total yearly consumption of oil (including oil for the machinery) for the years 1803-5 was as follows:

	£
1803	777
1804	857
1805	639

Oil lamps were in use at Holywell and James Watt jun. noted that the firm 'state their present consumption of oil to be £800', reckoning their consumption of oil to be around ten gallons per

Gas Lighting
Philips and Lee, Salford Twist Mill, 1806
'The Old Mill' illustrated here was lit throughout by gas, whereas the 'New Factory', the famous fire-proof factory was only partly gas-lit at first.
Portf: 242

Mess.rs Philips & Lee June 1806

Plan of Story K, Old Mill

Story L

Stories M. N. O.

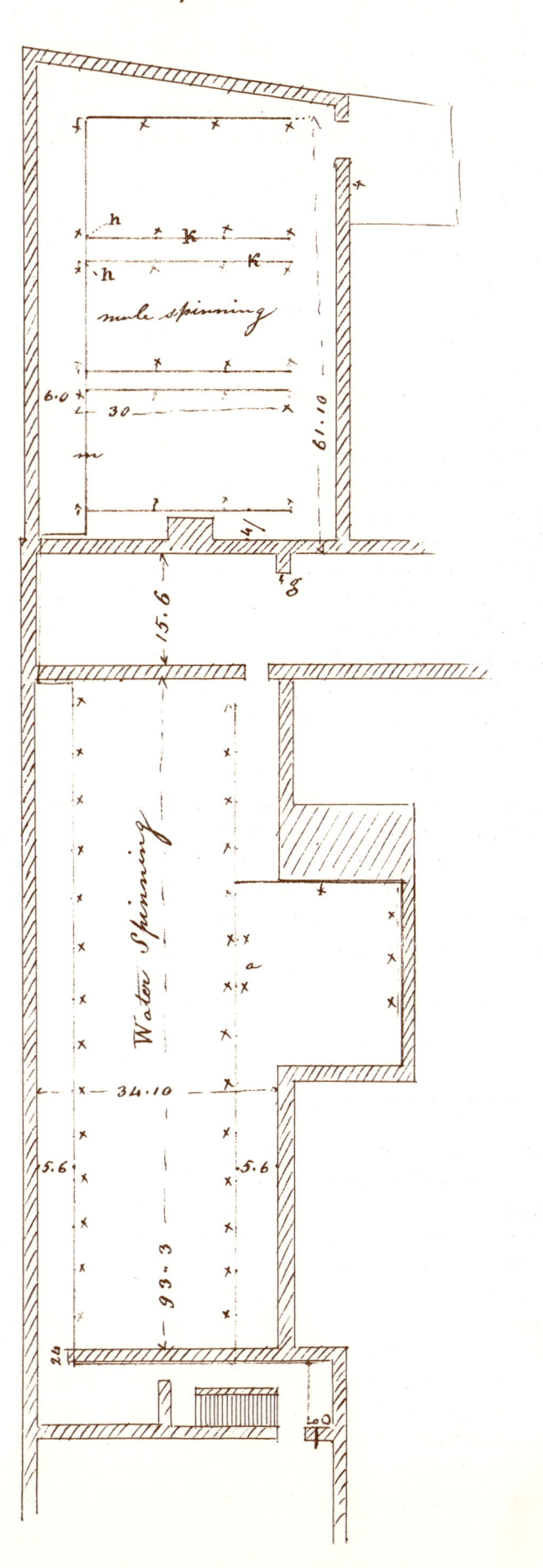

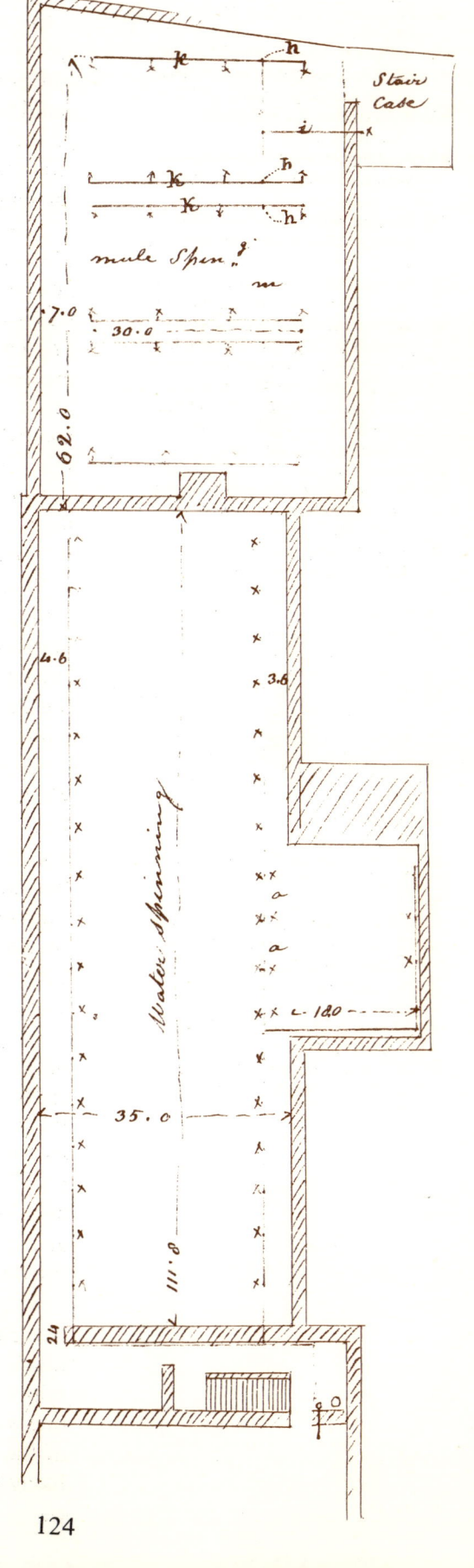

This mill is to be supplied with gas from a pipe already fixed in the staircase betwixt the old & new mills. The burner pipes for the mule spinning in all the stories will be similar — 18 inches —

Those for the water spinning in all the stories will be suspended from the horizontal pipes which will run close under the Beams but as the rooms are of different height their lengths will vary

For Stories 1 & 2 length 24 inches
3 —————— 25
4 —————— 24
5 —————— 14

Those for the water spinn.g are to have tube burners (argands) with glasses & reflectors, some of them a, a, are double the two lights b. b. in the engine house will hang down 5.0 below the horizontal pipe & f c, will be 18 inches below it

The pipe c, supplies the twist room mentioned hereafter — the light f is supplied from above

124

9. Ibid.
10. *Phil. Trans. Roy. Soc.* 1667.
11. Ibid 1733.
12. William Matthews, *An Historical Sketch of the Origin and Progress of Gas Lighting*, 2nd ed, 1823, p.11.
13. W.H. Chaloner, *People and Industries*, p.125.
14. W. Matthews op cit, p.22. T. Wilson to Boulton & Watt, Jan. 27, 1808, Box 30.
15. Charles Hunt, *A History of the Introduction of Gas Lighting*, 1907, p.42.
16. Minutes of Evidence Taken Before the Committee to whom the Bill to Incorporate Certain Persons for Procuring Coke, Oil, Tar, Pitch, Ammoniacal Liquor, Essential Oil, and Inflammable Air, from Coal; and for other Purposes was Committed. 1809 (hereinafter Mins of Evidence Coke etc. Bill).
17. Ibid.
18. Ibid. Evidence of James Watt jun.
19. Aris's *Birmingham Gazette*, 15. April, 1802.
20. Samuel Clegg jun. *Treatise on the Manufacturing of Coal Gas*, 1841 p.6.
21. Soho Box.
22. Mins of Evidence Coke etc. Bill, evidence James Watt jun.
23. W. Matthews op.cit. p.24.
24. Southern to James Watt jun., 9 Mar. 1806, Box 30.
25. Watt Notebook.
26. W. Matthews op cit pp.37-39, 52. Trans. Roy. Soc. Arts XXVI, p.202.
27. Boulton & Watt to G. Lee, 31 Oct., 1800, Office Letter Book.
28. W. Murdock to Boulton & Watt, 23 Dec., 1805, Box 30.
29. W. Murdock to Boulton & Watt, 20 Dec., 1805, Box 30.
30. W. Murdock to Boulton & Watt, 1 Jan., 1806, Box 30.
31. G. Lee to Southern, Aug. 11, 1806, Box 30.
32. W. Murdock to Boulton & Watt, 5 Feb, 1806, Box 30.
33. G. Lee to Southern, Aug. 11, 1806, Box 30. Watt Notebook.

day. The introduction of fish oil had reduced their costs slightly — this was now mixed with seed oil 'and we may reckon it at 4/6d per gallon'. Experiments at Pendleton showed that in addition to being more costly oil lamps gave out less light than candles, one lamp being equal to only one candle of eight in the pound.[9] In view of these costs it is not surprising that when an alternative means of lighting factories was found the owners of the large cotton factories were the first to adopt it.

The burning properties of natural gas were known in the mid 17th century[10] and in 1733 it was noticed that coal gas could not be set alight except by a flame,[11] hence the flint steel which was used as a light by miners. Sir James Lowther sank a pit to drain a colliery of his near Whitehaven in the 1730s and discovered extensive deposits of natural gas.[12] In 1765 Spedding lit his office at Whitehaven with natural gas.[13] But in spite of these and other discoveries of the properties of gas it was not until the 1790s that William Murdock, an agent of Boulton and Watt's in Cornwall, demonstrated the commercial possibilities of gas lighting. In 1792 Murdock lit his office at Redruth and in 1795-6 he lit the counting house at Neath Abbey Ironworks.[14] Between 1797 and 1798 he travelled a number of times to Birmingham finally settling there in 1798. In that year he built a larger gas apparatus 'which was applied during many successive nights to the lighting of the building', and various new methods of washing and purifying the gas were practised.[15] The experiments were continued with some interruptions until 1802.

On his return to Birmingham, Murdock approached James Watt jun. on the question of a patent but was discouraged by Watt: 'I told him that I was not quite certain if it were a proper object for a patent, and I was induced to be rather nice upon the subject of patents, from being at that time engaged in carrying on the defence of a patent which my father had obtained for improvements on the steam engine, and which had then occupied more than four or five years, and which had been attended with a great expenditure of money.'[16]

In spite of the lack of encouragement from James Watt jun., Murdock continued to carry out experiments on his lighting apparatus at Soho Foundry. Gregory Watt went to Paris in 1801. What he heard there and reported back to Soho forced Boulton and Watt's hand 'if we intended to do anything with Mr. Murdock's light no time should be lost because he [Gregory] had heard that a Frenchman of the name of Le Bond [Lebon] was at the same period endeavouring to apply the gas obtained from the distillation of wood to similar purposes.'[17]

Murdock began to receive encouragement and to celebrate the peace of 1802 'a public display of these lights was made in the illumination of Mr. Boulton's Manufactory'.[18] From this description by Watt jun. and from other contemporary accounts one would be justified in assuming that the whole of the front of the Soho Manufactory was lit by gas but this seems highly unlikely bearing in mind the stage in his experiments that Murdock had then reached. Considering the novelty of gas lighting the local press could be expected to have devoted some space to a description of the source of light had it been gas, yet it is silent on this point although there is a lyrical account of the illuminations[19] as a whole. Samuel Clegg jun.[20] states that gas lights known as Bengal lights were placed at the ends of the building and were supplied by a retort placed in the room below whilst the rest of the Manufactory was lit by conventional oil lamps. The existence of a sketch of a Bengal lamp amongst the papers on the Soho Manufactory[21] in the Boulton and Watt Collection supports Clegg's assertion.

In 1803 the lighting at Soho Foundry was extended and between then and 1805 Murdock continued to experiment on the use of different types of coal and on retort design. 'The experiments were very long and attended with a great deal of expense.' Watt estimated that the expense in Murdock's time and the cost of the equipment amounted to £5,000.[22]

Between 1802 and 1805 Murdock altered his apparatus considerably. His first retorts were similar to the glass retorts commonly used in chemical experiments, then he tried cast iron cylinders placed perpendicularly in a common portable furnace — these contained about fifteen pounds of coal. In 1802 he tried horizontal retorts. In 1804-5 he constructed retorts with an aperture at each end, one for introducing the coal and the other for removing the coke but these were not successful and he tried a large retort shaped like a bucket with a cover to it which contained a loose grate to hold 15 cwt of coal. Later still, smaller elliptical retorts were designed.[23]

At first Murdock only lit the retorts shortly before a supply of gas was required. This meant that the retorts were cold each time and fuel was wasted in heating them up before the manufacture of gas could begin. It was John Southern, the faithful draughtsman, clerk and eventually a minor partner in the Firm, who first pointed this out and suggested that a great saving could be made by using fewer retorts continuously and storing the gas rather than by using a larger number of retorts intermittently.[24] At Philips and Lee's factory 1,400 cu ft per hour were required during the hours in which lighting was used. The total time of lighting during the winter was eight hours so that 11,200 cu ft of gas had to be made daily. If no means of storage had been provided about nine retorts would have been required to produce the necessary quantity of gas at the times it was needed. But Lee had five working retorts and one spare and stored the surplus gas made during daylight hours in gasometers. At McConnel and Kennedy's factory fourteen retorts would have been needed to provide the 2,250 cu ft per hour required during lighting times in their mill had there been no gasometers. Instead McConnel and Kennedy had five retorts and one to spare and needed either six gasometers 16 ft square and 5 ft deep or 12 ft square and 5 ft deep.[25]

Interest was shown in gas lighting by the chemists of the day, and Dr. Stancliffe gave a course of chemical lectures in Birmingham in which he 'exhibited and explained' the inflammable properties of coal gas. At the end of 1804 Dr. Henry gave a course of lectures on chemistry at Manchester in which he showed how to produce gas from coal and demonstrated the burning of it by argand lamp in Murdock's manner. The results of Dr. Henry's experiments on gas were published in Nicholson's Journal in 1805. Murdock communicated his discoveries in a paper presented to the Royal Society in 1808 for which he was awarded the Rumford Medal and in the following year Samuel Clegg sent a paper to the Royal Society of Arts[26] for which he was awarded the silver medal.

George Lee became interested in Murdock's experiments at an early stage, whilst Boulton and Watt were still sceptical. 'Murdock's countenance brightened upon my mentioning to him that his various schemes were shortly to be inspected by tw such approved connisers [sic]',[27] they wrote in 1800. However, a firm order does not seem to have been placed until 1804 when Lee's house was lit by gas. The success of this experiment led Lee to place an order for a lighting apparatus for the Salford Twist Mill. Part of the apparatus was installed in late December 1805, but Murdock was obviously impatient with the slowness of both Soho Foundry and the communications system of the day for he wrote 'if materials cannot be forwarded in a more expeditious manner than they hitherto have done it is of no use to think of taking orders here, for your old servant Clegg is manufacturing them in a more speedy manner than it appears can be done at Soho.'[28]

From the start Lee took a keen interest in the installation of his lighting system. 'Mr. Lee enters into the spirit of the project and wishes much to have it lighted before Christmas but I think it impossible, however we may get it accomplished by New Years Day.'[29] On New Years Day fifty lamps were lit. Mrs. and Miss Lee visited the factory to inspect them 'and there [sic] delicate noses have not been offended'.[30] Lee carried out experiments on the efficiency of the lights, the relative brightness compared with candles and the cost of operating the apparatus. His was the final decision when the types of lamp were chosen: 'I propose to have Argands in all the water spinning rooms and no where else.'[31] Cockspurs were used elsewhere. But he left the design to Boulton and Watt: 'Mr. Lee thinks designing to be more in the way of the people at Soho than a cotton spinner therefore he hopes they will please their own fancy in his chandeleers.'[32] When he tried to estimate the gas consumption of the two types of lamp he found that argands exceeded cockspurs by about one third. In February 1806 James Watt jun. noted 'Mr. Lee deduces from his Experiments that one hours consumption of a candle of six in the pound produces light equal to two thirds of a cubic foot of gas or that two cubic feet equal one cockspur burner of three issues for one hour.' 'These experiments', said Lee 'were made under my direction and inspection.

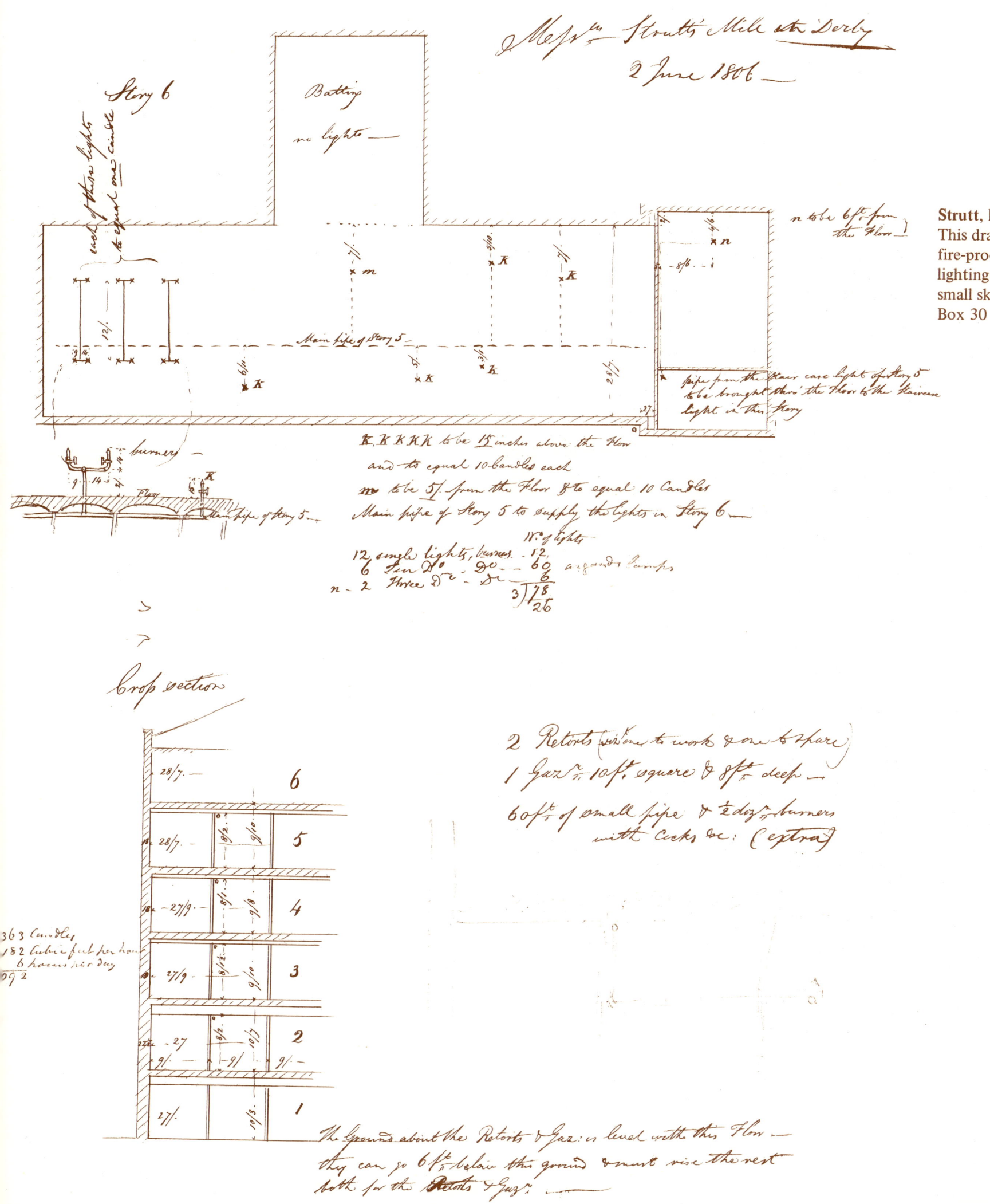

Strutt, Derby Calico Mill, 1806
This drawing is probably of greater interest in showing the first fire-proof factory in the world than in illustrating an early gas lighting installation. The brick arch floors can be seen in the small sketch beneath the main plan.
Box 30

34. Watt Notebook. Mins of Evidence Coke etc. Bill. W. Murdock, An Account of the Gas from Coal to Oeconomical Purposes.
35. Memorandum Concerning Lee's Photogenous Apparatus, 2 June, 1807, Box 1 L.
36. Nearly all Boulton and Watt's estimates for gas lighting are now illegible. Box 30.
37. W. Matthews op cit p.40. Clegg probably obtained Lodge's order when he installed Lodge's steam engine in 1803, Southern to Clegg, 1 Feb. 1803, Foundry Letter Book.
38. W. Matthews, op.cit. pp.52, 83.
39. Shrewsbury Library, Bage MSS Bage to Strutt, 5 March, 1808.

Each of Lee's retorts produced from 160 to 200 cu ft of gas per hour and one hundredweight of coal produced about 370 cu ft In March 1806 further experiments were conducted. There were at this time three gasometers of 7 ft diameter and 5 ft deep, and a fourth was to be installed. Lee estimated that he would require 1,400 cu ft of gas per hour. It was found that Cannel coal was by far the best type to use in a lighting system and it was given 'a decided preference . . . over every other coal, notwithstanding its higher price'. Best Wigan cannel coal cost 13½d per cwt, 22/6d per ton at the factory. The annual consumption for 313 working days was 110 tons costing £125. Common coal was used to heat the retorts and 40 tons were used each year which at a cost of 10/- per ton amounted to £20 per year. The 110 tons of Cannel coal produced 70 tons of good coke which was sold on the spot for 1/4 per cwt and this over the year produced £93. 11 to 12 ale gallons of tar per ton were produced — about 1,250 ale gallons per year. This could be sold for 1/- per ale gallon. Without allowing anything for the tar the annual cost of running the apparatus was £52. To this cost had to be added the interest on capital plus depreciation which Lee estimated at £550. He estimated that the annual cost of candles to give the same light would have been approximately double although on re-examination before the Committee on the Coke, Oil, Tar . . . Bill he said that the difference in expense between oil and gas lighting was 50% and between candles and gas 70%.[34] Even this figure was far lower than Boulton and Watt's estimate. In 1807 Boulton and Watt drew up a memorandum concerning Philips and Lee's 'Photogenous Apparatus'. In this it was noted that 550 cockspur burners and 300 argands had been supplied to Philips and Lee. These produced light equal to 3725 candles and had candles been used the annual cost would have been in the region of £2,600 per year. The memorandum concludes with a detailed costing of Philips and Lee's apparatus:[35]

	£
Cost of metal materials of apparatus from Soho	3,670
Cost of gasometer pits, retort houses, putting up, pipes etc.	1,330
	5,000
Say 6 retorts £100 each = £600 at 20%	120
8 gasometers £150 = £1200 at 15%	180
Pipes, burners, lamps = £1800 at 10%	180
Buildings, etc. = £1330 at 10%	133
	613

The high cost of Philips and Lee's lighting apparatus must be partly explained by the fact that when it was installed in the Salford Twist Mill Murdock's lighting system was still very much in the experimental stage. Later customers were to benefit from these experiments and their apparatus seems to have cost less. Estimates varied from about £800 for a medium sized order to £1,270 for 400 burners and 26 spare for Marshall Hives and Co in 1811. Two estimates were sent to Monteith Bogle and Co in 1814. In the first, the materials came to £1,161. 1. 0d and in the second they came to £1,710. 15. 11d.[36] Both James Watt jun. and George Lee gave evidence to the Committee on the Coke, Oil, Tar . . . Bill in which the cost of Philips and Lee's installation was set out and this may have deterred some smaller manufacturers from approaching Boulton and Watt to solve their problems of factory lighting.

Although Philips and Lee's factory was undoubtedly the first large factory to be lit by gas it was not the first factory (Soho apart) to be gas lit, for a former employee of Boulton and Watt, Samuel Clegg, successfully lit the cotton factory of H. Lodge near Halifax in 1805.[37] Clegg was one of a number of former employees of Boulton and Watt who branched out on their own and made valuable contributions to the design of machinery. In Clegg's case he was to improve gas lighting. In 1809 he introduced a paddle at the bottom of the tank to agitate the lime at a Coventry gas plant and in 1817 he invented the collapsing gas holder. Clegg was the first man to superintend the lighting of a town by gas. 'His active, ardent and enterprising disposition being united with many useful acquirements eminently qualified him for such large undertakings.'[38] Murdock was aware of Clegg's competition in Manchester as his letter to Boulton and Watt, quoted above, shows, and it was to Manchester that Bage went in 1808 when he wanted to see a gas-lit factory:[39]

'I am just returned from viewing some gas lights in a small factory near Manchester put up by Clegg. There are 28 splendid lights equal each to 10 or 12 candles of 6 in the pound; not nearly so offensive as candles or common oil; and what smell they have (in itself rather a pleasant one) might be diminished by ventilation, which seems very necessary where there is so much combustion. The expense according to the reports of Mr. Knight the proprietor and Mr. Clegg the Engineer is nearly as follows:—
Each retort holds 3 cwt of coals, and 2½ cwt is burnt under it. The gas supports the combustion of 28 lights 4 hours. The expense of an apparatus for 100 lights including fixing, brickwork and tin reflectors amounts to £500 which at 10 per cent is . . . £50 per an.

Strutt, Milford Mills, Derbyshire, 1806
The cruciform building shown on the right hand side of the drawing was Strutt's second fire-proof building known as Milford Warehouse.
Box 30

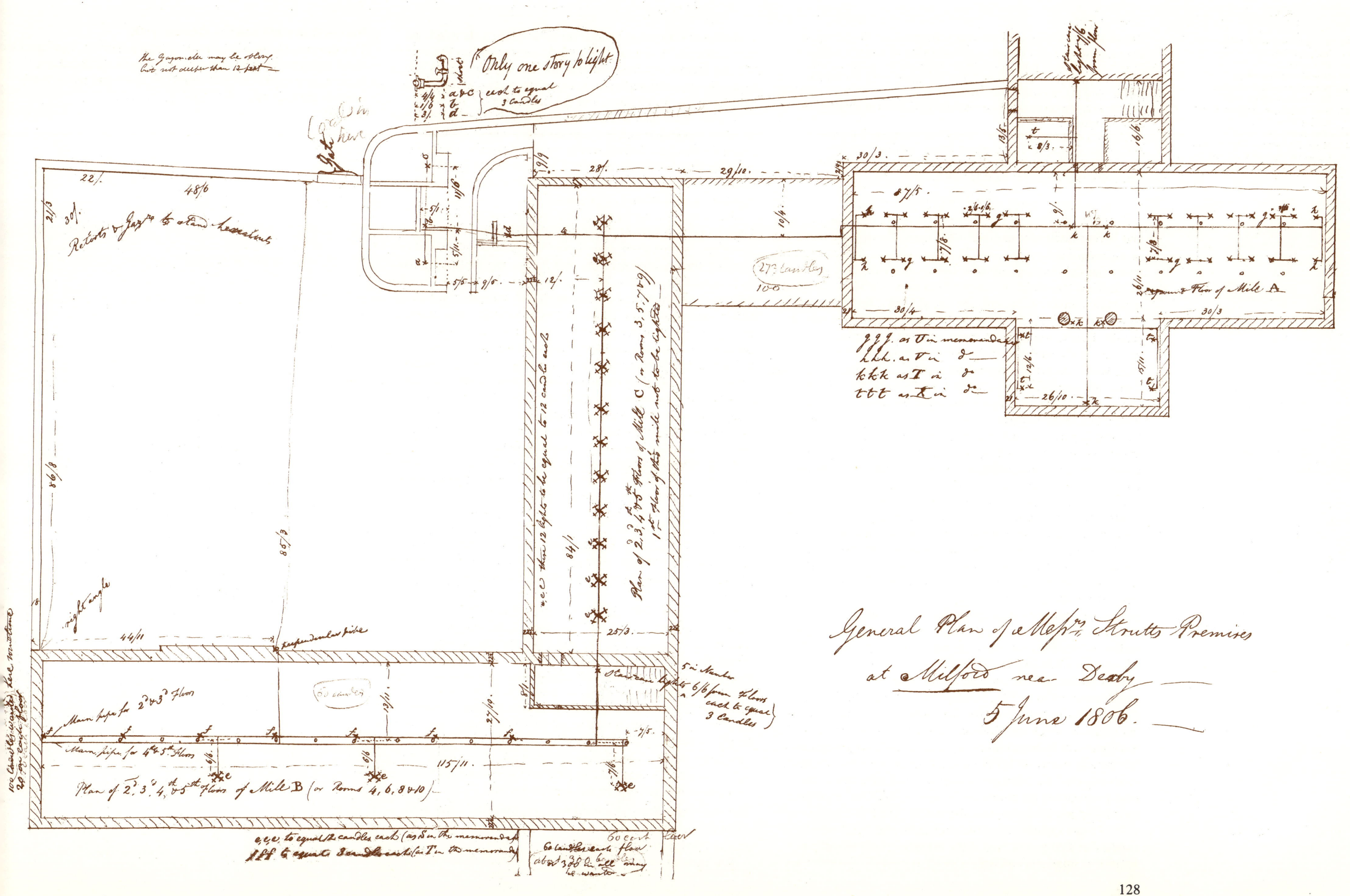

General Plan of Messrs Strutts Premises at Milford near Derby 5 June 1806.

40. W. Matthews op.cit. p.40.
41. Ibid. pp.39-42.
42. Ibid. p.43.
43. Charles Hunt op cit, pp.80-82.
44. W. Matthews op cit, 52, 83. Charles Hunt op cit, p.69.
45. Watt Notebook. Boulton & Watt Engine Book.
46. Watt Notebook.
47. Mins of Evidence Coke etc. Bill.
48. Cusworth Hall MSS, B. Gott to B. Gott jun. 9 Jun, 1809.

Repairs Clegg says would be over estimated at £40 per an. as the chief source of expense is the burning of an iron saddle interposed between the fire and the retort; serving at once to preserve the retort and to equalise the heat, without which different gases would come over   £ 40

Attendance one man & an assistant whilst discharging and charging say   £ 40

Coals 5 cwt per hour for 100 lights. Suppose the season 400 hours, equal to 100 tons; of these one half are improved into coak and cost nothing, and half the remainder (they say the whole) is paid for by the tar. Say 25 tons   £ 15

Total expense (per an.)   £145'

Early in 1806 Josiah Pemberton, a Birmingham manufacturer and 'a very intelligent and ingenious man',[40] exhibited a number of different lights outside his Birmingham factory and began to install gas lighting apparatus for the smaller workshop/factories of the West Midlands. He designed a scheme for Mark Saunders, a button manufacturer, whereby the gas not only lit the premises but produced jets that could be used for soldering button shanks. Pemberton is said to have installed gas lighting for a number of Midland manufacturers including Isaac Spooner of Park Mill.[41] Another Birmingham manufacturer, Benjamin Cook, a brass tube and toy manufacturer, having seen Murdock's paper to the Royal Society, wrote a letter to *Nicholson's Journal* in November 1808 in which he stressed the economies to be made by using gas light in preference to other methods of lighting. Matthews suggests[42] that Cook's apparatus was made by Pemberton but this is not mentioned by Cook in his letter to the Journal. Cook described how, because of 'the present rupture with Russia and the northern powers the want of importation of tallow has increased to a very considerable height the price of candles.' What Murdock did for the large-scale manufacturer Cook attempted for the small manufacturer. 'Mr. Murdock's paper is on the great scale, therefore far above the circulation of the simple mechanic, and it is to the great number of these that the thing ought to be made clear. To them a small saving of ten pounds per annum is of as great consequence as to the wealthy their thousands.' Cook's apparatus consisted of a small eight gallon pot for the coal; and a gasometer holding about 400 gallons. He made about 600 gallons of gas from twenty to twenty-five pounds of coal. His former annual expenditure on candles amounted to £18. In addition to this he used to spend £30 per year on oil and cotton for soldering but found that in 'all trades in which the blow-pipe is used with oil and cotton the gas flame will be found more superior, both as to quickness and neatness in the work; for the flame is sharper, and is constantly ready for use.' His annual expenditure, including the wages of a man to make the gas amounted to £18. 4. 0d. 'The expense of putting up my apparatus was about £50.' Cook estimated the cost of a gas apparatus 'for the poor man who lights only 6 candles' at £10 or £12 'and if the pipes are made of old gun barrels, as mine are, and once a year, or once in two years, are coated over with tar to keep them from rusting, they will last half a century.' Cook modified his opinion regarding the use of old gun barrels on account of the number of joints which had to be made, notwithstanding this they were in general use for gas lighting in the Birmingham area for many years.[43]

In 1807 Alderman Wood attempted to light the Golden Lane Brewery in London and in the same year a weaving shed at Pollokshaws near Glasgow was lit by gas, the retort being the cylinder of an old steam engine. Two years later Clegg lit Harris's factory in Coventry, and in 1817 he was responsible for lighting the Royal Mint in London.[44]

Meanwhile, following the successful lighting of Philips and Lee's factory in Salford, Boulton and Watt received enquiries about gas lighting from a number of important manufacturers, many of them cotton spinners. But it was only the largest manufacturers who applied to Boulton and Watt, the others, probably put off by the high cost of lighting Philips and Lee's factory, either continued to use oil or candles or went to Clegg or one of the other firms competing in the artificial lighting market. Strutts were amongst Boulton and Watt's earliest customers and in 1806 details were drawn up for lighting the Derby Calico Mill (page 126). T. Houldsworth required approximately 600 burners which were expected to consume £45. 15. 0d worth of gas per year. James Kennedy ordered a lighting apparatus in 1808, and in the following year Greg and Ewart ordered theirs.[45] Creighton recommended one retort and one gasometer of 1,000 cubic feet capacity for the latter's seven storey building. There were to be twenty-six lights per storey with additional lights in the roof, staircase and boiler house providing altogether light equal to 1,200 candles. One of the most ambitious lighting schemes was McConnel and Kennedy's. They required 1,500 burners producing light equal to 4,500 candles. James Watt jun. computed that the cost of coal per year would be £106 or 1/21 of the cost of tallow to produce an equivalent amount of light.[46]

Benjamin Gott, woollen manufacturer of Leeds, stayed with George Lee and had 'regular access to the Salford Twist manufactory at all times.... I was induced to order the apparatus for one of the mills and when the engineer came to take the dimensions of that mill to order a drawing of two or three intending to apply a smaller apparatus to them so soon as it was ascertained whether the lights in the first were as advantageously applied for the convenience of the workmen as possible.'[47] Gott sent an enthusiastic report of gas lighting to his son: 'We have had Burley Mill illuminated by gas from coal for ten days past and a beautiful and interesting object it is, the shade of the light is so pure, ye quantity so great and at so small a price I have ordered similar apparatus for Armley Mill.'[48]

The two Shrewsbury flax mills were gas lit in 1811 and in 1814 plans were drawn up for the lighting of John Maberley's mill in Aberdeen (page 133).

It is noticeable that the owners of fire-proof mills were

Watson Ainsworth and Co, Preston, 1806
This drawing shows a large mule spinning factory. The 'machine shop' indicates that the company employed clockmakers, turners, carpenters and millwrights to make and repair machinery in the factory.
Box 30

Messrs Watson Ainsworth & Co. Preston
June 1808.

New Mill Stories 1 & 2

Story 1

the pipe in the ground at x will be abt 1 foot under the surface, it will rise up on the outside of the Machine makers shop & come thro' the wall

49. Mins of Evidence Coke etc. Bill.

amongst Boulton and Watt's first customers for gas lighting. George Lee said that the Sun Fire Office offered to insure the Salford Twist Mill for a third the usual premium and Warner Phipps, the secretary of the Albion Fire and Life Insurance Co, on being asked if he thought gas lighting was safe, commented 'It appeared to me to diminish the risk of fire very greatly.'[49] But just as fire-proof buildings remained exceptional until the mid 19th century, so did the lighting of factories by gas. For although there is no doubt that once installed it gave a better light than oil or candles and was also cheaper to run, with the exception of the cheap Midland type installations, the initial capital outlay was too great for all but the largest firms.

APPENDIX

Gas plants supplied by Boulton and Watt

No. on Portf.	Earliest date on Drawings	FIRM
804	1805	Philips and Lee, Salford.
818	1806	Strutt, Milford Mills.
816	1808	H. H. Birley & Co Manchester.
817	1808	James Kennedy, Manchester.
812	1808	Wormald, Gott and Wormald, Park Mill, near Leeds.
819	1809	Richard Gillespie & Co
809	1809	Neilson & Co
820	1809	Benjamin Gott, Armley Mills, Yorks.
811	1809	Marshall Hives & Co, Shrewsbury.
813	1810	Lister Ellis & Co, Burley near Otley, Yorks.
814	1810	T. Coupland & Sons, Leeds.
815	1810	John Thomas & Edward Lewis, Manchester.
810	1811	Marshall Hutton & Co, Shrewsbury.
805	1811	Benyon, Benyon & Bage, Near Shrewsbury.
806	1811	Huddart & Co Rope Manufactory, Limehouse.
808	1814	John Maberley, Broadford Mill, Aberdeen.
807	1814	Williams, Jones & Co, Birmingham.

The following drawings of crank engines show gas plants.

Greg & Ewart, Birley & Hornby, Huddart & Co, Heathfield, The Britannia Nail Co, McConnel & Kennedy.

James Watt jun. attempted to obtain orders from the following cotton spinners:

Houldsworth, Bury & Co, Radcliffe & Ross, Douglas (Pendleton), Douglas (Holywell), P. Marsland, J. Lees (Oldham), Pooley, Pollard, Wood, Horrocks, Peel (Burton).

Source: Engine Book, James Watt Notebook, Box 30.

Greg and Ewart, Manchester, 1809
This was one of the new mule spinning cotton factories. It may have been iron framed for the pillars appear to be cast iron. Creighton's calculations indicate the hours during which light was required.
Portf: 307

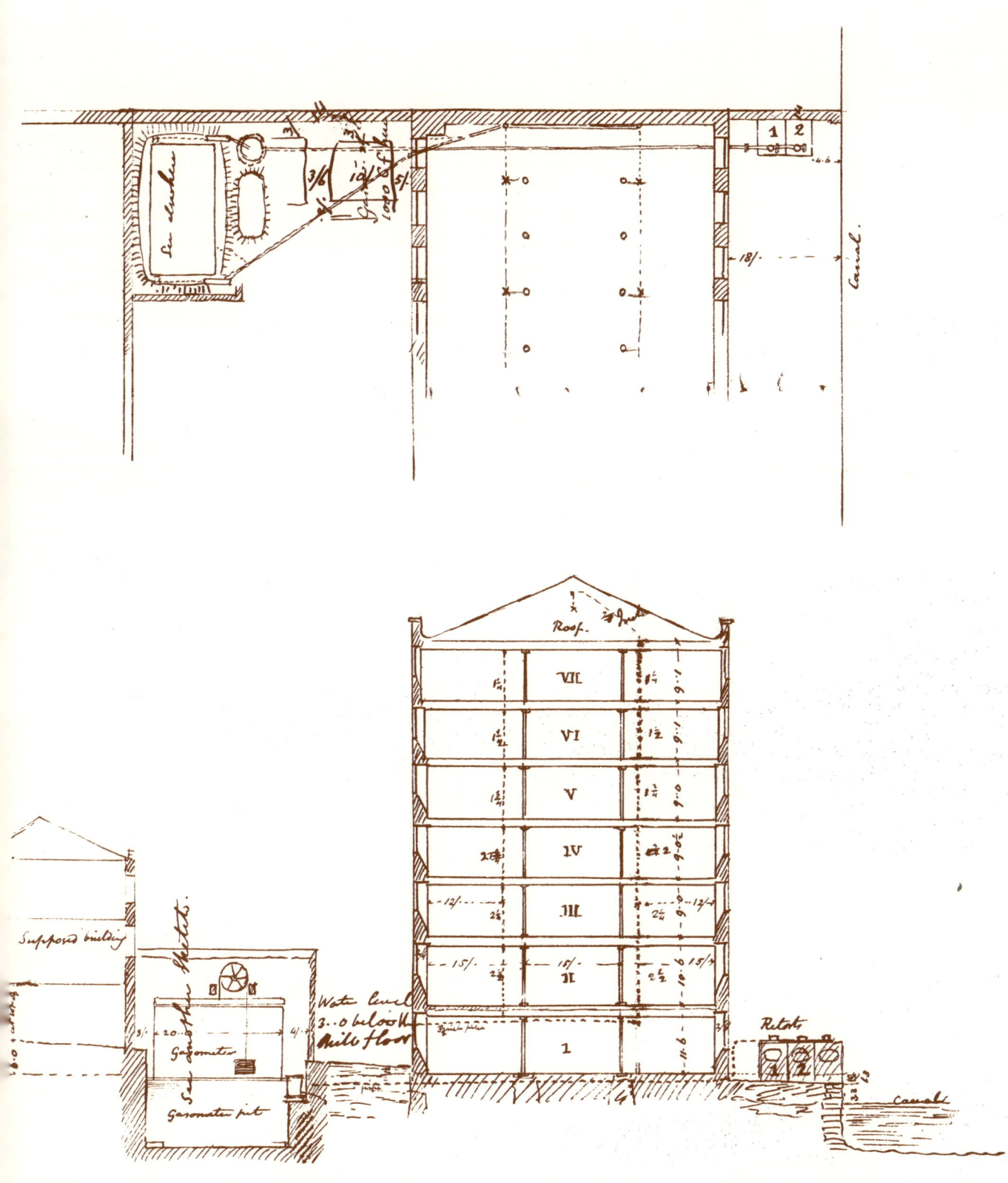

Messrs. Greg & Ewart. Manchester 30 June 1809.

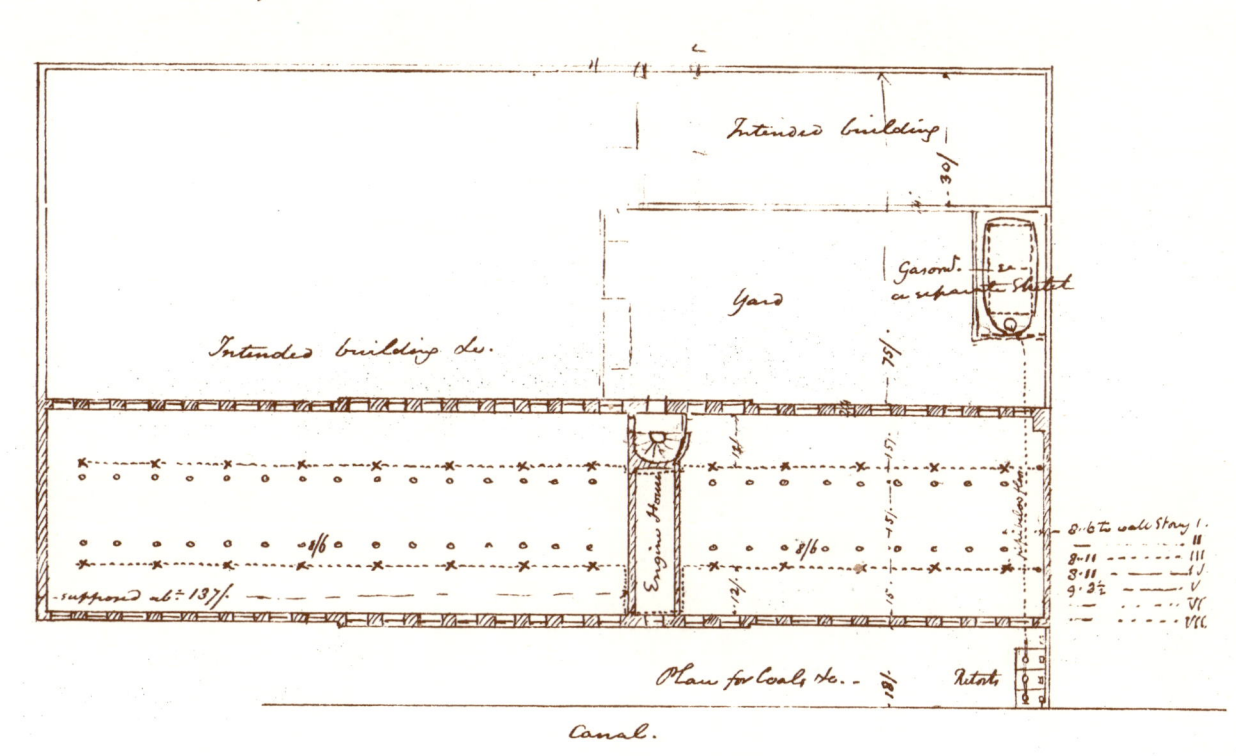

For a building of the above size there are supposed 26 lights in each story & 7 stories = (26×7) = 182 + 13 for roof = 195 & 4 over the Engine house in Stories IV. V. VI. VII. = 199. call 200 at 6 Candles each = 1200 C.
Staircase 7 lights = 14 C. E & Boile H say 6 Candles = 20
                                                        1220 Candles

which would require 600 Cf: of Gas pr hour & 7 hours pr day = 4200 C:f: which 2 retorts wo.d supply working constantly — in 4½ hours (the longest time of burning at once) 2700 Cf: gas will be used & if the Gasom: contains 2000 Cf: the two retorts will make out the rest. — the total quantity of lights may however be = to 1600 Candles at some period or other but at present the following will be a statement of what is required —

One retort & one Gasom: of 1000 Cf will be suff.t for the present build.g (taking the lights = 6 Candles) (79×6=) say 80×6 = 480+20 = 500 Candles or 250 Cf Gas p.h.s = 1750 Cf: p. day —

level of Canal is 4.4 below ground floor of Mill or 3..3 below the ground where the retorts stand.

132

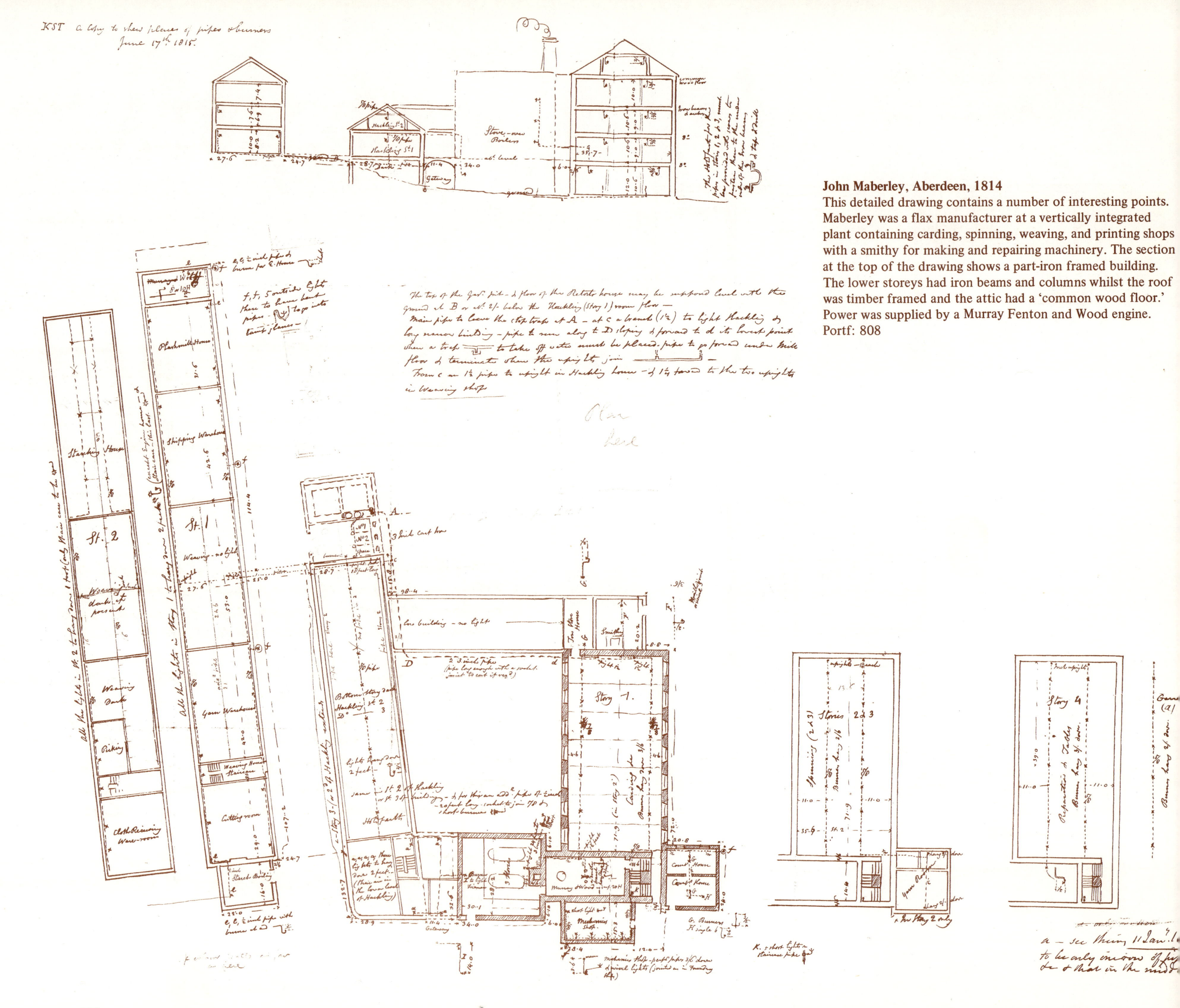

John Maberley, Aberdeen, 1814
This detailed drawing contains a number of interesting points. Maberley was a flax manufacturer at a vertically integrated plant containing carding, spinning, weaving, and printing shops with a smithy for making and repairing machinery. The section at the top of the drawing shows a part-iron framed building. The lower storeys had iron beams and columns whilst the roof was timber framed and the attic had a 'common wood floor.' Power was supplied by a Murray Fenton and Wood engine.
Portf: 808

Wormald, Gott and Wormald, Leeds, 1808
Murdock used vertical retorts in his earlier gas experiments but by 1808 had developed the horizontal retort as shown in this drawing. A similar retort was installed at the Salford Twist Mill in the same year.
Portf: 812

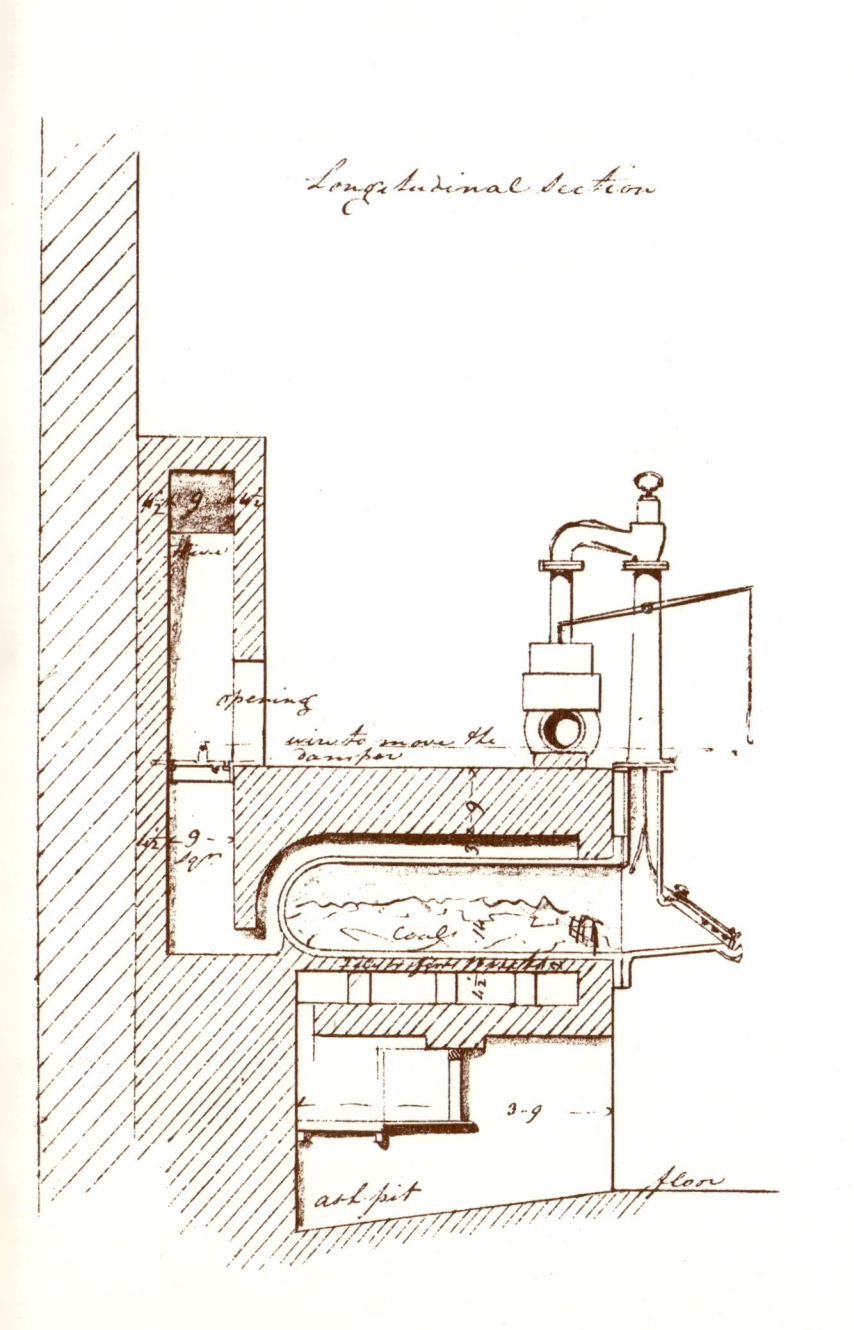

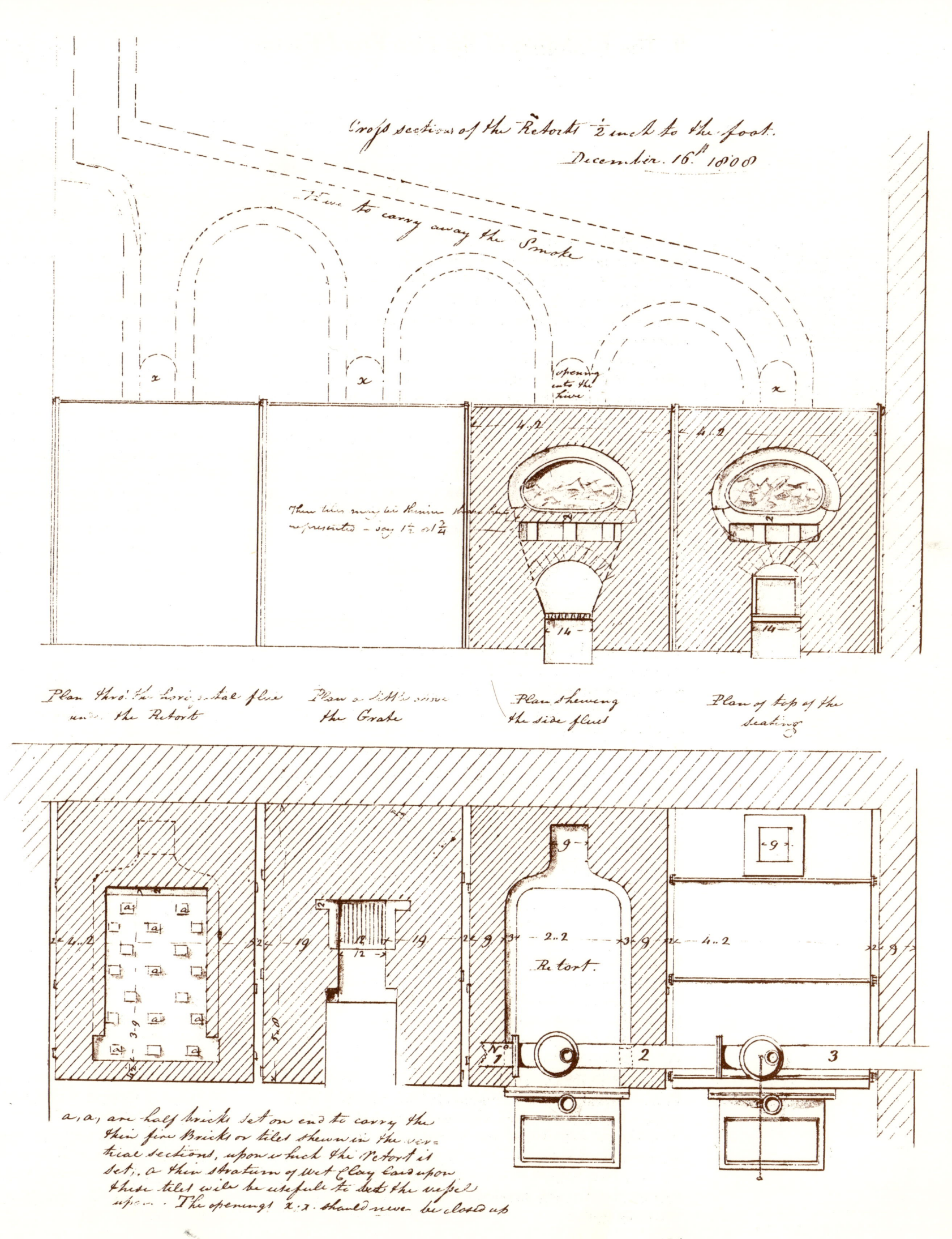

9. The Evolution of the Fire Proof Factory

1. Turpin Bannister, 'The First Iron-Framed Buildings,' *Architectural Review*, 107, 1950; A. W. Skempton & H. R. Johnson, 'The First Iron Frames,' *Architectural Review*, 131, 1962; H. R. Johnson & A. W. Skempton, 'William Strutt's Cotton Mills 1793-1850,' *Trans. Newcomen Soc*, 30, 1955-7.
2. T. Bannister op cit pp 233.234
3. C. L. Hacker, 'William Strutt of Derby,' *Journal Derbyshire Arch. & Nat. Hist. Soc*, 80, 1960.
4. *Life of William Strutt*, Derby Public Library. Quoted T. Bannister op cit p.235.
5. M. Boulton to W. Strutt 8 May 1793 A. O.
6. H. R. Johnson & A. W. Skempton, 'William Strutt's Cotton Mills' op cit pp 180-183.
7. *Derby Mercury* 20 Jul 1853, quoted in ibid p.184.
8. A. W. Skempton & H. R. Johnson 'First Iron Frames' op cit p.178.
9. Quoted R. S. Fitton & A. P. Wadsworth, 'The Strutts and the Arkwrights' p.205 S. Slater left Strutts employment to begin cotton spinning in New England.

A chronology of early fire-proof factories was established in three important articles by Professor Turpin Bannister, Professor A.W. Skempton and Mr. H.R. Johnson.[1] Little new material has come to light since they wrote and subsequent work has revolved around some of the details of the building of fire-proof factories.

The risk of factory fires was only too apparent in the 18th century. Multi-storey buildings with wooden beams and floors containing combustible materials such as cotton or wool; the floors saturated with the oil which dripped off wool and machinery; these hazards were made all the worse by the systems of artificial lighting adopted—candles and oil lamps. Understandably the premiums charged by fire insurance companies rose. In 1791 the complete destruction of the Albion Mills, the largest steam driven flour mill in the country, concentrated attention on the need to find a solution to the problem of the fire risk in existing mills and factories.

Early architects had considered the possibility of the replacement of inflammable timber by iron, but the high cost of iron had limited its use to tie rods, cramps and other small items. It was the higher cost of iron rather than any fears of breakage which deterred Boulton and Watt from using cast iron beams on their steam engines until 1800. Nevertheless, the use of iron for structural purposes increased during the 18th century. Two European iron bridges were proposed in 1719 and an 80 ft chain footbridge was built over the Tees in County Durham in 1741. In 1777-9 the first successful cast iron bridge was thrown across the Severn at Coalbrookdale. Iron gradually became a structural element in buildings from the 1750s. The first experiments had a wide geographical dispersal. Columns were used to support a hearth at a Cistercian monastery in Portugal in 1752, iron beams were introduced at St. Petersburgh in 1768-72, while St. Annes Church, Liverpool was built with iron columns in 1770-2. Vital experiments were made in France in the late 1770s to 1780s. A wrought iron roof was built over the stairway at the Louvre and M. Ango a Parisian architect used wrought iron to carry a floor in a private house. In 1785 de St. Fart, an architect specialising in hospitals, introduced the use of hollow pots in vaults. The culmination of these experiments was the building of the new Palais Royale in Paris in 1790 in which stone vaults, hollow pots, and wrought iron replaced inflammable materials. In all these buildings in Paris wrought iron was used rather than cast iron. But development was halted by the Revolution and the next steps towards the completely fire-proof factory were taken in Britain. News of the new buildings in Paris reached Britain quickly and in 1788 hollow pots were used in Dover House, Whitehall, Sir John Soane used hollow pot vaults in re-building the Bank Stock Office in 1792-3 at the Bank of England and at the same time William Strutt, son of Jedediah Strutt, a former partner of Arkwright, began investigating the possibilities of fire-proof mills.[2]

William Strutt had a keen and original mind. Friend of Erasmus Darwin, R.L. Edgeworth, Tom Moore and the Benthams, he was the man chiefly responsible for the designs of the Derby Infirmary (1805-10), the most advanced hospital of its day. He made important contributions to the theory and construction of water wheels, designed a central heating system, experimented on the strength of cast iron beams, investigated the most advanced methods of chemical bleaching and devised a domestic cooking stove.[3] His factories and the machines in them were considered amongst the most advanced in England and he was a model employer for the times, housing his employees in dwellings that were superior to much of the late 18th century and early 19th century industrial housing. The Strutt empire had been expanding since the dissolution of partnership with Arkwright in 1781 and in 1792 they began to build a calico mill in Derby. William Strutt had obviously heard of the hollow pot ceiling construction of the Palais Royale for he wrote to enquire of John Walker whether he could obtain samples of the pots. Walker replied from Ashbourne:[4]

'Dear Sir,
I am afraid I shall have more difficulty in getting you the drawings than I was aware of. As soon as I received your letter I wrote to an English architect in Paris for them. But three days before I left London a Frenchman who had made his escape informed me that the massacres of the 2nd September had driven him to England, as they did all ye English and notwithstanding my enquiries I cannot learn where he is.... Previous to your letter I had ordered one of each sort of the hollow bricks, of which the Arches are composed, to be sent to me, and I expect soon to hear of their being arrived in London.... Unluckily I only saw the building the evening before I left Paris, at a time when I was unwell, so that I have not so perfect a recollection of the Plan as I should have had, had I reviewed it at my leisure. However, perhaps I will give you as good a description as I can, least perchance I should not be able to obtain drawings at all.

The Building of the Palais Royale seems to me to be about 24 feet wide, the iron bars supporting the Arches are about four feet from each other.... The roof of the Palais Royale is of framed Iron, with the larger sort of hollow bricks to fill up the panes.'

It seems improbable that Strutt received the plans or samples of the hollow bricks but he contacted people whom he thought would be able to offer advice, including Matthew Boulton who wrote:[5] 'I understand you have some thoughts of adopting the invention of forming Arches by means of hollow pots and thereby saving the use of Timber in making Floors and guarding against Fire. Allow me to say that I have seen at Paris floors so constructed, and likewise at Mr. George Saunders' in Oxford St, London, who is an eminent Architect.... I have therefore no doubt but it might be applyed also with great success and security to a cotton mill.'

The construction of the Derby calico mill began in 1792 and the building was completed in 1793. The site was 150 yds south east of the Market Place in Derby. The mill no longer exists but by comparing their work on Milford warehouse with drawings of the Derby mill Johnson and Skempton have suggested that it had brick arches for the first five floors and hollow pots for the ceiling of the sixth.[6] The use of brick arches and hollow pots was, to contemporaries, the most striking feature of the mill yet the method of supporting the arches by cast iron columns and wrought iron tie rods in equally interesting, for this structural form was the antecedent of the modern framed building. The Derby mill was 115 ft long and 31 ft wide with an internal width of 27 ft (page126). There were five central bays with arch spans of 8 ft and three bays at either side of the central section with spans of 9 ft. There was a two bay wing on the west side. Drawings show that beams of wood, not iron, were about 12 ft wide and a report of 1853[7] indicates that they were, at least in part, encased in sheet iron. Because of the lightness of the ceiling of the top storey there was no need for columns on the top floor. The mill was not absolutely fire-proof for a fire broke out in the loft in 1853 but practically the whole of the rest of the building survived which testifies to the success of the construction

While the Derby mill was under construction a cruciform-shaped warehouse was being built at Milford. This was completed in 1793 and had brick and pot arches springing from wooden skewbacks covered with sheet iron. The soffit of the beams was plastered. The first storey consisted of stone piers and brick arch floors and on the other storeys the brick arches sprang from 12 in square pine beams supported by two rows of cruciform cast iron columns. The columns divided the internal width into three equal spans of 9 ft. In the outermost arches the brickwork was replaced by hollow pots made at Smalley Pottery. The pots were 5 in high and 4⅓ in diameter. A hole in the bottom of each pot provided a key for plaster and flat covers were placed in the open ends to form a base for the sand infilling on which the brick paving was laid. Skempton and Johnson note two drawbacks to Strutt's design: the roof trusses were timber and the tie rods were visible.[8]

Before either the Milford or Derby building was completed Strutt began a factory at Belper. On 11 June 1793 Samuel Slater's old schoolmaster was writing to him in New England, U.S.A.[9] 'Messrs Strutts go on swimmingly — they are erecting a very large mill at Belper; and Mr. George is beginning to build himself a noble house on the bridge Hill, just above the watering troughs.' The mill, begun in 1793, was not completed until 1795. It was 190 ft long, 31 ft wide and 6 storeys high and like the buildings at Derby and Milford constructed of brick arches paved with brick. The first storey, like Milford warehouse, consisted of brick arches springing from brick piers and the other storeys were built with brick arch floors springing from 13in timber beams supported by cruciform cast iron columns. The

COTTON MANUFACTURE.

Sections of one of Messrs Strutt's COTTON MILLS at Belper in Derbyshire.

Longitudinal Section. Fig. 1.

a School Room

Section of the Wing. Fig. 3.

Fig. 2. Cross Section

Belper North Mill

10. H. R. Johnson & A. W. Skempton, 'William Strutt's Cotton Mills' op cit pp 189-193.
11. R. S. Fitton & A. P. Wadsworth op cit pp 207-210.
12. J. Southern to J. Lawson 17 Feb 1796, Foundry Letter Book.
13. Science Museum, Goodrich Papers.
14. W. Fairbairn, *On the Application of Cast and Wrought Iron to Building Purposes*, p.3.
15. T. Bannister op cit, A. W. Skempton & H. R. Johnson op cit.
16. W. G. Rimmer, *Marshalls of Leeds, Flax Spinners 1777-1886*; W. G. Rimmer, 'Castle Foregate Flax Mill, Shrewsbury, 1797-1886,' *Trans Salop Arch.Soc.* LVI, 1957-8.
17. Shrewsbury Public Library, Bage MSS, undated letter C. Bage to W. Strutt 1796.
18. C. Bage to W. Strutt 28 Oct 1811.

ceiling of the sixth floor was made of hollow pots.[10]

Building accounts for Belper West Mill have been analysed by Dr. Fitton.[11] The cost of cutting and laying the foundations was £8. 8. 6d. It took seven men two days to lay the beams for each floor 'exclusive of the other area which worked regularly at the building'. Lime was obtained from Crich. Pots cost 52/6d per thousand and 35,609 were used in the top storey at a cost of £93. 9. 5½d. Some of the cast iron columns came from Ebenezer Smith and Co of Chesterfield, the second largest iron foundry in Derbyshire, and much respected by Boulton and Watt for the quality of their castings. Much of the stone, timber and other building materials would have come from the Strutt's estate which had several quarries and plantations on it but they were unable to produce all the timber that was needed. The mill cost £4,688. 19. 1¾d to build, excluding the water wheel, wheel house, bridge and cut.

Strutt's first method of fire-proofing impressed Boulton and Watt and when Marshall of Leeds lost a large mill by fire in 1796 Southern sent some papers which had been drawn up by James Watt after the fire at Albion Mill to their agent Lawson[12] asking him to distribute them as he saw fit. Southern said that if either of the ways suggested in the papers was put into practice there would be a great reduction in the fire risk. Watt suggested replacing floor boards by iron plates. This method would be better for mills already built as the floor boards could be quickly removed and the plates fixed in their place. Watt's other suggested method was to cover exposed timber with a layer of plaster but the disadvantage of this was that it would take some time to dry. Southern added 'Messrs Strutts of Derby have adopted one which promises well and in new buildings might be adopted generally. If Messrs Marshall and Co would wish to see it, or know what it is I daresay Messrs Strutt would with pleasure shew it. It consists of throwing brick arches betwixt beam and beam ... & tying the beams together by iron bolts to prevent them from yielding to the spur.' Watt's paper is of great interest for it appears to contain the substance of Simon Goodrich's winning design of 1802 for 'Rebuilding the Albion Mills, Fire-Proof'[13]:

'The under side of the beam is to be protected from the action of fire by being covered with a double coating of lath and plaster or with plate iron: the other three sides of the beam are sufficiently protected, being buried in brickwork. The beams might be of cast iron, but the preference is given to wood, since being fully secured against fire, it is cheaper and more to be depended upon for strength in this position, than cast iron.'

Wooden beams had the advantage of being extremely slow to burn, smouldering for hours without collapsing. So far no example of Watt's and Goodrich's 'intermediate' type of fire-proof building has been found.

The next stage in the development of the fire-proof factory was the substitution of cast iron for the wooden beams formerly used. For over a century Fairbairn's[14] statement that the Salford Twist Mill was 'the first instance on record of the successful application of cast iron beams for the purposes of building' remained unchallenged. But it has since been shown[15] that the first factory to have iron beams as well as iron columns was a flax mill in Shrewsbury. John Marshall of Leeds obtained a licence to use Kendrew's flax spinning frame and in 1788 he began flax spinning at a water mill four miles outside Leeds. In 1791 Marshall moved to Water Lane in Leeds where a four storey steam powered mill was built. Two years later he took the brothers Thomas and Benjamin Benyon, merchants, of Shrewsbury into partnership and in 1794 work began on a larger mill at Water Lane. In 1796 a mill was begun at Shrewsbury. Marshall opposed the scheme and invested no capital in the mill until 1800 although he had a quarter share in the new concern.[16]

Marshall had depended on the skill of Matthew Murray for advice on engineering matters. Similarly the Benyon brothers sought a man who could provide them with the necessary advice on buildings and machinery. The man they chose was Charles Bage (1752-1822), son of Robert Bage of Derby who had established a paper factory at Elford near Lichfield, Staffs. Bage moved to Shrewsbury in *c.* 1780 and became a wine merchant. Whereas Murray, a mechanic, could be offered a wage, Bage required more and became a junior partner holding an eighth share in the Shrewsbury concern. The choice of Bage, a man known to be interested in the principles of engineering, was probably prompted by the burning of Mill B at Water Lane in 1796 (referred to by Lawson above). Before May 1796[17] Bage was writing to William Strutt asking for comments on his ideas for building a factory in Shrewsbury: 'inverted arches we had had under consideration & I have no doubt your opinion will be decisive in their favour.' Bage raised several important questions 'respecting the expansion of iron by heat and its effect on our buildings'. He mentioned Smeaton's experiments on the expansion of iron by heat and asked whether cast iron expanded as much as wrought iron. Bage was obviously seriously considering using cast iron beams by this date for he asked Strutt about the extremes of temperature that the beams would be subjected to in the year 'suppose no precautions used, would it injure the walls to force them in & out as the beams contract and expand?' He commented that a beam of 37.6 in, 'our length', would expand 0.377 in for every addition of 10°F. Bage later acknowledged his great indebtedness to Strutt: 'How much we are obliged to you for teaching us how to make buildings Fire Proof';[18] but his originality has been overshadowed by Strutt's. Previous writers have given most of the credit for the fire-proof factory to Strutt yet the Bage correspondence shows the lively, original mind of a person who understood the strength of cast iron better than his contemporaries, who told Strutt when he

Belper North Mill
Sketches from Goodrich's journals showing the structure

Creighton's drawing of the Salford Twist Mill, 1801

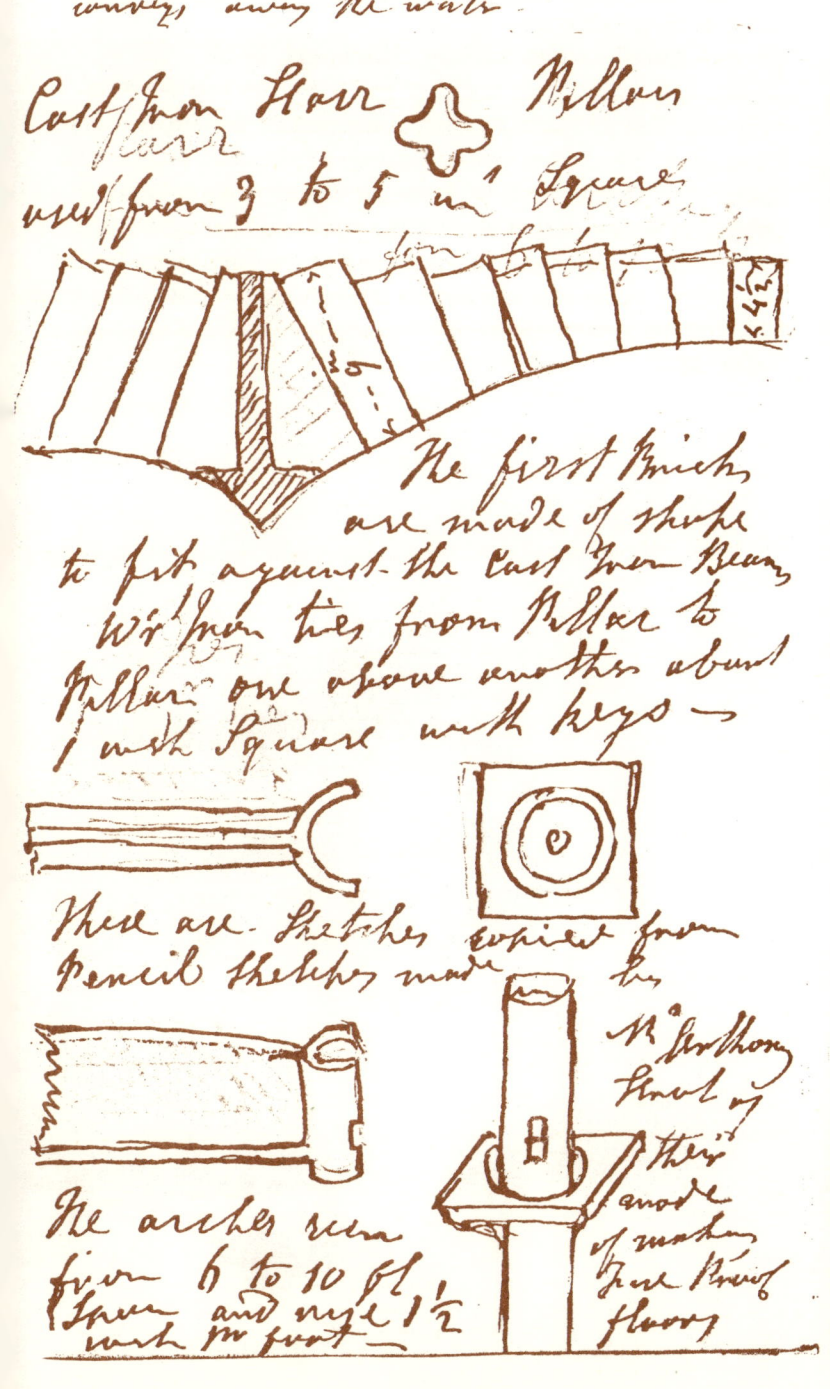

138

19. C. Bage to W. Strutt 1796 undated.
20. C. Bage to W. Strutt ibid.
21. Brotherton Library, Marshall MSS 57, p.44.
22. Shrewsbury Chronicle 1 Sept 1797, quoted A. W. Skempton & H. R. Johnson, 'The First Iron Frames,' op cit. p.180.
23. ibid p.180.
24. C. Bage to W. Strutt 15 Oct 1802.
25. C. Bage to W. Strutt 11 Nov 1802.
26. C. Bage to W. Strutt undated letter 1802-3.
27. C. Bage to W. Strutt 19 May 1803.
28. Ibid. For Hazledine's obituary see Bodl. Lib. Top-Salop. C.1 f 699. I am indebted to Mr. R.A. Chaplin for this reference.
29. C. Bage to W. Strutt 29 Aug 1803.
30. G. A. Lee to J. Watt jun 10 Mar 1798.

had made mistakes in his arithmetic and who by no means copied Strutt's designs without question. He considered Strutt's pillars of 50 ft, the total height of the Derby mill columns, too long as a difference of 30° would push the floor of the upper storey 1½ in into an arch.[19]. On the question of columns he sought advice from a number of people. Joseph Reynolds of Ketley gave Bage the results of an experiment on the weight bearing properties of cast iron which he had made in 1795 'We are differently advised about the strength of pillars. This shape [cruciform] which nearly resembles yours is doubtless the strongest.'[20] A similar shape of pillar was adopted in a later mill at Leeds and John Marshall noted that 'the strongest shape for a prop is [cruciform] because it cannot fail by bending and the strength to resist bending will be as the square of the distance [from one side to the other].'[21] Cast iron columns were thus designed in a rational manner before the end of the 18th century.

The mill took less than a year to build for in September 1797 the Shrewsbury Chronicle[22] reported that Benyon and Bage 'have just finished a spacious Flax spinning mill which is fire-proof. The materials consist wholly of brick and iron; the floors being arched and the beams and pillars being formed of cast iron.' The mill is 177 ft long and 39 ft 6 in wide with an internal width of 36 ft. The castings were made at Hazledine's foundry. Bage may have considered cast iron roof frames for this building. If he did, for some reason he decided against them but being opposed to the idea of using timber he adopted a system of brick arches carried on beams spanning 18 ft from the walls to the central pillar, with slating almost immediately above the arches. This produced an unusual saw-tooth profile.

Shrewsbury was the centre of some of the most interesting developments in the use of structural cast iron. The first iron bridge, erected in 1779, was only twelve miles away. William Hazledine, a millwright, started his foundry in Shrewsbury and in 1795 Telford, county surveyor of Shropshire, directed the work on two iron structures, Buildwas Bridge and a cast iron aqueduct at Longdon on Tern. As Skempton and Johnson said 'No other region in England, or in the world, could show so much work in structural iron, and there cannot be the slightest doubt that Bage was aware of this activity.'[23]

After the completion of the flax mill at Shrewsbury Bage began a series of experiments on the strength of arches and cast iron beams. In particular he was much occupied with the idea of spherical arches and wrote 'I have little doubt of spherical arches answering perfectly.'[24] In November 1802 he wrote to Strutt having just completed an experiment. He had let four posts, 5 ft long, into the ground so that a parallelogram 10 ft by 9 ft was formed. An arch was turned from post to post, bevelled on the inside, and the space between the arches was filled with hexagon bricks. The courses were wedged and the centres removed and the arch was weighted with 2,000 bricks and some barrow loads of soil, a weight of c. 6 tons altogether. The centre of the dome dropped by nearly 2 in. 'Thus I feel perfectly satisfied that brick beams will answer as well as Wood or iron ones and be cheaper.' He added 'Do you think that Government would not do something for the encouragement of Fire-proof buildings? and what is the best method of setting such a plan on foot?'[25]

Later Bage wrote again with information on the latest improvements that had occurred to him. He proposed connecting the pillars by arched plates of cast iron ⅝ in thick. The under edges would have projections from which the spherical arches would spring.[26] In May 1803 Bage asked Strutt for the result of his experiments on arches 'the subject being somewhat complex and not obvious'. He mentioned that when they had last met they had discussed whether the failure of one arch would affect the rest and said that in the case of spherical arches he imagined that it would.[27] Strutt was involved in the building of Belper North Mill – the original mill built by Jedediah Strutt had been burnt in January. Strutt was evidently still undecided about cast iron beams for Bage wrote 'If on a balance of advantages and disadvantages you should at length prefer iron beams I should be glad to submit my reasoning on the strength and shape of them to your examination.'[28]

By August 1803 Bage was carrying out experiments on roof frames 'which have cost no little time or money, but which on our plan were essentially necessary to be ascertained'. He sent Strutt three pages of diagrams showing a near triangular roof frame with tables showing the loads born by the frames and the amount that they had bent. Attempting to discover the effect of the centre pillar sinking more than the walls Bage had lowered it gradually 'on dropping ⅛ in the frames broke in pieces' and he concluded with mathematical proof of his theory of the strength of cast iron beams. He also set out his ideas on the tension flange on beams, a system that is very reliable in practice.[29] These were matters of immediate concern because Bage was involved in building two more mills, one at Leeds and the other at Castle Fields, Shrewsbury. But before either of these two mills was built Bage's influence was seen in another factory.

This was the Salford Twist Mill, the cotton spinning factory of Philips and Lee. There are several drawings of this mill amongst the Boulton and Watt papers and these might seem to support Fairbairn's claim that Boulton and Watt played an important role in the design of the building. They were accepted as such by Giedion and by Skempton and Johnson; however most of the drawings relate to the installation of the steam engine supplied in 1801 or the gas plant ordered in 1804. None of them are working drawings to guide the erection of the mill. G.A. Lee, the active partner in the firm was clearly the key figure in the execution of matters of policy once they had been agreed on by the partners. In March 1798[30] Lee said that at a recent partnership meeting it had been agreed 'that I should immediately erect

Fire Extinguishing Apparatus

The frequent occurrence of factory fires in the 18th and early 19th centuries prompted some entrepreneurs to build fire-proof factories. But these were beyond the means of all but a minority and the installation of fire extinguishing apparatus provided a cheaper, although less effective, means of dealing with the fire risk. However in practice the majority of manufacturers who purchased fire extinguishing equipment from Boulton and Watt had fire-proof mills. This is another example of the shortage of capital experienced by many manufacturers who found difficulty in recruiting capital for the basic equipment in their factories.

Thomas Orrell, cotton factory, Stockport, 1819
Orrell's equipment was installed in 1819. The top drawing shows in plan and section the connection of the apparatus to the engine. The lower drawing shows in plan and section the staircase of Orrell's mill with the air vessels at each landing level. The water cistern was on the roof and there were pipes at each floor which could be connected to the main pipe by the stairs. The pump was worked by a beam connected to the main beam of the steam engine.
Portf: 466

140

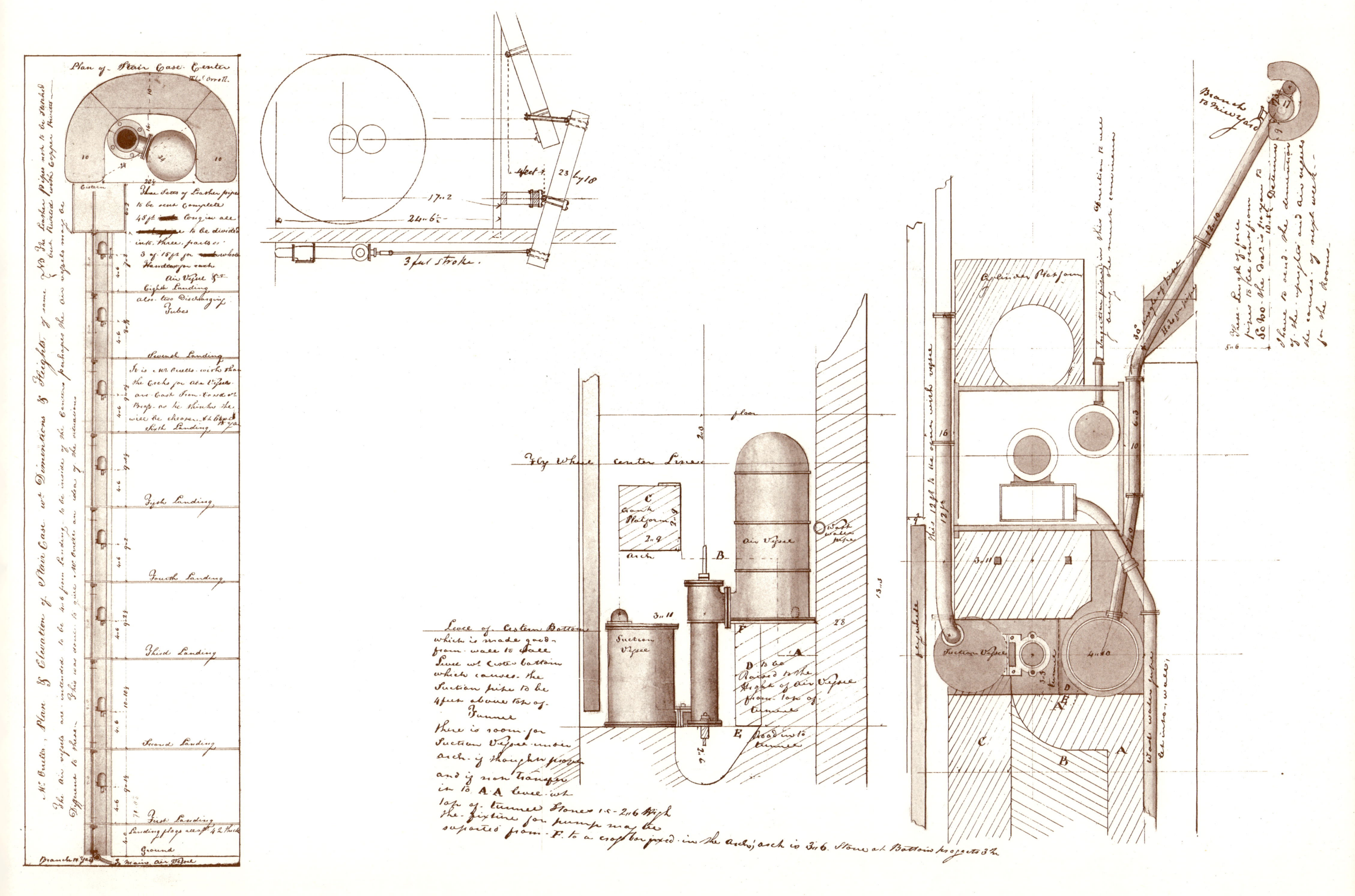

31. A. J. Pacey, 'Earliest Cast Iron Beams,' *Architectural Review* 145, 1696.
32. G. A. Lee to J. Lawson, 26 June 1797.
33. J. Watt jun to G. A. Lee 2 Jan 1799.
34. G. A. Lee to J. Watt jun 8 Jan 1800.
35. Ibid.
36. Ibid.
37. Ibid.
38. J. Watt jun to G. A. Lee 11 Jan 1800 Office Letter Book.
39. J. Watt jun to G. A. Lee 25 March 1800, Office Letter Book.
40. G. A. Lee to J. Watt jun 8 Jan 1800.

another mill as the Situation of Affairs was favourable to economy and good workmanship. But, he added, 'their hearts have since fail'd [and] I have neither proposed or press'd the undertaking, and the knowledge of it may remain amongst ourselves.' The decision against building must have been reversed for in 1799 foundations were laid and the walls were built up to ceiling level on the first storey. Thereafter there was little progress until mid 1800. A.J. Pacey[31] suggests that the time interval between the mill being first proposed and its being built were of great importance from a design point of view and that it was only in those months that Lee decided to use iron beams. This is possible but there is no evidence to show that Lee had not intended using cast iron beams from the outset. The delay in building was probably caused by economic, not technical, problems.

As early as June 1797[32] Lee, with his eyes trained keenly on Marshall's success in flax spinning, was considering embarking on a similar venture. He wrote to Lawson at Soho asking him if, when he went to Leeds, he would 'make any further Enquiries about the Flax spinning there without implicating yourself more than you would wish. I should be glad to hear and compare it.' Lee, apparently half jokingly had put a proposition to his friend James Watt jun. and in January 1799[33] Watt wrote 'You may recollect that we have sometimes talked jocosely of the erection of a flax mill and we now begin to consider it *seriously.*' He asked for Lee's advice on the profits to be made in flax spinning 'and whether your engagements and inclinations would permit you to take an active share in the planning of such an establishment'. Watt mentioned that a capital of £30,000 could probably be raised 'by ourselves and friends'. Lee broached the subject to his partners sometime in 1799 and they agreed to appropriate £20,000 to begin the undertaking 'and confided to me the selection of the situation and the choice of the manager'.[34] Then the derangement of American and West Indies trade 'locked up so much of the manufacturers property' that the idea was temporarily suspended. One question needs to be raised. Did Philips and Lee consider putting up a flax factory rather than a cotton factory on the site of the new Salford Twist Mill? They may done so for a few months but in a later letter to Watt, Lee implied that after the idea of going into flax spinning was 'temporarily suspended' by his partners he had considered the possibility of going into flax spinning in another partnership whilst remaining a partner and manager in the firm of Philips and Lee. However, as he pointed out 'There are many general objections justly exist against multiplying partners.'[35] When previously in Leeds Lee had inspected Matthew Murray's workshops where he was shown the machinery that Murray was preparing for Marshall 'I avail'd myself of a pretended difference of opinion about the form of wheels to desire him to call upon me afterwards and then I learn'd that he was under articles not to make machinery for any other person, but he added slyly there was no restraint against giving instructions or drawings. From that moment I saw the palm of his hand was open to corruption.'[36]

Perhaps Marshall had an inkling of what was in the wind for he called on Lee, a situation which Lee turned to his own advantage: 'In consequence of Marshall calling and asking to see our Mill I placed myself in a situation that he could not decline introducing me into his and I found him on the whole as as communicative as I expected. I have now the same claim upon Bage who paid me a similar visit a few weeks since which I shall certainly return the earliest opportunity.'[37] Three days later Watt replied 'Let me recommend it to you to lose no time in getting the information you wish from Murray and Bage. If once our intention transpires it will become very difficult to gain further intelligence.'[38] He added that 'the Old Gentlemen' (Boulton and Watt senior) might be against the scheme, having suffered from having a stake in the Albion Mill Company. Shortly afterwards Watt declined to take the project further saying that he felt that if Boulton and Watt were to invest in other manufacturing concerns it would be wiser for them to invest in the iron trade.[39]

There can be no doubt that Lee and Watt seriously considered going into flax spinning. Lee's desire to see the Leeds and Shrewsbury mills was prompted by the wish to see Murray's machinery at work. Nevertheless, he must also have been aware of the improvements in factory design effected by Bage. Lawson had probably reported on this as early as 1797 when he went to Leeds for he must have heard about the Leeds firm's new factory at Shrewsbury. The delay between 1799 and 1800 in the building operations at the Salford Twist Mill, added to the fact that building stopped at the position of the first storey beams could indicate that operations were halted pending a decision on what materials the beams were to be made of. But it seems more likely that the delay was due to economic factors rather than technical ones. The 'derangement' of overseas trade and the shortage of capital added to the fact that the partnership agreement expired in 1799 and was not renewed, the partnership continuing 'only by mutual convenience and upon Honour' would have delayed matters. The delay in renewing the partnership agreement may have been due to the indecision of the partners over whether to embark in flax spinning or not. Moreover, Lee remarked in January 1800 on 'The substantial and expensive manner in which everything has been done in the mill,'[40] which might indicate that he had decided to use cast iron from the outset. Nevertheless, the delay in building enabled Lee, whilst acquiring knowledge of flax spinning, to gain information on Bage's iron framed buildings.

Birley and Marsland, cotton spinners, Manchester, 1804
This drawing shows the cotton mill in plan with the position of the main water pipes.
Portf: 310. Reverse. Scale approx 1in : 10ft.

Sketch of pipes for Extinguishing Apparatus &c

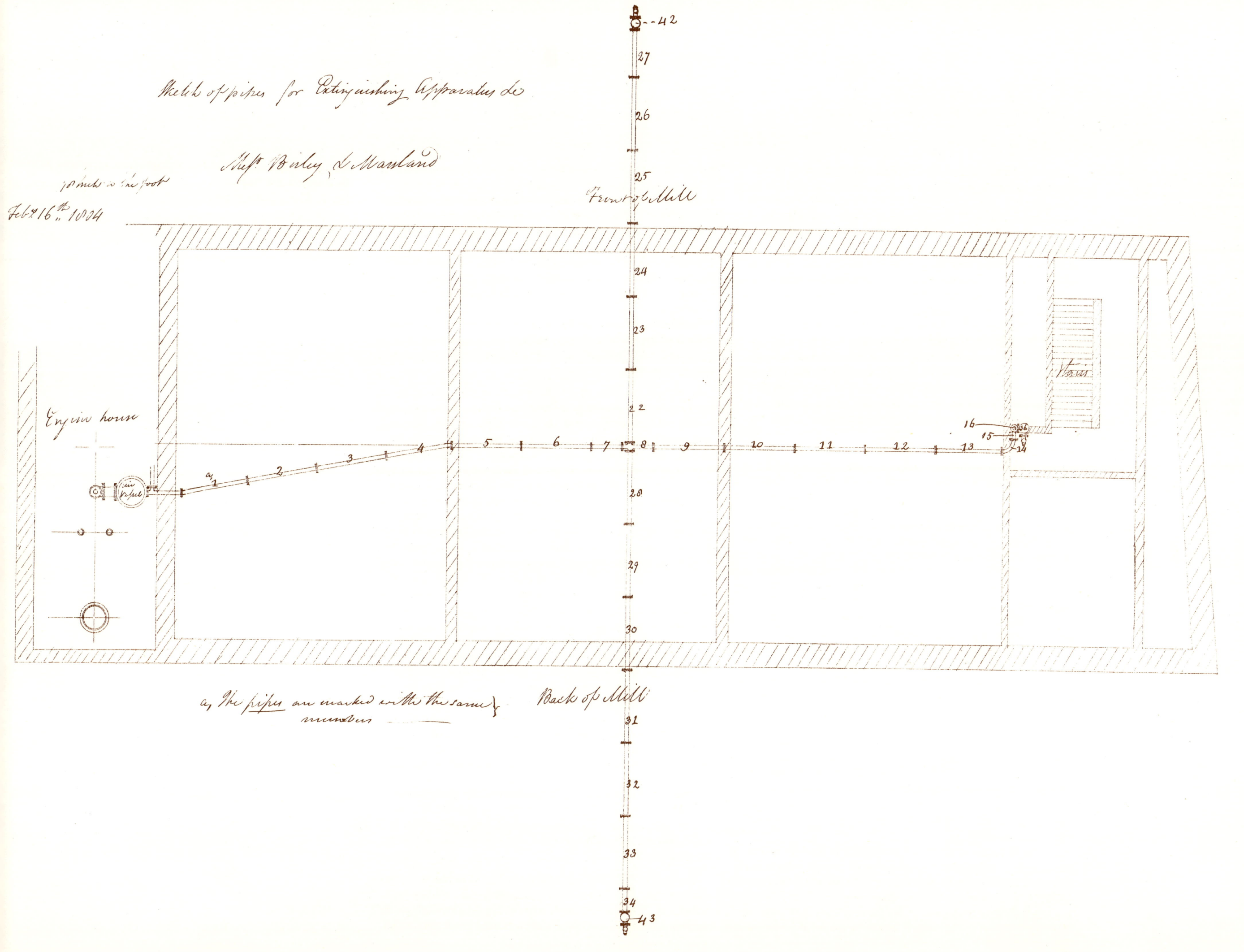

Messrs Birley & Marsland

10 inch to the foot

Feby 16th 1804

Front of Mill

Engine house

a, The pipes are marked with the same numbers

Back of Mill

142

41. J. Southern to G. A. Lee 29 May 1800 Foundry Letter Book; J. Southern to G. A. Lee 7 Nov 1798, Foundry Letter Book.
42. G. A. Lee to J. Watt jun 16 Jul 1800.
43. J. Southern to G. A. Lee 30 July 1800 Foundry Letter Book.
44. G. A. Lee to J. Watt jun 25 Nov 1800.
45. G. A. Lee to J. Watt jun 23 Mar 1801.
46. G. A. Lee to J. Watt jun 28 Jul 1801.
47. A. J. Pacey op cit.
48. A. W. Skempton & H. R. Johnson, 'The First Iron Frames' op cit p.184.
49. Science Museum, Goodrich Papers, Memoranda Book.
50. C. Bage to W. Strutt 29 Aug 1803.
51. *The Leeds Guide,* Printed by Edmund Baines for the Author, Leeds, 1806 p.103.

By May 1800 Lee had sent drawings of the factory to Boulton and Watt, apparently inviting comments, and Southern returned them with sketches of what he and Watt considered were improvements.[41] Watt suggested modifications to the section of the beams: 'Instead of two beams three are drawn across the mill supported at the pillars. The pillar to go through the beam, and the joint to be so that it may be got out in case of breakage .... the ends of the beams to have circular eyes to clasp the top of the pillar.' He continued 'I am aware of the advantages you expect to derive from making two beams only and screwing them together (at mid span) but neither Mr. W. nor myself think it safe to trust...such screws for the purpose'. Watt thought tie irons essential 'Mr. W. recommends wrought iron bars from end to end of the mill to keep the beams from springing out, and each to be independent of any others that if necessary any arch may be removed without endangering the rest.' Lee rejected Watt's advice on the number of beams to span the mill and Creighton's drawing of 1801 shows two beams connected at mid span. It is not clear at what date Lee decided to use hollow circular pillars rather than adopt the cruciform ones used by Bage and Strutt but as early as November 1798, probably in connection with the engine house for the earlier mill at the site Southern wrote 'We cannot cast hollow pipes or columns more than 9 ft in length without a very considerable expense.'

The building progressed and in July Lee reported 'The mill is larger than Marshalls at Shrewsbury'.[42] He requested a sketch of a 50 or 60 hp engine to be sent 'that we may fix the bearings in the walls as we proceed, a precaution which an arch'd building absolutely demands as it would be highly dangerous to derange them.' Later the same month, perhaps in answer to a question from Lee Southern wrote 'We have no opinion of frame work cast in one piece — and wrought iron stays the whole length will come very expensive.'[43] By November the iron framing was complete and the walls probably finished, for Lee requested Watt to allow Creighton 'at leisure hours to make a Drawing of our Building with the Beams and Pillars in perspective before they are conceal'd by the Arches and gearing.'[44] By March 1801 the floor arches had nearly been completed 'and the effect perfectly corresponds with my expectations'.[45] But there was a set-back in July when one of the brick piers which supported a pillar suddenly gave way killing a man, 'everything else stood bolted and firm into place after a strain which is scarcely calculable'.[46] After this accident the spaces between the piers were filled in.

Fairbairn seems to have been inaccurate in his description on three main counts. In the cross section of the mill which he reproduces he shows three beams resting on the pillars as suggested by Watt but as rejected by Lee. There is no evidence to support his assertion that the beams at the Salford Twist Mill were an inverted T shape, for as Pacey[47] has shown what Fairbairn would have seen in the mill would have agreed with his experience of T beams, the web of Lee's beams being obscured by the brick arches. Although Boulton and Watt's opinions were valued by Lee, on major points their advice was rejected, and on the available evidence it must be said that Boulton and Watt did not play a major role in the design of the Salford Twist Mill. The improvements in the design at Salford compared with Shrewsbury were in the slimmer, more economical beam sections, not T shaped but with less web above the exposed base, and in the concealing of the tie rods. The hollow cylindrical pillars were, theoretically, stronger for the same weight than the cruciform shaped pillars in use at Derby, Milford and Shrewsbury. But, as Skempton and Johnson have said, it is strange that Lee accepted timber roof frames.[48]

It will be remembered that Bage had been experimenting on the load bearing properties of cast iron and on the construction of cast roof trusses between 1802 and 1803. In 1804 Marshall separated from the Benyons and Bage, taking the Leeds mill and the mill at Castle Foregate, Shrewsbury. The Benyons and Bage took their share in capital and machinery and started again both in Leeds and Shrewsbury building a factory at Meadow Lane, Leeds (1802-3) and at Castle Fields, Shrewsbury (1803-4). The Leeds mill was begun late in 1802 and Bage was there from January to March 1803. He put into practice his tension flange idea and made the beams simply supported at the columns by a flexible joint. Simon Goodrich visited the mill in August 1804[49] and noted that the mill was on the same plan as Marshalls. He described the arched floors and iron beams *c.* 12 in deep in the middle with 1½ in flanges supported on 'star pillars'. With a flexible connection at each end the beam was in simple bending tension in the lower portion only. It will be remembered that Bage had also experimented on a triangular shaped cast iron roof truss of 38 ft span, noting in a letter to Strutt that the experiments 'on our plan were essentially necessary to be ascertained'.[50] He added that the frames would be spaced 10 ft apart. These dimensions agree with the ground plan of the mill and it seems probable that this mill had the first cast iron frame roof. A contemporary writer describes the buildings as 'completely fire-proof, no timber whatever being used in the building'.[51]

The mill at Castle Fields, Shrewsbury was begun in about July 1803 and was in operation by the end of 1804. It was 208 ft long and had Bage's standard internal width of 36 ft. It was 5 storeys high with a pitched roof probably supported by iron roof trusses.

In January 1803 the original North Mill at Belper was totally destroyed by fire. The loss was immense as the building was uninsured. Strutt almost immediately began to rebuild the mill but his decision to use iron beams was only made after a visit

William Creighton's list of fire extinguishing apparatus supplied by Boulton and Watt, n.d.
Portf: Wheel Carriage Box.

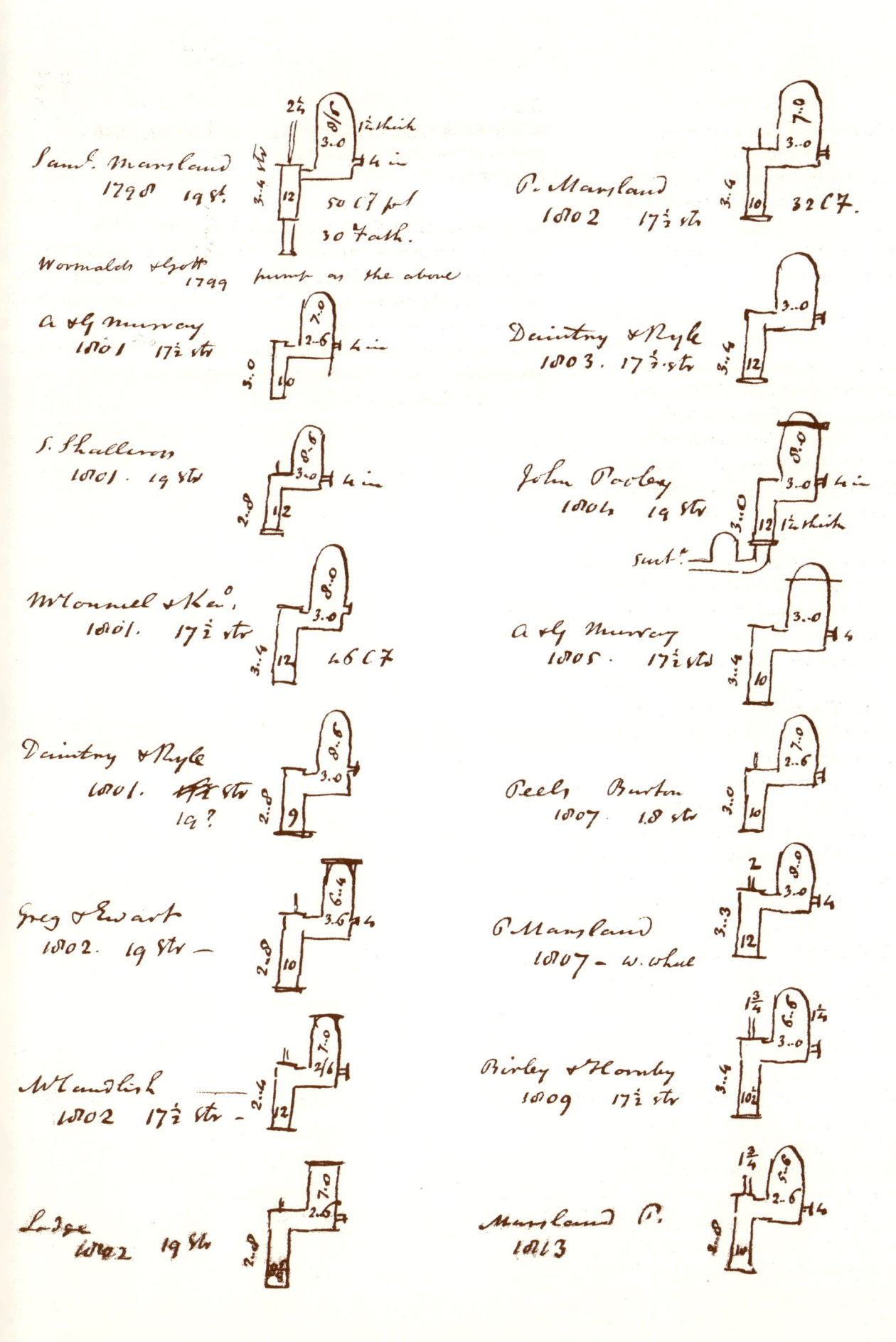

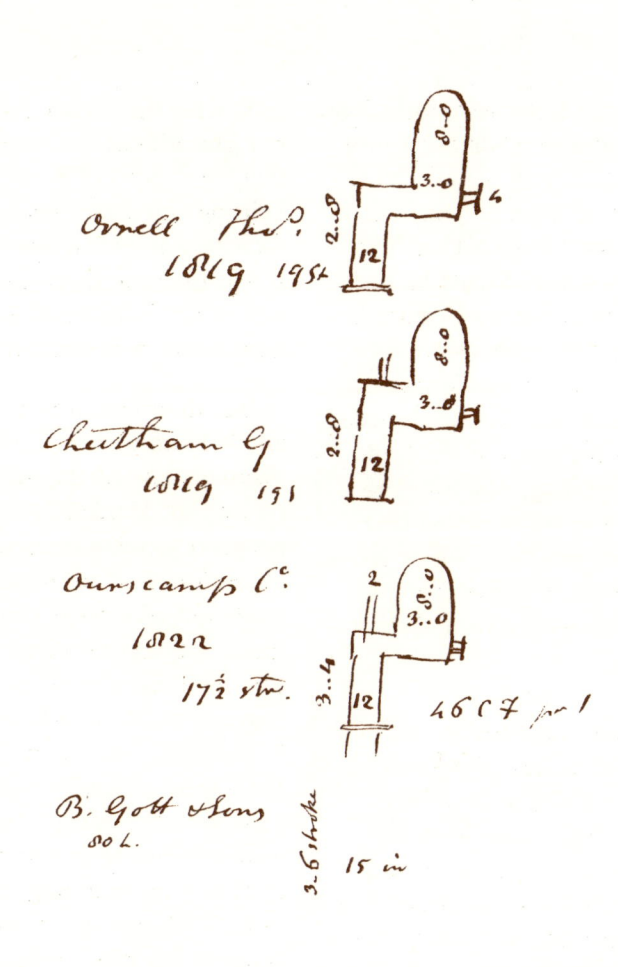

Saml. Marsland
1798   19 str

Wormalds & Gott
1799   pump as the above

A & G Murray
1801   17½ str

S. Challen
1801   19 str

McConnell & Ken.
1801   17½ str   46 C 7

Daintry & Ryle
1801   str
        19?

Greg & Ewart
1802   19 str —

Mc Candlish
1802   17½ str —

Lodge
1802   19 str

P. Marsland
1802   17½ str   32 C 7.

Daintry & Ryle
1803   17½ str

John Pooley
1804   19 str

A & G Murray
1805   17½ str

Peels   Burton
1807   18 str

P Marsland
1807 — W. whl

Birley & Hornby
1809   17½ str

Marsland P.
1813

Orrell Thos.
1819   19 str

Chetham G
1819   19 s

Ourscamp Co.
1822
17½ str   46 C 7 p 1

B. Gott & Sons
80 L.
               15 in

52. *Rees Cyclopaedia*, Article 'Manufacture of Cotton.'
53. J. Sutcliffe, *Treatise on Canals and Reservoirs...* 1816, p.33.
54. William Brown, 'Information Regarding Flax Spinning in Leeds,' 1821, p.3, Typescript in Leeds City Library.
55. J. & F. Naismith, *Recent Cotton Mill Construction and Engineering*, quoted in Owen Ashmore, *Industrial Archaeology of Lancashire* p.48.
56. S. Andrew, *Fifty Years Cotton Trade*, Oldham 1887, p.1.
57. Owen Ashmore op cit p.49.
58. Brotherton Library Gott MSS, 152, Fairbairn & Lillie to B. Gott & Sons, 21 Oct 1824.

paid to Derby by Bage in May 1803. In his letter of thanks Bage said that 'If on a balance of advantages and disadvantages you should at length prefer Iron beams I should be glad to submit my reasoning on the strength and shape of them to your examination.' In his next letter written in August Bage set out his theory and it was perhaps this which persuaded Strutt to use iron beams. The internal structure was sketched by Goodrich probably in August 1804. He noted that the beams were 9 ft long and that the roof truss was made of cast iron. Concealed wrought iron tie bars connected the pillars and Goodrich noted that ties were used to join the beams in the walls of the building. Farey was enthusiastic about this building, describing in detail the pot arches of the wing above the water wheel that were 'light but sufficiently strong to bear anything which is ever required to be loaded upon them'.[52] The mill is five storeys high with an attic, 127 ft long, 31 ft wide with a wing 41½ ft by 34 ft.

For the next quarter of a century there was little advance in the design of framed buildings. In 1805-6 an east wing was added to Milford Old Mill and between 1807-8 a six storey structure known as the Reeling Mill was built near West Mill, Belper. The Evans at Darley Abbey, related by marriage to the Strutts, built two of their mills on Strutt's model at about this period. In 1804 Henry Houldsworth's iron framed cotton spinning factory at Glasgow was built. This factory was 224 ft long with an internal width of 39 ft and was six storeys high plus an attic. There were iron beams spanning 14 ft and the pillars were cylindrical. Between 1805 and 1807 James Kennedy's mill was built in Great Ancoats St, Manchester. It was of immense size, being 370 ft long with an internal width of 40 ft 8 in and six storeys high. South Mill, Belper built between 1811 and 1812 marks the culmination of Strutt's work on iron framed buildings. A five storey building 118 ft long by 40 ft wide, it had 12 ft span beams in the main part of the mill but in the east wing the beams spanned 17 ft.

In the south of England one of the first framed buildings and perhaps the finest of all, certainly the most elaborate, is Stanley Mill, Gloucestershire, built in 1813. The designer is unknown but the ironwork was obtained from the Level Ironworks near Dudley. Traceried iron arches supported by cast iron pillars support the brick arch floors. Between 1813 and 1820 several other important iron framed buildings were erected. Sedgwick mill was built for McConnel and Kennedy in Union St, Manchester in 1818. This was an eight storey mill with beam to pillar connections similar to those used by Bage at Leeds.

There is no doubt about the enthusiasm of the designers and builders of the early fire-proof factories for this type of construction. John Sutcliffe,[53] the Halifax millwright, writing of an ideal cotton mill said 'whether the building is made fire-proof rests with the owner though it would be much better if it was so, but,' he added, 'certainly the Manchester plan of making a factory fire-proof is capable of much improvement.' Like gas lighting and steam heating in cotton factories the fire-proof factory was a structure that only the largest firms contemplated. It was manufacturers like William Strutt, Charles Bage and George Lee or McConnel and Kennedy, people who had capital and were keen to experiment, who led the way.

But their lead was not quickly followed by other manufacturers. William Brown[54] noted in his tour of the Leeds flax mills 'Scarcely any of the mills are fire proof and excepting one or two of Mr. Marshall's and one of Mr. Benyon's, they are old irregular looking houses seemingly much disfigured with alterations and additions. Some of the smaller ones are even made out of a range of old dwelling houses and are extremely mean and inconvenient.... The warehouses and heckling houses like the mills are by no means well arranged or commodious, they are built of brick and tile and of all shapes and sizes.' J. and F. Nasmith[55] wrote in 1909 that 'English mills between 1825 and 1865 were generally constructed with wooden floors on transverse wooden beams, crossed by longitudinal joists with cast-iron columns to support the beams.' Samuel Andrew[56] remarked that mule spinning mills were generally non-fireproof and other 19th century writers state that until the mid to late 19th century fire-proof mills were rare. This is also implied by the absence of drawings of fireproof mills in the Boulton and Watt collection. A survey of all the surviving textile mills in Preston, made in the early 1950s showed that the majority were constructed with wooden beams and floors until the last two decades of the 19th century.[57] Several important fire-proof factories were built in the 1820s and 30s but as a percentage of the total factory building of the time the proportion of new fire-proof factories was extremely small. All these factories were erected by large scale manufacturers who were by no means typical. Thomas Harrison of Stalybridge built a large fire-proof factory for spinning and weaving in 1823 (page 24) and George Cheetham of Stalybridge and John Wood of Bradford (page 41) both built new iron-framed mills in 1833. Wood's mill was designed by Fairbairn. In 1824 Fairbairn and Lillie designed a four storey mill for Gott of Leeds. The pillars were to be cast in Leeds but Fairbairn and Lillie made the iron roof trusses. The beam span was 18ft which caused Gott some alarm after a 20ft beam had broken locally. Fairbairn and Lillie assured him: 'As respects the stability of your beams we have no hesitation in pronouncing them perfectly safe and would beg you to remain free from all suspense on the score of strength'.[58] Strutt had used a 17ft span at the beginning of the 19th century. In 1820 G.A. Lee of the Salford Twist Mill designed a fire-proof mill for J. and N. Philips of Tean Hall, Checkley, Staffs. The Company decided to build another fire-proof factory at Cheadle at the same time. Lee set

Glass

British Plate Glass Co, Ravenshead, Lancashire, 1846
A plan of the works dated 1846
Portf: 537

British Plate Glass Co, Ravenshead, Lancashire, 1788
A steam engine was ordered in 1788 when Boulton and Watt noted 'The boiler shed and polishing shops may be built in the spring, but the engine house, chimney and grinding shop should be built up as soon as possible.' Henry Gardner of Liverpool was millwright but James Watt was much concerned in the design of the polishers. This drawing shows the polishing shop.
Portf: 39. Scale approx 1in : 7ft 3in.

Plan shewing the proposed position of the 2-60H. Engines & Boilers.   **BRITISH PLATE GLASS Cº**

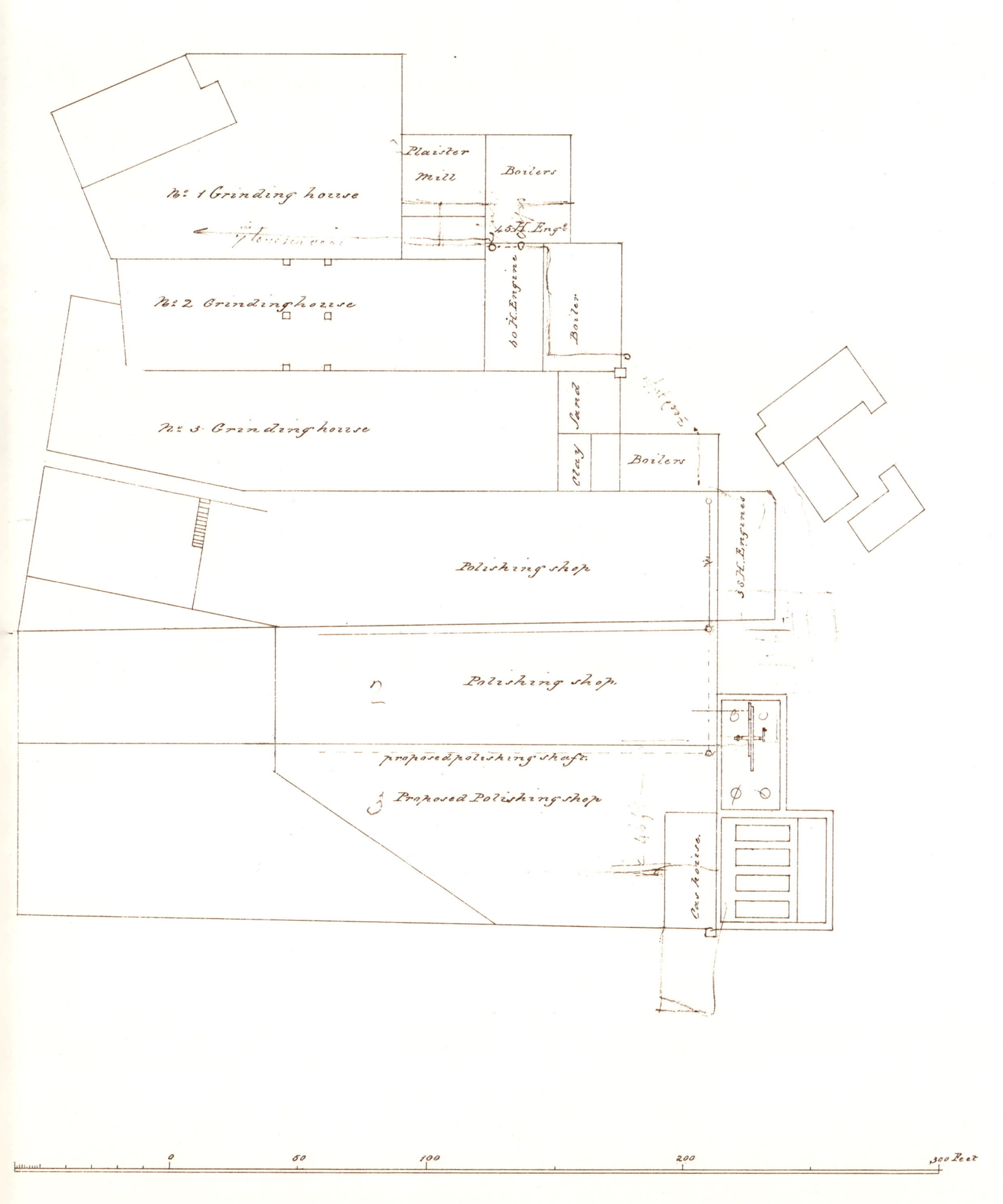

59. Staffs R. O. D644/8/1. I am indebted to Mr. Robert Sherlock for this reference.
60. A. Ure, *The Philosophy of Manufactures*, p.33.
61. M. Davies-Shiel & J. D. Marshall, *Industrial Archaeology of the Lake Counties* p.97.
62. H. R. Johnson & A. W. Skempton, William Strutt's Cotton Mills, op cit p.193.
63. Marshall MSS 37, p.1.
64. S. D. Chapman, 'Fixed Capital Formation in the British Cotton Industry, 1770-1830,' pt IV. I am indebted to Dr. Chapman for showing me the typescript before publication; James Montgomery, *The Theory and Practice of Cotton Spinning*, Glasgow, 1836, pp 248-55; Andrew Ure, *The Cotton Manufacture of Great Britain*, 1836, vol 1, pp 297-313.

out the costs and advantages of bringing weavers into the mill, halving the labour force by using power looms.[59] In 1834-5 Orrell's factory at Stockport was built, designed by Fairbairn,[60] and in 1836 Fairbairn designed Shaddon Mill, Carlisle, a seven storey mill 224ft by 58ft.[61]

One reason why the fire-proof mill was only slowly adopted was that the cost of a fire-proof mill was considerably greater than the cost of a traditional timber framed building. Johnson and Skempton have shown that in the 1790s fire-proof construction increased the cost of building a factory by about 25%. Belper West Mill cost £4,689 to build excluding the wheel and wheelhouse. With a total floor area of 33,500 sq ft the cost of the building was just under 3/- per sq ft. In comparison a four storey mill put up by John Marshall at Water Lane, Leeds in 1791 cost £1,600, or about 2/2 per sq ft.[62] Bage acknowledged the higher cost of fire-proof factories. Another supposed disadvantage of the fire-proof factory was noted by Marshall: 'John Kennedy said that for some time after fire-proof mills were used in cotton spinning it is found that equally good work could not be done in them as in the old wood mills'.[63] The few large steam mills that were built in the first major expansion phase of the cotton industry produced less in proportion to the investment than did the smaller mills. Bage's Shrewsbury factory was similarly over capitalised. The application of power to the mule necessitated the widening of mills but until the 1830s the mule was not entirely satisfactory. When mills were widened, in terms of total floor space, the factories were very little larger. The smaller unit of production had economic advantages over the larger. In 1836 it was shown that the cost of a cotton factory of traditional construction was *c.* 1/8d per sq ft whereas the cost of a Fairbairn iron framed building, the most advanced of the day, was in the region of 4/8 per sq ft. It seems probable that manufacturers delayed major re-equipment and building programmes until the self-acting mule and the power loom were perfected in the 1830s. Then, the larger fire-proof factory became capable of producing an adequate return on the capital invested.[64]

Chance Bros, Birmingham, 1846
A plan of the works dated 1846
Portf: 535

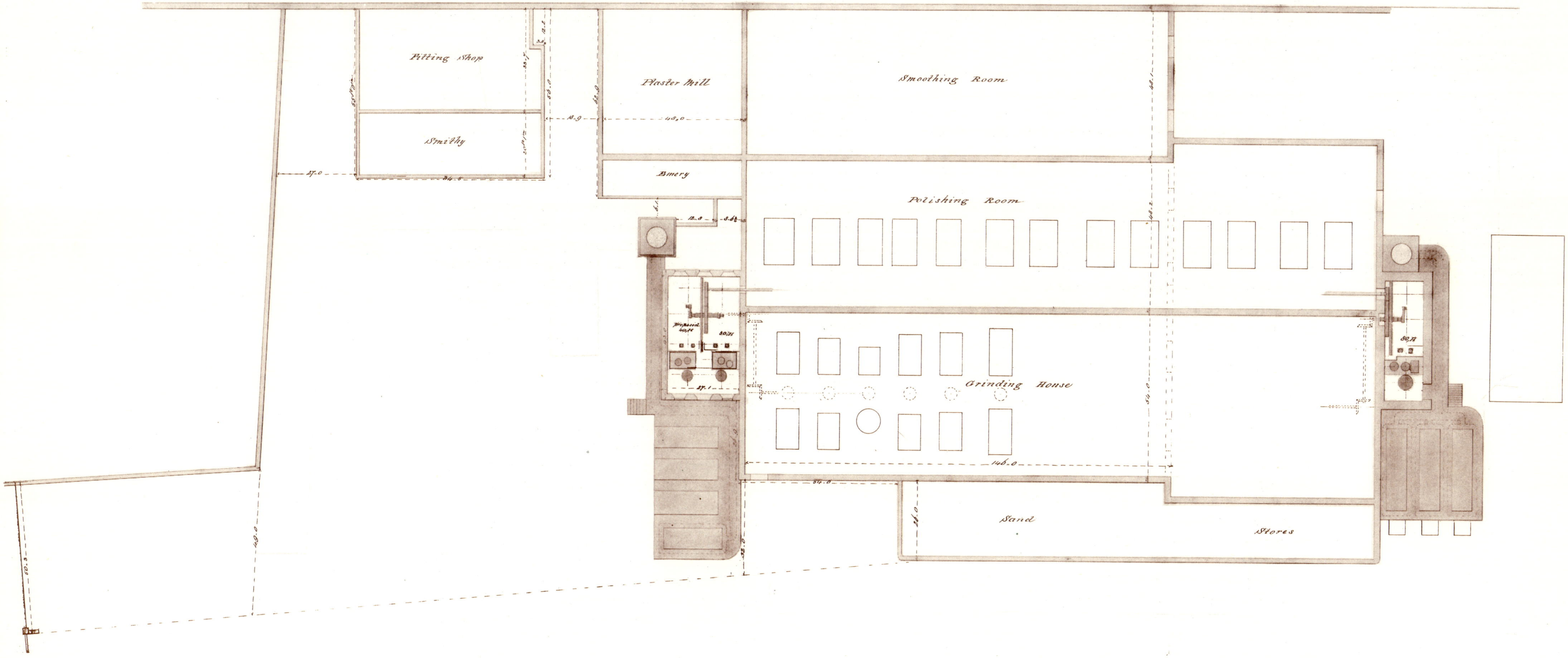
Fitting Shop
Smithy
Plaster Mill
Emery
Smoothing Room
Polishing Room
Grinding House
Sand
Stores

10. Factory Architecture

1. John Gloag, *The English Tradition in Architecture.*
2. K. G. Klingender, *Art and the Industrial Revolution.*
3. Turpin Bannister, 'The First Iron Framed Buildings,' *Architectural Review.* 107, 1950.
4. A. W. Skempton & H. R. Johnson, 'The First Iron Frames,' *Architectural Review* 131, 1962.
5. J. M. Richards, *The Functional Tradition in Early Industrial Buildings.*
6. William Harvey Pierson jun. 'Notes on Early Industrial Architecture in England,' *Journal of the Society of Architectural Historians*, 8, 1949.
7. S. Giedion, *Space, Time and Architecture*, pp 118-119.
8. A. Yarranton, *England's Improvement by sea and land*, 1677.
9. Pierson op cit p.7.

18th century factory architecture, like domestic architecture must be considered in the context of the thoughts, fears, aspirations and political views of the person who was responsible for the building in question. To ignore the social context of factory architecture, or to study a building in isolation, is meaningless.

Factory buildings have received attention from a number of noted architectural historians. John Gloag[1] has discussed the 'external urbanity' of early industrial buildings; Klingender[2] has explored the relationship between art, science and industry. The study of the history of metal framed buildings has been pioneered by Turpin Bannister[3] and A. W. Skempton and H. R. Johnson.[4] J. M. Richards and Eric de Mare[5] in a well known study have considered the theme of functionalism in early industrial architecture and one author, William Harvey Pierson jun.[6] has investigated, in an all too little known article, the architectural style of the 18th century factory.

In the first waves of the 18th century factory movement many entrepreneurs moved not to newly built factories but to converted workshops, dwellings, sheds, barns, warehouses and old water mills. Because of their diverse origins these were necessarily makeshift factories in many cases. They had no unity of architectural expression. The first purpose-built factories on the other hand, were erected to fulfil a specific function and it is possible to detect certain underlying themes in the buildings. Once the manufacturer had decided on the scale of his operations the millwright/engineer was able to calculate the quantity of machinery required. The main structural problem was to provide adequate space to accommodate the machinery which would be gradually installed and to arrange the machinery in the most satisfactory way in relation to the source of power. A rectangular multi-storey form was the most satisfactory unit for the textile and corn milling industries. Its development was not accidental; the necessary space could be achieved more cheaply in a multi-storey block than in a spread of single storey buildings. Spinning machinery was light and well suited to this layout. In corn milling the grinding floor required strengthening but the dressing machinery on the upper floors was light and abundant storage space was necessary. The need for regular motion and the desire to avoid loss of power by friction imposed limitations on the relationship between the machines and the source of power. The width of the factory was determined mainly by the dimensions of the machinery and the desirability of keeping the machines as close as possible to the main shaft. The amount of light required was also a consideration since, except for the top floor, it was impossible to have overhead lighting. It was therefore necessary to have windows fairly close to the machines. But the great variation in the quality of natural light in the early cotton spinning factories suggests that this point has been given too much prominence by some writers.

In certain industries the dimensions of the early factories were remarkably constant, in others there was a greater variation, but in those industries in which factory dimensions varied considerably there was often a characteristic shape to the plan of the works. In cotton spinning, for example, the factories built to house machinery on the Arkwright pattern were usually about 30 ft wide for two rows of frames could be installed within this width. The length varied from approximately 60ft to over 170ft (pages 12,15). With the introduction of the mule at the end of the 18th century the width of the spinning factory was increased to about 45ft. Thus in cotton spinning, the width of the factory was primarily a function of the machinery inside it. The corn mill varied in size but was generally square in plan for the grinding stones were placed around the circumference of a large spur wheel (page 64).The brewery, printworks, mint, glassworks some ironworks and the pottery were generally planned around a courtyard. Two or three storey buildings predominated and the multi-storey building was uncommon until well into the 19th century. In some industries, minting for example, a bastille-like plan was adopted for security. In all these industries the power requirements were generally low and power was needed in small quantities in several separate workshops. In a brewery extensive stabling had to be provided for the dray horses and there had to be room for carts to be loaded in the yard.

The materials used and the methods employed in the construction of the early factories were traditional. It has been suggested that the introduction of the multi-storey building depended on the introduction of the cast iron pillar in the 1780s.[7] But multi-storey warehouses were known in the 17th century and Andrew Yarranton planned a granary of seven storeys in his projected industrial village of New Harlem, Milcote near Stratford-on-Avon.[8] Lombe's mill at Derby was five storeys high. The heavy wooden beams were capable of sustaining the weight of the small, light machinery in use in the 18th century. The 30ft wide cotton mill could be spanned by a single unsupported beam and if additional width was required timber pillars were used. The heavy wooden beam continued to be used long after the introduction of cast iron into factory buildings in New England U.S.A. Pierson comments 'In these interiors some of the heaviest spinning frames ever built were carried on a minimum of wooden columnar support. Iron began to replace wood in the last decade of the century not because of any structural weakness in wood but because iron was less combustible.'[9]

The basic factory form was a response to the function that the building was to serve. In some cases a new or unusual form developed because the mill builder could find no other way of overcoming an unfamiliar problem. Yarranton's proposed flax mill and granary had saw tooth roofs; so did Marshall's Castle Foregate mill at Shrewsbury because, in the latter case Bage wished to avoid using timber in the roof and slated over an 18ft cast iron beam carrying brick arches. In Yarranton's case the

Saw Mills
Don Fernando de Torres, Spain, 1790
This was one of Boulton and Watt's early foreign orders. The steam engine was ordered in 1789. There were to be eight frames for sawing wood and nine for sawing marble.
Portf: 58

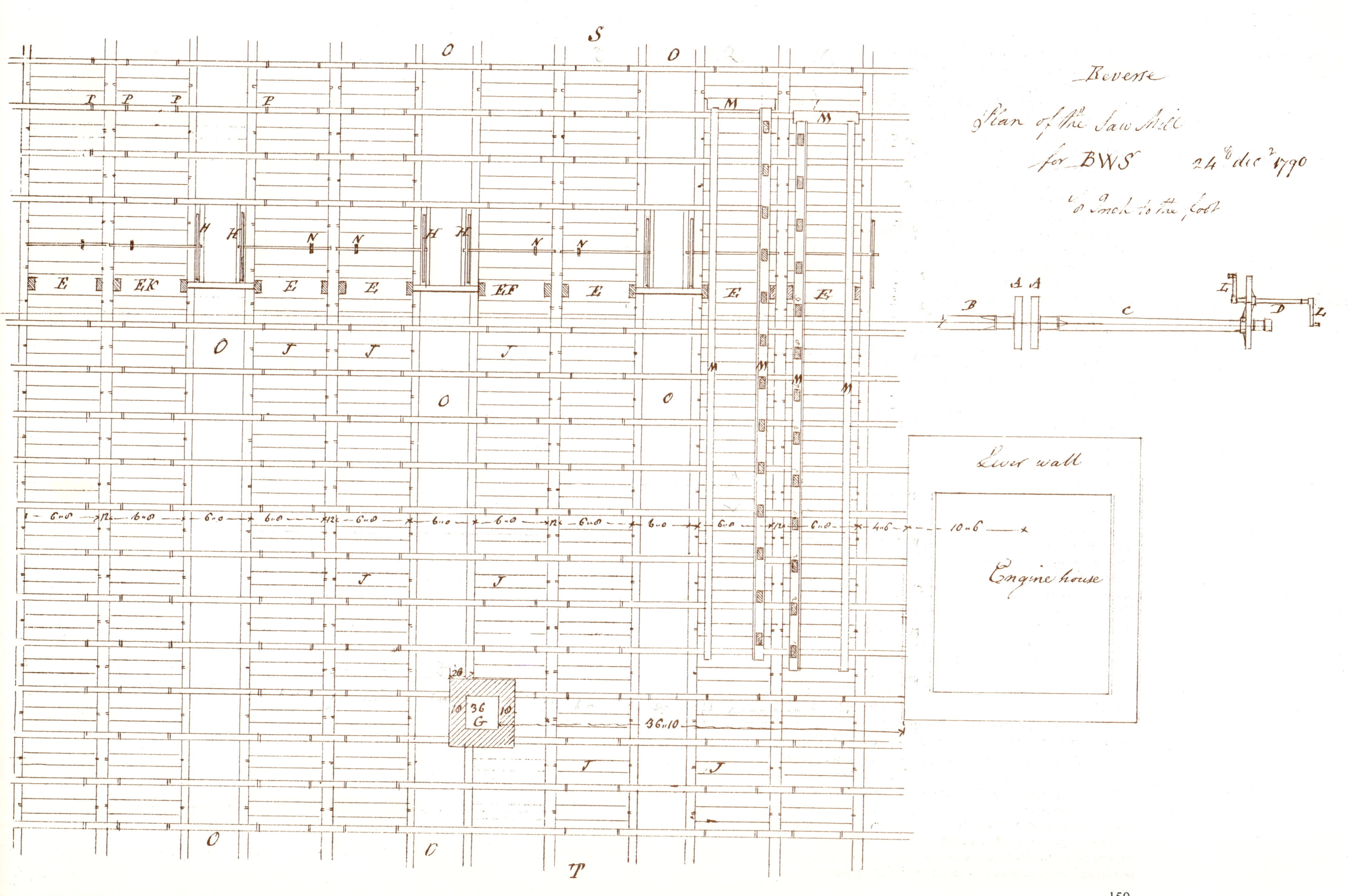

Reverse
Plan of the Saw Mill
for BWS 24th decr 1790
⅛ Inch to the foot

Lower wall

Engine house

10. J. M. Richards op cit p.14.
11. Ibid. pp. 17-18.

novelty of a seven storey block may have prompted him to suggest a roof composed of what amounted to a series of domestic gables. Functionalism extended beyond the basic dimensions of a building to the peculiar adaptations necessary for certain industries. The long attic dormer workshop windows so characteristic of the West of England textile mills were required for burlers who inspected and repaired cloth; the iron plates noticeable on the walls of some metal working factories were not the terminating point of a tie iron but were the means by which work benches for heavy machines were fixed to the building. The lucam, housing a hoist is a characteristic of the corn mill; the louvred window of tanneries and maltings.

The functional building, as J. M. Richards writes,[10] has 'as its object the fulfillment of such needs as logically and economically as possible by taking full advantage of the means and materials available'. Nevertheless, factory buildings were to some extent interchangeable. The Birmingham nail factory (p 86) was, as the inscription says, 'an ale manufactory now a nale manufactory'. Many Yorkshire cotton factories were converted to woollen or worsted mills in the early 19th century and in the West of England fulling mills became paper or silk throwing mills.

Few architectural historians have considered factory architecture beyond the idea of functionalism. But there is another important aspect of factory architecture; the question of aesthetics. Although the factory was often a rectangular multi-storey block or a series of two storey buildings around a yard the front elevation frequently did not conform to the plain elevations of the other sides of the buildings and there was a noticeable movement towards a conscious architectural style. The 'remarkable architectural qualities' of early industrial buildings isolated by J. M. Richards are strength and majesty, their 'breath-taking scale and power'. The factory is linked with the massive civil engineering works of canal and railway and Richards speaks of 'the vigorous and forthright qualities shared by all the artefacts of the Industrial Revolution — engineering structures and industrial buildings alike.' He adds that in many instances the functional character 'is partly overlaid but not disguised by embellishments in Georgian style as it was thought proper at the time to give them'.[11] The decorated front elevation of a factory however was more than a mere overlay for it represented a conscious desire to be accepted, it epitomised the social aspirations of many 18th century entrepreneurs. This should perhaps be called the non-functional aspect of early industrial architecture. Many buildings with a distinctive front elevation are portrayed in *The Functional Tradition in Early Industrial Architecture* but they are beyond the authors' definition of functionalism. The only justifiable inclusion of these buildings as part of the functional tradition is if the need to impress either the local gentry or a prospective customer can be considered a function of a factory building. If this is the

A wind driven saw mill designed by Rennie, n.d.
It is not known if the mill was ever erected but a Mr. Osborn is known to have had a wind saw mill at Hull. Saw frames as shown remained in common use until superceded by the circular saw. A 6 hp engine working three saw frames such as these would cut about 100 ft of 1 ft thick deal timber per hour.
Misc. Mills
Portf: 218

152

12. Sir John Summerson, 'The Classical Country House in 18th Century England,' *Jour. Roy.Soc.Arts*, 107, 1959.
13. Soho Box.
14. Matthew Boulton Notebook 8, pp. 18-19 A.O.
15. E. Sitwell, *English Eccentrics*, 1960 ed, pp. 33-4.
16. Matthew Boulton Notebook 8, p.18. A.O.
17. Ibid, 1, 0.21. A.O.
18. Pierson op cit p.9.
19. W. Fairbairn, *Mills and Millwork*, vol 2, p.113.
20. S. D. Chapman, *The Early Factory Masters*, p.67.
21. R. D. Owen, *Threading My Way* p.9.

case these factories are more successfully functional than the unadorned plain brick rectangular facades of which many 18th century examples can be found. At all events when the front elevation of the factory, adorned or not, is considered in relation to the person who built it and the prevailing taste of the times interesting light is thrown on the entrepreneur.

Sir John Summerson[12] in three illuminating articles on country house architecture traces the development of the Palladian movement in the 18th century and its implications for the classical country house and the smaller house or villa. The large, classical country house was beyond the means of most 18th century entrepreneurs although in the 19th century the Arkwrights, for example, purchased Sutton Scarsdale, Derbyshire. But the villa type house was copied extensively by wealthy manufacturers conscious of their rising position in society. Matthew Boulton expended much time in discussing the building of Soho House with the architect brothers James and Samuel Wyatt. Early drawings illustrate how the architects modified their designs in accordance with Boulton's wishes, the final version being based on a modified form of the Palladian villa. Boulton was concerned that his garden should be pleasingly landscaped in accordance with 18th century practice. He took an inventory of the fruit trees in his garden[13] and visited country houses noting improvements. He was much impressed by the Hon Charles Hamilton's gardens and especially by his grotto: 'The Grotto past description....There are many winding cells to pray in whilst water falls raise ye odour of coolness both to ye ear and feel.'[14] The Hon Charles even imported a hermit to give the grotto greater authenticity.[15] Boulton's improvements were more modest. 'Improve Soho a bridg opposite Eggintons for company to walk ye round so that a one horse chair may go round ye brow of ye hill...plant ye bottom by ye side of ye great pool low trees.'[16] This absorbtion in the ways of the nobility extended beyond their houses and gardens to more personal matters. Boulton noted with scrupulous accuracy, amongst jottings on Newton, 'The Rt Hon the Earl of Hopetown informs me that from many years experience he finds new laid eggs to be very wholesome and more nutritious food than almost anything else in use so he thinks two with salt and Bread only are sufficient for a Meal and that hath been his constant Breakfast for many years'.[17]

Boulton characteristically paid considerable attention to the exterior of Soho Manufactory. The main building was begun in 1762 and finished in 1764. It was a three storey complex consisting of three rectangles, a large block forming the main section and two smaller ones at right angles at either end. The two end blocks projected slightly towards the front as shallow pavilions, extending as wings to the rear. There was a central pediment over three bays breaking forward and the hipped roof was crowned by a cupola. Doors, windows and chimneys were domestic in proportion but the third storey windows were approximately half the size of those of the two lower storeys. The finished facade was a classic piece of English Palladianism. Boulton was a practical industrialist. He housed workers in the upper floors of the wings and at the rear of the main block were the forges and melting furnaces and the huge water wheel which provided the motive power. But he was also as Pierson has said 'an influential citizen in the new social order'[18] who recognised in the architecture of aristocratic England the elements of formal elegance and applied them to his factory.

Some interesting regional variations in factory architecture can be observed in the textile industries. The Midlands cotton factories in particular exhibit great variety and it is worthwhile to bear in mind Fairbairn's[19] assertion that 'At first these mills were square brick buildings, without any pretensions to architectural form....This description of building with bare walls was for many years the distinguishing feature of a cotton mill'. Arkwright's first Cromford mill, built in 1771 made few concessions to architectural style. The hipped roof had a square cupola at one end of it rather than in the middle. But at some of his non-factory building at Cromford and his later mills elsewhere Arkwright displayed a growing awareness of the Palladian idiom. The Greyhound Inn at Cromford shows this well. Built in stone with prominent quoins, it is three storeys high with a pediment containing a clock. The windows on the first two storeys have prominent keystones and the centred doorway has a small pediment over it echoing the main one. At Rocester Mill, Staffordshire, an unusually low pitched valley roof is a striking feature and prominent quoins, keystones and string courses relieve the brickwork. Arkwright's finest Midland factory is undoubtedly Masson Mill at Matlock Bath, built in 1784. This is a red painted brick factory with three bays breaking forward containing a central row of lunettes flanked by venetian windows. There was originally a wooden cupola containing a bell over this central portion. It has been suggested that Cromford was chosen by Arkwright not primarily because of good water power facilities, but as part of a plan to gratify Arkwright's social ambitions.[20] The change in style from the rather bleak early mills at Cromford to the fine inn, school and superior housing there and Masson mill about ¾ mile from Cromford undoubtedly reflects this desire to live and behave like a country gentleman. His preoccupation with the aesthetics of factory building is well illustrated by what is probably an apocryphal tale concerning his partnership with David Dale at New Lanark.[21] Four almost identical factories were built in a line along the bank of the Clyde. On each factory there was a central pediment with a lunette over three bays breaking forward. The central bays had a long narrow window at each storey flanked on either side by venetian windows. On one of the mills there was a fine tall cupola containing a bell. It is said that Arkwright and

Soho Manufactory

East Front of Enoch Wood's factory at Burslem

Mellor Mill, Derbyshire

22. G. Unwin, *Samuel Oldknow and the Arkwrights*, p.124.
23. Catherine Hutton, *Reminiscences of a Gentlewoman of the last century*, 1891, p.23.
24. Illustrated in S. D. Chapman op cit p.132.
25. D. Stroud, *Humphry Repton*, pp.39-40, pl.45.
26. Schinkel, quoted Pierson op cit p.14.
27. Both near Stroud. Woodchester Mill demolished.
28. John Mosse, 'The Albion Mills,' paper read to Newcomen Society, Jan. 1968.
29. John Gloag op cit p.213.
30. Illustrated in Ward, *History of Stoke-on-Trent*, p.266.

Dale quarelled over the positioning of the cupola and agreed to dissolve their partnership. Arkwright's social ambitions were fulfilled for he was knighted during his term of office as High Sheriff of Derbyshire. His mansion house, Willersley Castle, was built in 1788 overlooking the Derwent between Cromford and Masson Mill. Although Byng scathingly called it 'an effort of inconvenient ill taste' Willersley Castle is the embodiment of Arkwright's social aspirations; it was built in the gothic idiom then gaining favour amongst the local aristocracy.

One of the most outstanding cotton mills was Mellor Mill, Derbyshire, (page 154) built by Samuel Oldknow in 1790. This was built on classical country house lines. A main block with a central bay breaking forward contained venetian windows and was crowned by a cupola. A single bay projected at each end forming small wings. The roof was hipped and partly obscured by a parapet. Flanking the main block on either side were three storey pedimented pavilions. Mellor Mill adhered to the Palladian plan with central block and linked pavilions but while the pavilions might have belonged to some gracious country seat the centre block shows the style stretched to breaking point to encompass the requirements of the factory. Mellor Mill was admired by contemporaries, amongst them Robert Owen:[22] 'being ambitious he [Oldknow] desired to become a great cotton spinner as well as the greatest muslin manufacturer. He built a large, handsome and imposing cotton mill, amidst grounds well laid out, and the mill was beautifully situated, for he possessed good taste in these matters.' Oldknow obviously felt keenly the importance of aesthetic considerations in planning his Mellor estate but the cost was high and he had to mortgage the estate almost immediately the mill was built. Ironically the mortgagee was Arkwright.

Wye Mill at Cressbrook, Derbyshire, built in 1815 to replace a smaller mill is of the classic three and a half storeys and has a pediment over four bays breaking forward on both the front and rear elevations. A cupola crowns the hipped roof.

Some entrepreneurs favoured the gothic style. Samuel Unwin, a hosier, went into cotton spinning in the 1770s and built himself a pretentious mansion house in Sutton Park, Sutton-in-Ashfield, Notts, which a visitor described in 1779:[23] 'Mr Unwins house is built of stone and on the outside seems fit for a nobleman, but the best rooms are occupied as warehouses for the cotton manufactory.' He mingled with the gentry and the nobility and it was reported that his carriage was received at Welbeck Abbey, the Nottinghamshire home of the Duke of Portland. Before the end of the 18th century Unwin had built four mills and his Sutton factory[24], a fantastic four storey crenellated building with a chateau-like turret crowning the roof of the wing and with a windmill on the roof of the main block, exhibited some of the wilder fantasies of the entrepreneur with social ambitions. Humphry Repton the landscape gardner and architect was employed at Welbeck Abbey for about ten years after 1789[25] turning a pedimented classical house into 'a large mass of gothic building' a change based, according to Repton, on the Duke of Portland's own ideas. Sutton mill was therefore a direct and intended compliment to the Duke. Castle Mill, built by the Robinsons in the Upper Leen Valley, Notts, is also a gothic crenellated factory with towers and pointed windows.

In the West of England textile area many of the factories have great aesthetic appeal. They were built in the vernacular Cotswold tradition rather than showing a positive feeling for Palladianism or gothic and it was by a bold use of the traditional building materials combined with unusual fenestration that the buildings succeeded. A German architect visiting the area in 1826 noted:[26] 'The factory buildings...lie hidden at a distance under tall lindens, elms and larches and mingle with small churches, which are equally picturesque in their setting.' Two buildings, Woodchester Mill and Hope Mill, Gloucestershire,[27] had the classic circular cupola supported on six columns and Woodchester Mill had a pediment. Both mills had the continuous dormer windows so characteristic of the region. Stanley Mill, Gloucestershire, built in 1813, is an exceptional factory not only for its unique traceried internal cast iron framing but for the fine use of venetian windows in which the dividing members are doric pillars. Venetian windows are generally absent in the West of England mills although one small late 18th century mill at Bowbridge contains some.

The Albion flour mill in London was remarkable in its day and much visited by London tourists. The opening ceremony was attended by Sir Joseph Banks and masques and balls of which James Watt strongly disapproved were later held at the mill. The guests at these events included the Duke of Beaulieu, Duke of Roxburghe, Lord Lansdowne, Sir Robert Lawley and the Directors of the East India Company.[28] Albion Mill displayed what Gloag called 'the same external urbanity as other Georgian structures'.[29] An almost square building with a single bay breaking forward at each end, the building was remarkable for, amongst other things, an extraordinary variety of fenestration; venetian windows on the third storey with long flat headed windows between them, semi-circular headed windows on the second storey and three large lunettes on the first. Other flour mills demonstrate a similar awareness of Palladian motifs. Thompson and Baxter's flour mill at Hull (p 62) had the classic pediment with roundel over three bays breaking forward and two string courses and Stonard and Curtis's small London starch mill had a pediment with lunette (p 55).

An investigation of the facades of some 18th century potteries suggests that Etruria had some serious rivals. Enoch Wood's factory at Burslem[30] had a V-shaped block with a pedimented bay at the base of the V containing a venetian window and crowned by a cupola. To the left of this block was a gothic crenellated range terminating in a tower behind which the kilns

Masson Mill, Matlock Bath, Derbyshire
Scale approx 1in : 12ft 6in.

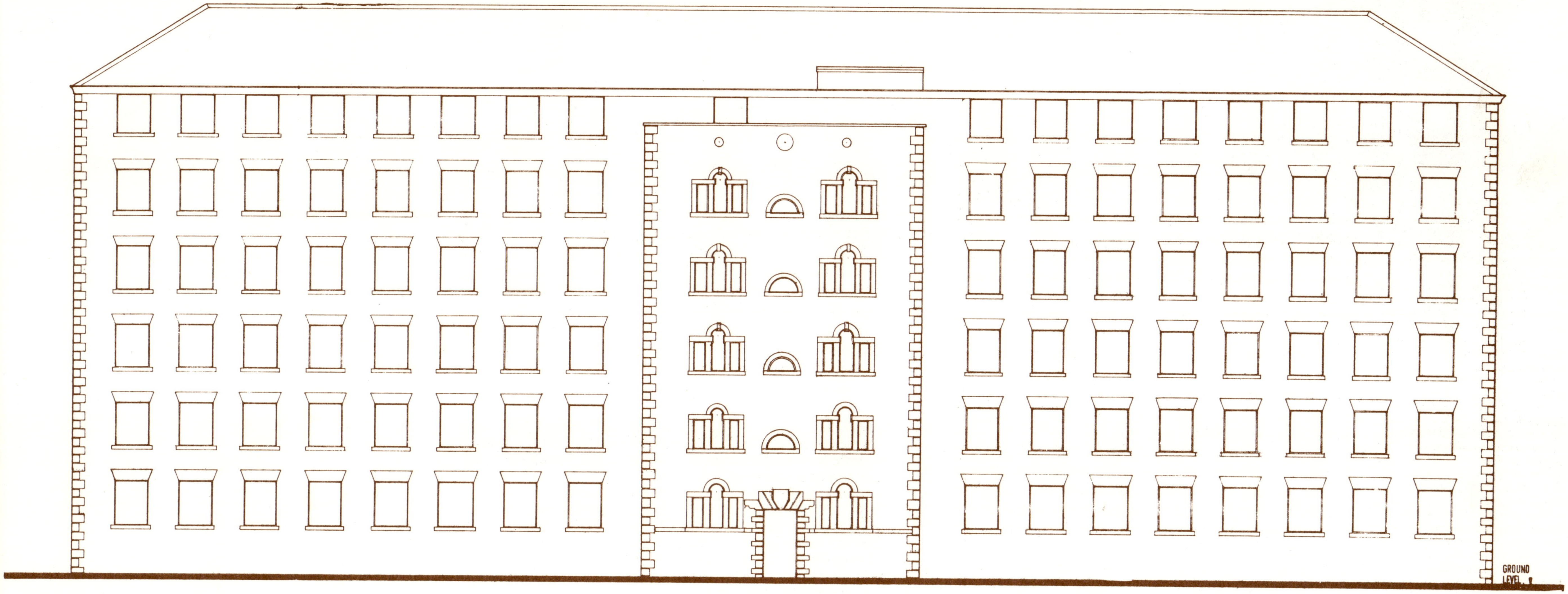
GROUND
LEVEL

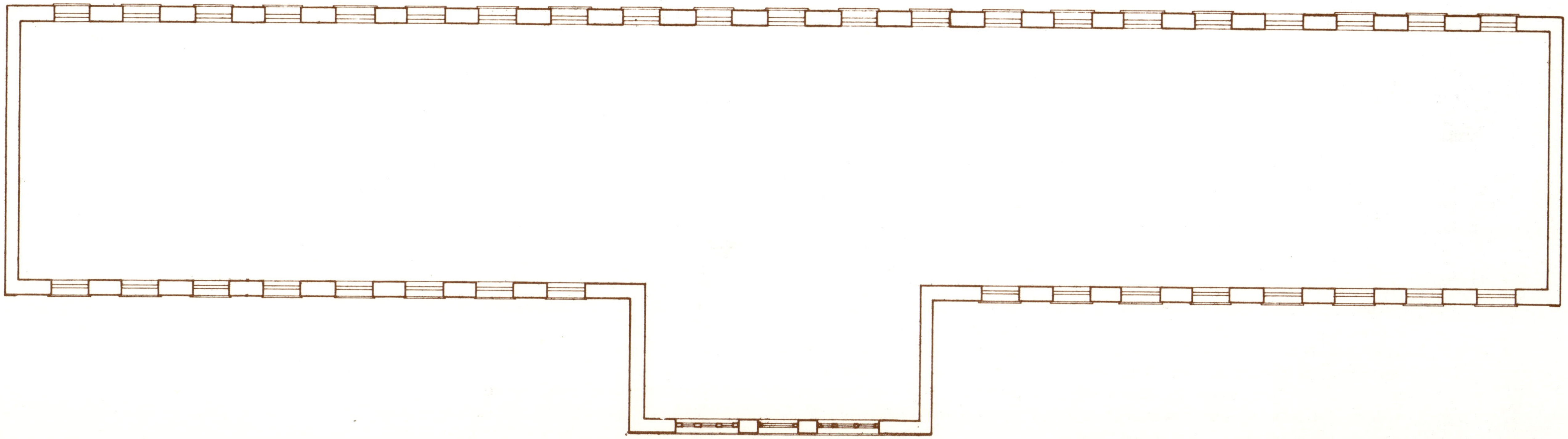

31. Illustrated in W. H. Chaloner & A. E. Musson, *Industry and Technology*, plate 55.
32. S. D. Chapman, 'The Peels in the Early English Cotton Industry', *Business History*, II, 1969.
33. Sir Charles Mordaunt went on a tour of the North in 1788, failed to get into Arkwright's factory 'He knew none of us, and had not time to talk.' E. Hamilton, *The Mordaunts*, p.232.
34. Esther Moir, *The Discovery of Britain*, p.91.
35. Rev. Richard Warner, *Excursions from Bath*, 1801, p.333.
36. Sir G. Head, *A Home Tour through the Manufacturing Districts of England*, p.184.
37. W. Cooke Taylor, *Notes on a Tour of the Manufacturing Districts of Lancashire*, p.22.
38. S. Shaw, *History of Staffordshire*, quoted with general description of Birmingham trade in *Universal Magazine*, August 1802.
39. T. Baines, *Lancashire and Cheshire Past and Present*, 1867.

could be seen, each one having a crenellated top. Wedgwood's factory facade at Etruria was a classic three storey range with a central pediment over three bays breaking forward crowned by a large octagonal cupola. The office and warehouse at the English glasshouse illustrated in Diderot's Encyclopaedia is a handsomely proportioned building with domestic scale doors and windows and vertical bands of rustication on the walls. A row of kilns was almost obscured at the cement works of I. C. Johnson and Co, Gateshead,[31] by an imposing block facing the river; a large coat of arms crowned the parapet over the central doorway and a pavilion at either end of the range completed the block in Palladian style.

It would be wrong to suggest that all 18th century manufacturers were concerned about the external appearance of their factory buildings, probably only a minority were. Robert Peel and his son (Sir) Robert, for example, built small mills of red brick with not a cupola, pediment, decorative quoin or string course to relieve the monotony. This was in keeping with their frugal habits which Sir Lawrence Peel noted, but they were also the wealthiest cotton spinners in the country; Sir Robert Peel built Draycot Manor and his daughter married a viscount.[32] It is interesting that (Sir) Robert Peel did not feel it necessary to adorn the mills or build housing of the standard of Arkwright's at Cromford although he had political ambitions and was, when elected M.P. for Tamworth, responsible for the introduction of the Health and Morals of Apprentices Act of 1802, the first factory Act. The majority of small Yorkshire textile mills, both the water driven and the early steam mills were plain. Benjamin Gott's factory at Leeds had only a small cupola to relieve the endless brick facade of the factory block facing the road.

The desire to embellish the 18th century factory was mirrored in the attitudes of 18th century 'gentlemen of taste' on tour in Britain. Many a traveller to the Lakes, such as Charles Mordaunt, 8th Baronet,[33] would visit a cotton factory on the way north or, after a visit to the great country houses of the Midlands, he might round off his tour at the Potteries or the Derby silk mills. Esther Moir[34] has captured the spirit of the 18th century traveller whose motives 'were at the same time romantic, economic, aesthetic and practical. A patriotism which delighted in the sight of the country's manufacturing and engineering achievements went hand in hand with an imagination which was nurtured on discussions of the distinction between romantic and picturesque and sought for gloom and terror as conductive to true emotions.' The tourist in his 'mood of enchantment' often showed a considerable knowledge of technical processes. One visitor was overwhelmed by the wonder of the machinery in a Gloucestershire woollen mill:[35] 'curious complicated machines above, moving with a velocity that defies the nicest vision to detect their motions, the ponderous engines down below, astonishing the mind in an equal degree by their simplicity and gigantic powers'. Sir George Head was awestruck by a steam engine in a Leeds cloth mill:[36] 'The brilliancy of polish bestowed on many of its parts so lustrous and the care and attention paid to the whole so apparent that imagination might readily have transferred the edifice to a temple.' W. Cooke Taylor[37] noted that Turton Mill, Lancs, was 'not without some pretensions to architectural beauty'. Stebbing Shaw devoted a long passage to Soho Manufactory where it is clear that he appreciated it not only as an example of great industrial organisation but as a piece of landscaping:[38]

'Soho is the name of a hill in the county of Stafford, about two miles from Birmingham; which, a very few years ago was a barren heath, on the bleak summit of which stood a naked hut, the habitation of a warrener.
The transformation of this place is a recent monument to the effects of trade on population. A beautiful garden, with wood, lawn and water, now covers one side of this hill; five spacious squares of building, erected on the other side, supply workshops or houses for about six hundred people.....Mr. Boulton, who has established this great work, has joined taste and philosophy with manufacture and commerce.'

By the mid 19th century the percentage of factories where the manufacturer had expended thought, time and capital on the adornment of the facade had fallen sharply. The unrelieved square multi-storey brick factories of Lancashire are characteristic of this period. Well known engravings of cotton factories in Oxford Street and Union Street Manchester illustrate the point. This was the age of the large urban steam mill, only rarely were rural sites chosen. Once removed from the obvious visual attractions of a water powered site in a rural environment, removed too from contact with the aristocracy and the desire to impress, the stimulus to adorn the factory facade vanished. The 19th century entrepreneur was an extremely practical man. He had often risen from the ranks of the artisan classes and, unlike his merchant forbears, acceptance by the aristocracy was relatively unimportant to him. The changing face of the factory in the mid 19th century is again mirrored in the attitude of the visitors to industrial towns. The rash of mill chimneys, the squalor of speculative builders' housing and the unsanitary conditions were offensive. One historian[39] of Lancashire and Cheshire writing in 1869 apparently found the reality of Liverpool, Preston and Macclesfield all too painful and he reproduced engravings of Liverpool seen from the opposite side of the Mersey, Preston with a Byronic sky and rainbow almost obscuring the town and Macclesfield from high up amongst the crags with 'peasants' in the foreground detracting the eye from the factory chimneys on the horizon.

As the size of the factory increased it was naturally more difficult to fit it into the Palladian framework which was so successfully used in smaller 18th and early 19th century factories.

Cromford Mill, Derbyshire

158

40. W. Fairbairn, *Mills and Millwork*, vol 2, p.114.
41. A. Ure, *The Philosophy of Manufacturers* p.33.
42. W. G. Rimmer, *Marshalls of Leeds, Flax Spinners* pp. 202-206.
43. A. Welby Pugin, *An Apology for the Revival of Christian Architecture*, p.12.
44. W. Fairbairn, *On the Application of Cast and Wrought Iron to Building Purposes* p.146
45. Derbyshire Record Office, Cressbrook Mill Building Accounts.
46. H.M. Colvin, *Biographical Dictionary of English Architects 1660-1840*.

In the cotton industry this was particularly difficult as factories reached immense proportions. Nevertheless, a minority of 19th century manufacturers was concerned to present an attractive facade to their buildings. Swainson, Birley and Co built a huge seven storey factory 157yds long and 18yds wide at Preston. An engraving shows two central crenellated turrets joined by a crenellated parapet and the roof crowned by a cupola. Symmetry is further displayed by one pedimented bay breaking forward four bays from each end of the building. Characteristically Fairbairn claims the credit for introducing in *c.*1827 'a new mill of a different class' being under the impression that all factories built before that date were plain square brick boxes. Fairbairn claimed to have 'persuaded the proprietor to allow some deviation from the monotonous forms then in general use.' The improvement 'consisted chiefly in forming the corners of the building into pilasters, and a slight cornice round the building... it was speedily copied in all directions with exceedingly slight modifications, but always with effect, as it generally improved the appearance of the buildings and produced in the minds of the mill owners and the public a higher standard of taste.'[40] It is impossible to estimate how widely Fairbairn's 'new Mill' architecture was copied but one suspects that it was less common than Fairbairn suggests. Ure[41] acclaimed Orrell's factory at Stockport designed by Fairbairn in the early 1830s: 'In beauty of architectural design it will yield to no analgous edifice, and may indeed bear a comparison in respect of grandeur, elegance and simplicity, with many aristocratic mansions.' In Orrell's factory Fairbairn used pilasters, cornice and hipped roof to adorn a pleasingly well proportioned plain building consisting of a main range with a wing at either end, a form so favoured by 18th century factory builders. The decorated chimney was set on a small hill nearby.

In 1836 Marshalls of Leeds began to consider weaving flax. James Marshall obtained estimates and plans for a multi-storey mill and warehouse and for a single storey mill and favouring the latter set about converting his father. He detailed the technical advantages but also considered it worth making the point that its 'external appearance may be made to look extremely well quite as much so altogether as the six storey plan'. In June 1840 a temperance tea marked the opening of what came to be known as Temple Mill. It was 132yds long and 72yds wide. Half way up the stone faced wall sturdy columns which divided the windows supported a massive entablature giving the whole the appearance of an Egyptian temple. The chimney was disguised as Cleopatra's needle. Skylights lit the building 'like cucumber frames in a garden' and on the immense roof 'a layer of earth, sown with grass, flourishes so well that sheep are occasionally sent to feed on it.' Matching offices were added a year or two later and the whole was said to be 'an exact copy of the temple of the pharoahs at Philae'.[42] This was truly a temple of industry marking the peak of Marshalls' achievement as spinners. Sir Robert Smirke encouraged the Egyptian idiom at the Mansion House, London in 1836 and this may have inspired Marshall. Certainly the growing vogue for exotic schemes such as these prompted A. W. Pugin's blistering attack in the late thirties.[43] He ridiculed the architects for showing off what they could do instead of carrying out what was required and described the facade to an imaginary works, The New Economical Compressed Grave Cemetry Company, which had along one wall 'a cement caricature of the entrance to an Egyptian temple 2½ inches to the foot...with convenient lodges for the policeman and his wife and a neat pair of cast iron hieroglyphical gates, which would puzzle the most learned to decipher.'

Saltaire marks the apogee of industrial architecture in the 19th century. In order to avoid monotony in a 550 ft by 50 ft building 'a bold Italian style' was adopted.[44] Pilasters, turrets, pediments and gracefully arched windows combined to produce a magnificent spectacle. But mills of the scale or magnificence of Orrell's, Marshall's or Salt's were exceptional. They were far too costly for the average manufacturer.

By the middle of the 19th century the adornment of factory buildings was probably due to two stimuli. No longer was there the desire to emulate the aristocracy, although it is interesting that Ure likened Orrell's factory to a stately home. Instead, the decorated facade represented the physical embodiment of the paternalistic attitude of a minority of entrepreneurs — Titus Salt's factory and model community at Saltaire is an example. There was also the desire to impress. With ever increasing competition between firms the potential customer could be attracted by the firm which appeared by its decorated facade to be the most successful. This is particularly noticeable in the brewing industry and later in banking and other buildings of commerce. The Lion Brewery in London, for example, presented a riotous Baroque facade to the river.

The changing attitude of the entrepreneur to the appearance of his factory has been examined but so far there has been no discussion of who was ultimately responsible for the design of the exterior of the building. Was an architect consulted or was the whole design the work of a millwright or engineer? The question is difficult to evaluate because of the paucity of records. Few factory building accounts have survived and in those that have, the accounts for Cressbrook Mill, Derbyshire,[45] for example, it is difficult to detect a minor architect in the list of names of masons, carpenters and sawyers who worked on the building. The architects' papers that exist belonged for the most part to the architects who did not design factories. Nevertheless it is possible from Mr. H.M. Colvin's[46] important work and from contemporary secondary sources to attribute a number of factory buildings to architects. It is almost certain that Boulton's Soho Manufactory was designed by either James or Samuel

Miller's Dale Mill, Derbyshire

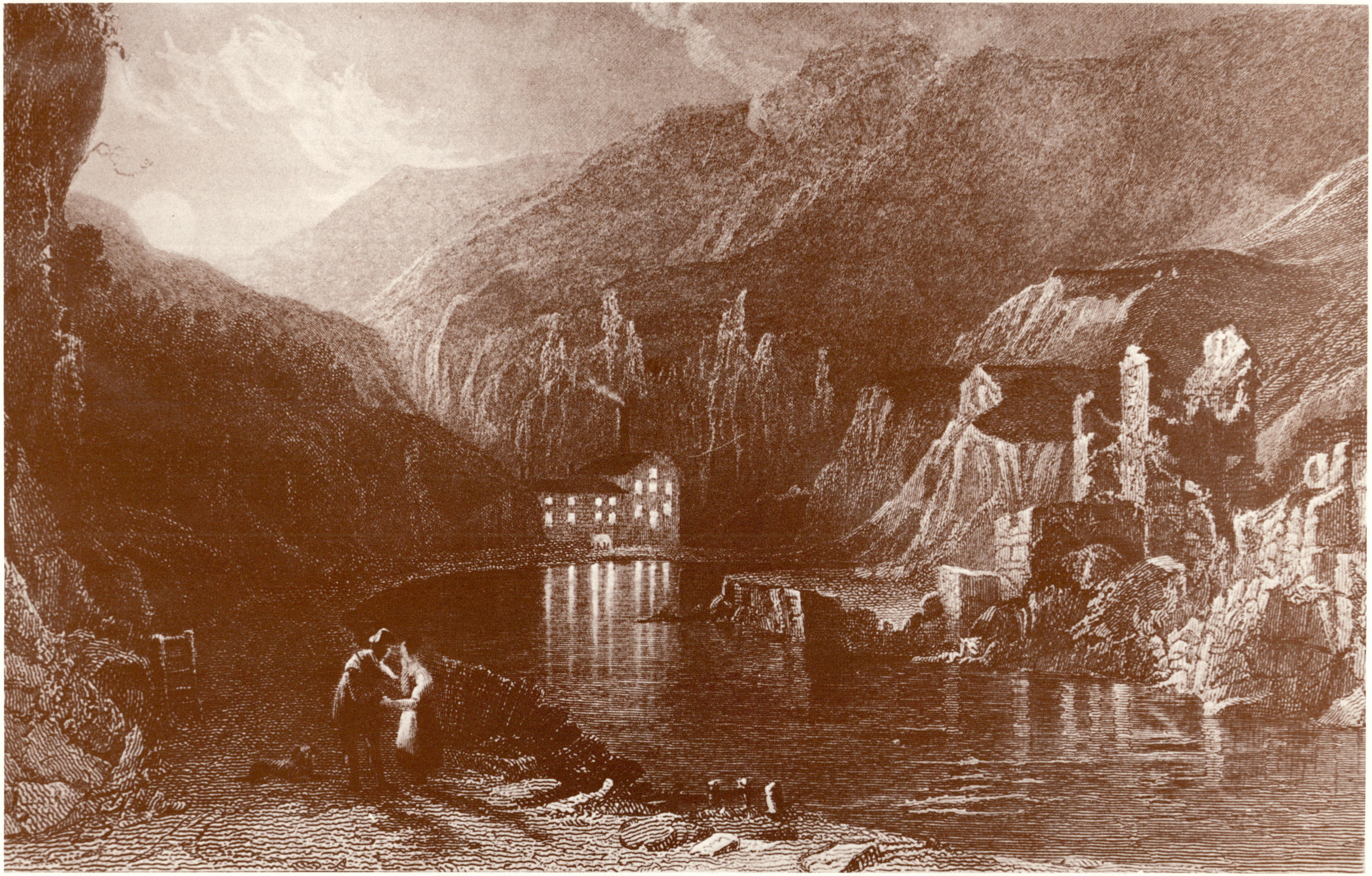

47. John Mosse op. cit. Albion Mills Box, A.O.
48. E. Meteyard, *The Life of Josiah Wedgwood*, 1865, II, pp.37, 82-3, 126-9.
49. H.M. Colvin op. cit. pp.190, 357.
50. G.C. Robertson, Ed., *The Stretton Manuscripts*, Nottingham 1910, pp. 179-81.
51. H.M. Colvin op. cit. see under architects' names.
52. M. Boulton to J. Rennie, 15 Nov. 1804, Rennie Box A.O.
53. Lord Chesterfield to his son, letter CXLVIII; 17 Oct 1749
54. Sir John Craig, *The Mint*, p.270.
55. Cusworth Hall, Sutcliffe MSS.
56. Isaac Ware, *A Complete Body of Architecture*, 1756.
57. J. Southern to J. Lawson, 15 Mar. 1802, Foundry Letter Book.
58. J. Southern to P. Ewart, 31 May 1802, Foundry Letter Book.
59. W. Fairbairn, *On the Application of Cast and Wrought Iron to Building Purposes* p.146.

Wyatt or both. Their friendship with Boulton and the fact that they designed Soho House which has affinities with the Manufactory supports the premise. Samuel Wyatt[47] designed the Albion Mill in which he and Boulton and Watt, amongst others, were partners. Joseph Pickford, the architect/builder of Derby, designed Josiah Wedgwood's Etruria pottery in 1767-73 and Etruria Hall in 1767-9.[48] Several brewery architects are known; William Robert Laxton designed three London breweries including Meux and Co's, and Francis Edwards designed the Lion Brewery, Golden Square and its namesake in Lambeth. George Byfield designed a brewery in Knightsbridge.[49] In 1796 Samuel and William Stretton designed and built Arkwright's first cotton spinning mill in Goosegate, Nottingham. In 1791 they designed Dawson's lace factory and in the following year Alderman Green's cotton factory in Broadmarsh, Nottingham. Between 1792 and 1794 they were concerned with two breweries, one of which was Evans, Storer and Green's in Poplar Place, Nottingham which was built at a cost of £12,000.[50] John Clark, the Leeds architect is known to have designed a flax mill for Hives and Atkinson and a cotton mill for John Wilkinson. Richard Tattersall, born in Burnley, is said to have designed 'many cotton mills' including one for Samuel Brewis at Goulbourne, Lancashire and Peter Dixon and Sons' fireproof Shaddon Mill at Carlisle. A list of architects known to have designed factories forms the appendix to this chapter.[51] The list is not long and it is reasonable to suppose that architects, even architect/builders were not commonly engaged to produce plans for factories although they were often employed to design entrepreneurs' houses. Boulton clearly saw the role of the architect as a secondary one in factory design. He made it clear that the architect was required merely to execute the facade when he discussed the new Royal Mint with Rennie:[52] 'The buildings in general should be plain simple and strong and all the operative buildings need not be more than two storeys high, many of them one at most, except the front which may be simply elegant in the Wyattistic style . . . Mr. Wyatt may design the ornamental part but I must sketch the useful.' Boulton was not being as condescending as he at first appears. Many an 18th century architect would not have wished to demean himself by being involved with the useful; as Lord Chesterfield wrote to his son: 'You may soon be acquainted with the considerable parts of Civil Architecture; and for the minute and mechanical parts of it, leave them to the masons, bricklayers, and Lord Burlington, who has, to a certain extent, lessened himself by knowing them too well.'[53] Samuel Wyatt was discharged 'for total neglect of Mint duties' and James Johnson completed the designs for the central pedimented building at the Royal Mint. After Johnson's death in 1807 Sir Robert Smirke was appointed master architect and the buildings were completed in 1809.[54]

The basic plan of a factory was generally decided by the millwright or engineer in consultation with the manufacturer and he employed local masons or carpenters and glaziers. The Sutcliffe/Gott correspondence makes it clear that the millwright's responsibility often extended to the exterior of the building for Gott employed Sutcliffe to plan the whole factory 'on such part of Bean Ing as Mr. Sutcliffe shall think most eligible' with houses for the master dresser, and the dyer and four other cottages.[55] Sutcliffe checked with Gott the dimensions of the windows suggesting that a 4 ft opening height would be required. He was aware of the acceptable rules adding that if stone cills were required 'they ought to be as broad as the brick walls are thick.' Sutcliffe also kept a keen eye on other factory buildings in the area: 'if your dyehouse windows were made 2ft 4in wide and 5ft 2in high they would be as large as those in Swinegate.'

There were numerous builders' manuals containing details of the acceptable forms for doors, windows, pediments, cupolas etc, as well as a growing volume of works on country houses. And if the manufacturer wanted a decorative building with words like Isaac Ware's[56] ringing in his ears a builder or millwright could adapt the facade of factory or workshop to the familiar idiom: 'If the plain decoration of architrave, and freeze and cornice, the addition of the pediment, or the ornaments of sculpture do not give satisfaction, let no false, foolish and fantastic decorations be added but at once admit an order.' The prevailing style was Palladian, until the 1780s when Wyatt and Nash led the gothic revival, and there was no reason for the millwright or builder to seek beyond these styles when designing a factory. Boulton and Watt's draughtsmen were conscious of the aesthetic element in factory building for there exists amongst their papers a series of small sketches of industrial buildings or parts of buildings in which the draughtsman has obviously experimented with the addition of the standard venetian window, pediment or portico. Although Pope was thinking of country gentlemen and local masons when he wrote the following lines they might equally apply to the millwrights and engineers who were responsible for the design of many factory buildings:[57]

'Yet shall (my Lord) your just, your noble rules
Fill half the land with imitating fools;
Who random drawings from your sheets shall take,
And of one beauty many blunders make;'

Southern was concerned about the appearance of the new fireproof mills.[58] 'I think the principal of erecting a separate staircase from the main building of these inflammable cotton mills and such like connecting by iron gangways is very good but will they not in most cases be unseemly?' Southern was well acquainted with architectural rules and orders and was not averse to satirising the architect, albeit rather heavy handedly:

'We have rummaged all the ancients and moderns from Palladio to that great aristocrat and Knight Sir William Chambers in quest of tables and propriety and having altered and altered and altered again and again various ingenious and highly meritorious designs have at last concentrated the essence of our 'labours' in one which for taste, beauty, magnificance, and utility will vie with the most renowned products of any genius of any age or nation before the conquest of Egypt.'

The discussion concerned the 'entablature' which in the case of a steam engine was part of the cast iron framing.

In the 19th century with the tendency for the factory to look more utilitarian there was even less need for the architect to be involved in its design. Fairbairn clearly considered himself equal to the task of relieving the endless brick, saying with honesty 'This alteration had no pretension to architectural design.' It was rare even in the late 19th century for the architect automatically to be employed to design the facade of a factory; Saltaire was an exception for although Fairbairn designed the functional part the 'architectural features, to avoid monotony in so large a dead surface have been most skilfully treated by the Architects Messrs Lockwood and Mawson of Bradford.'[59]

Tower Mint, London

Pottery

There was one process in pottery manufacture for which power was essential; this was flint grinding. The lack of water power in the Potteries was a severe disadvantage and although several wind flint mills were built in and near Burslem the majority of the flint mills were situated on the River Churnet some six or seven miles away. The flint was ground by stampers and by vertical edge runner stones. The potter who had water power near his pottery had an advantage over others who did not for not only were heavy transport costs cut and delays eliminated but the potter could supervise the mixing of his materials.

Hamilton's flint mill, Fenton Low, 1807

This flint mill originally belonged to Thomas Whieldon and it has been said that the proximity of the mill to Whieldon's pottery explains why Whieldon ware become famous in the early 18th century. Whieldon was in partnership with Josiah Wedgwood between 1754 and 1759. The 25 ft wheel drove stampers and edge runners in pans of varying diameters.

Portf: 400

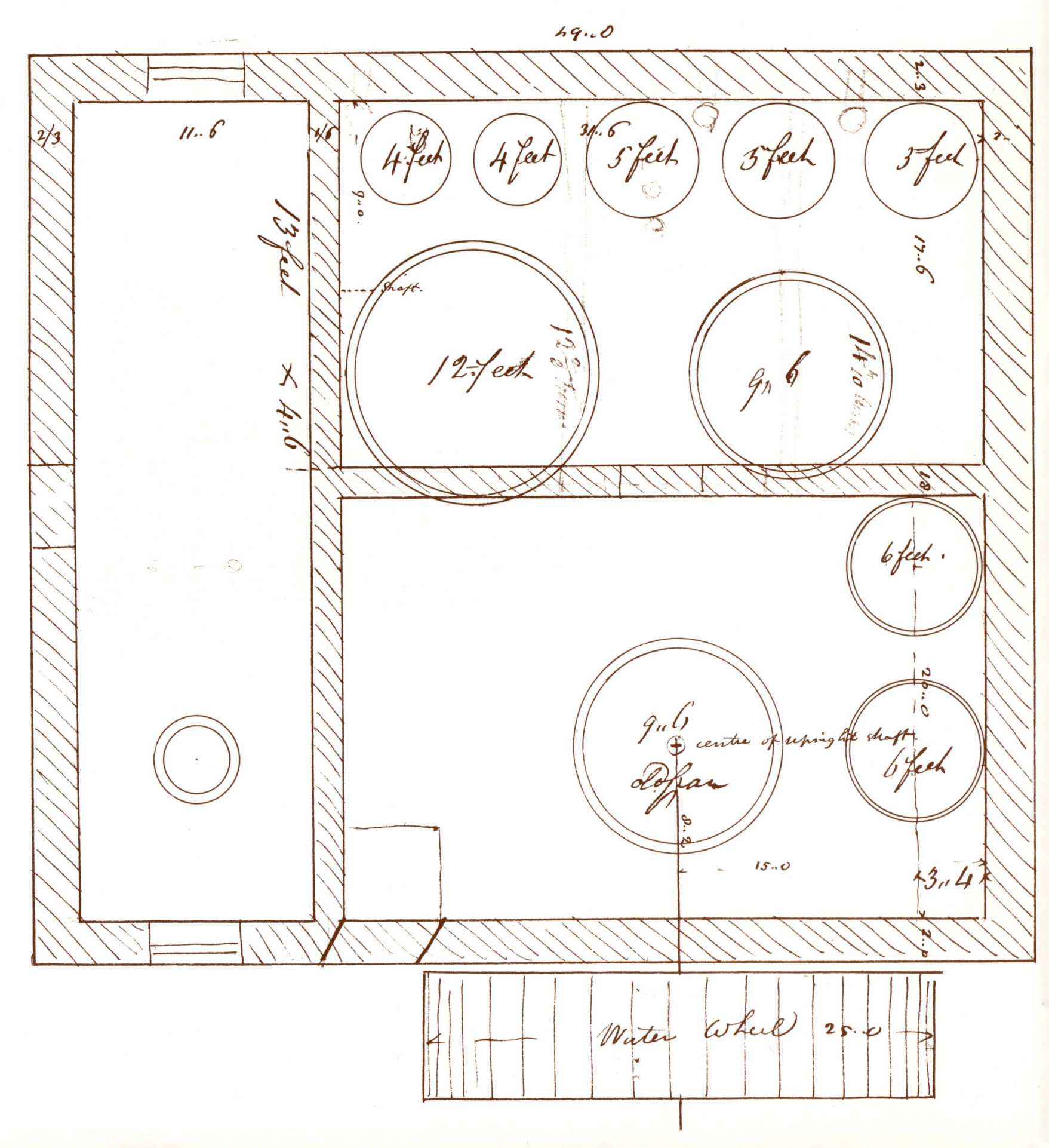

Wedgwood and Byerley, Etruria, 1800
Wedgwood was aware of the advantages to be derived from installing a steam engine at his works and he noted in 1793-4 that an engine could grind flints, grind enamel colours, operate a sagger crusher and mix clays.
Portf: 218. Reverse.

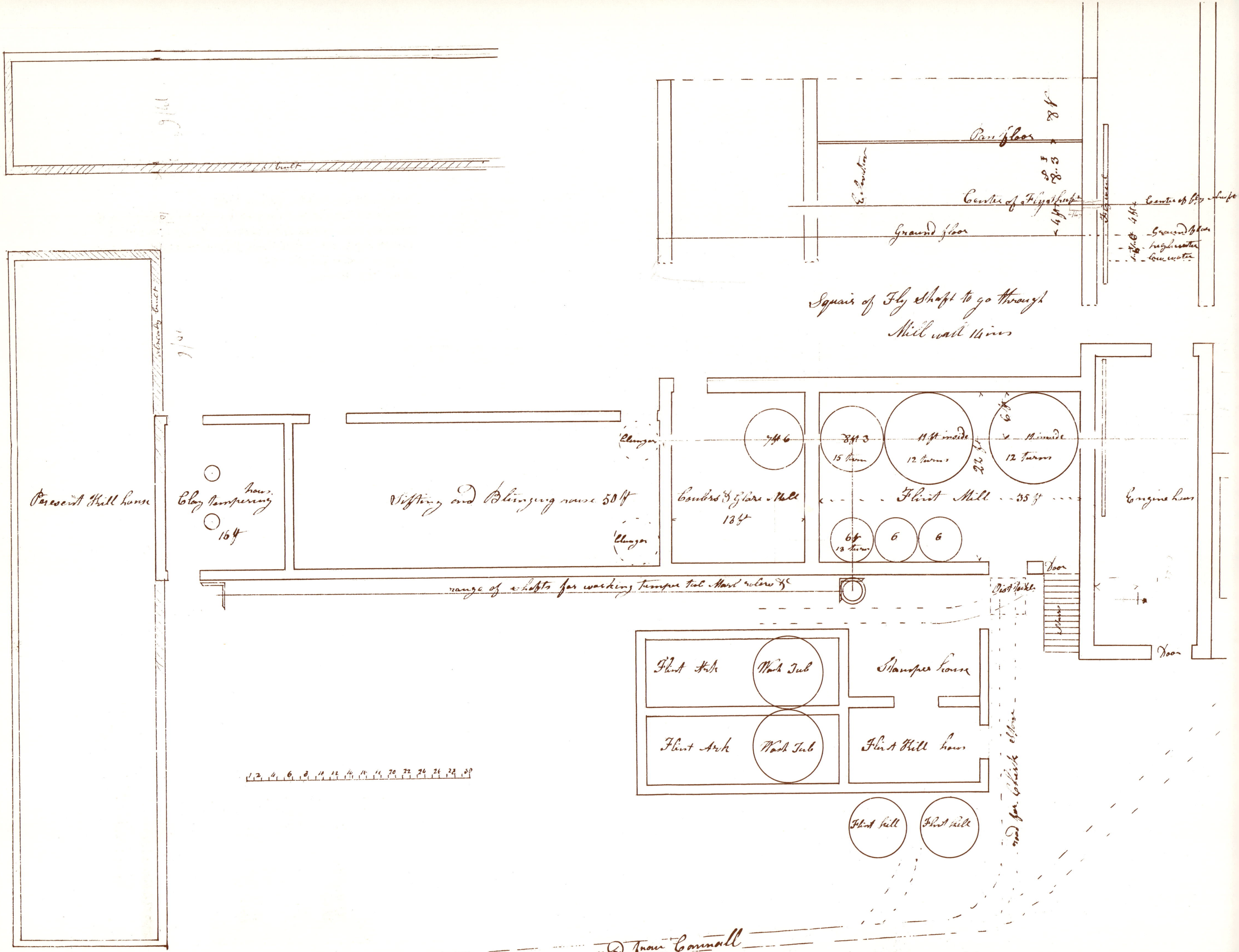
Pan floor
Centre of Fly shaft
Ground floor
Square of Fly Shaft to go through
Mill wall 14 ins
Present Mill house
Clay tempering house
16 ft
Sifting and Blunging room 50 ft
Blunger
Blunger
Colours & Glaze Mill
13 ft
Flint Mill 35 ft
11 ft inside
12 turns
15 turns
12 turns
Engine house
Door
Door
Flint Ark
Wash Tub
Flint Ark
Wash Tub
Stamper house
Flint Kiln house
Flint kiln
Flint kiln
road from Cornwall

Josiah Spode, Stoke-on-Trent, 1810
This interesting drawing shows the layout of Spode's works with the flint mill containing pans, flint stampers, the sifting and blinging house, a clay tempering house and the long kill (kiln) house. Boulton and Watt also supplied steam heating for his slip drying house.
Portf: 430

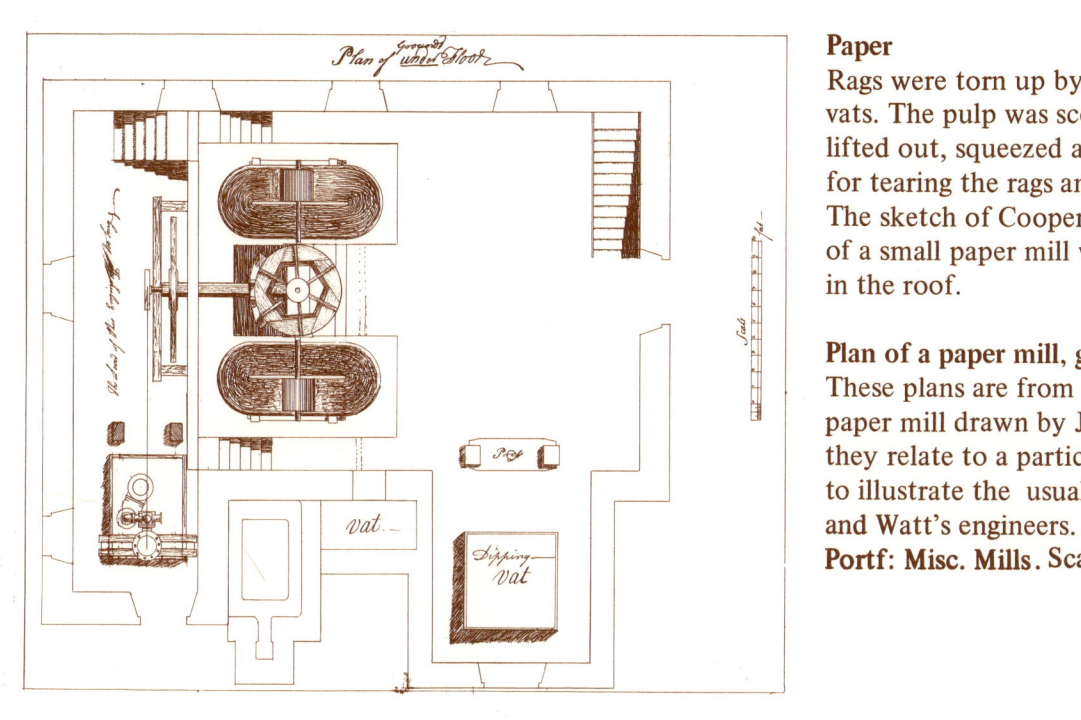

Paper
Rags were torn up by iron spiked cylinders and put into vats. The pulp was scooped out into a wire-mesh mould, lifted out, squeezed and dried. Power was required chiefly for tearing the rags and heat was needed to dry the paper. The sketch of Coopers and Boyle's mill shows the dimensions of a small paper mill with the characteristic louvred openings in the roof.

Plan of a paper mill, ground floor and upper floor. n.d.
These plans are from a small folder of drawings of an unknown paper mill drawn by John Southern. It is not known whether they relate to a particular mill or whether they were drawn to illustrate the usual layout of a paper mill to help Boulton and Watt's engineers.
Portf: Misc. Mills. Scale approx 1 in : 11 ft 6 in.

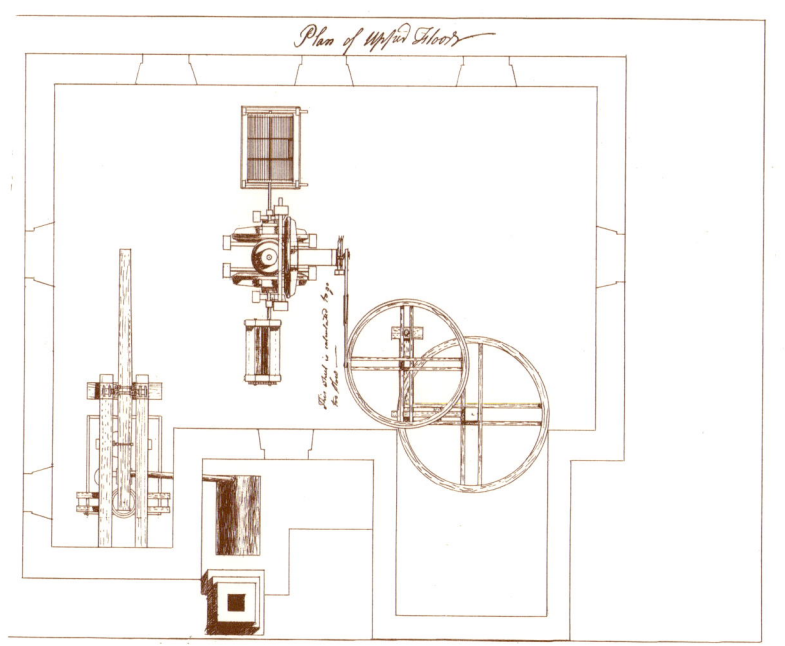

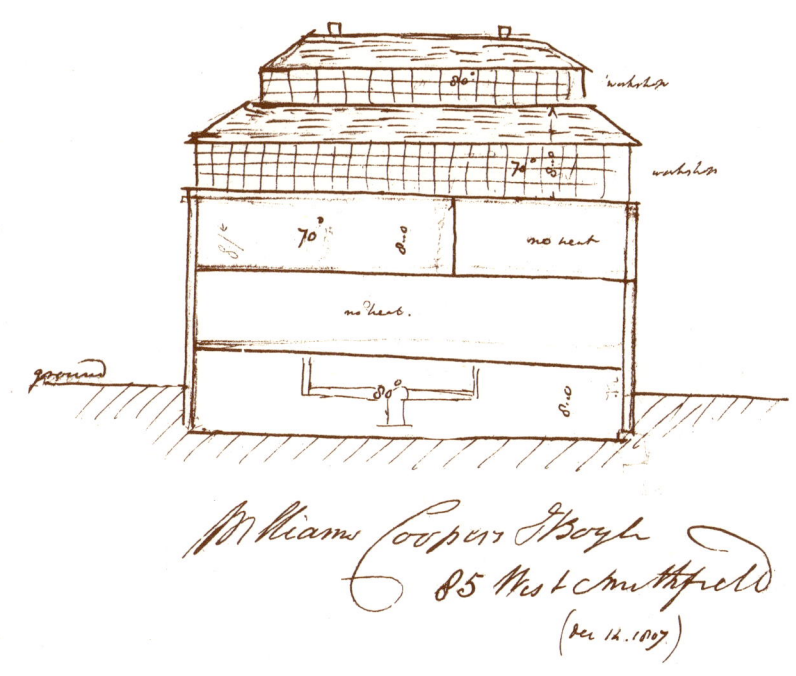

W & R Balstone, Maidstone, 1806
Steam heating was installed in this factory at the same time as a steam engine but a note records that experiments were to be carried out before the heating system was installed in the whole factory.
Portf: 372

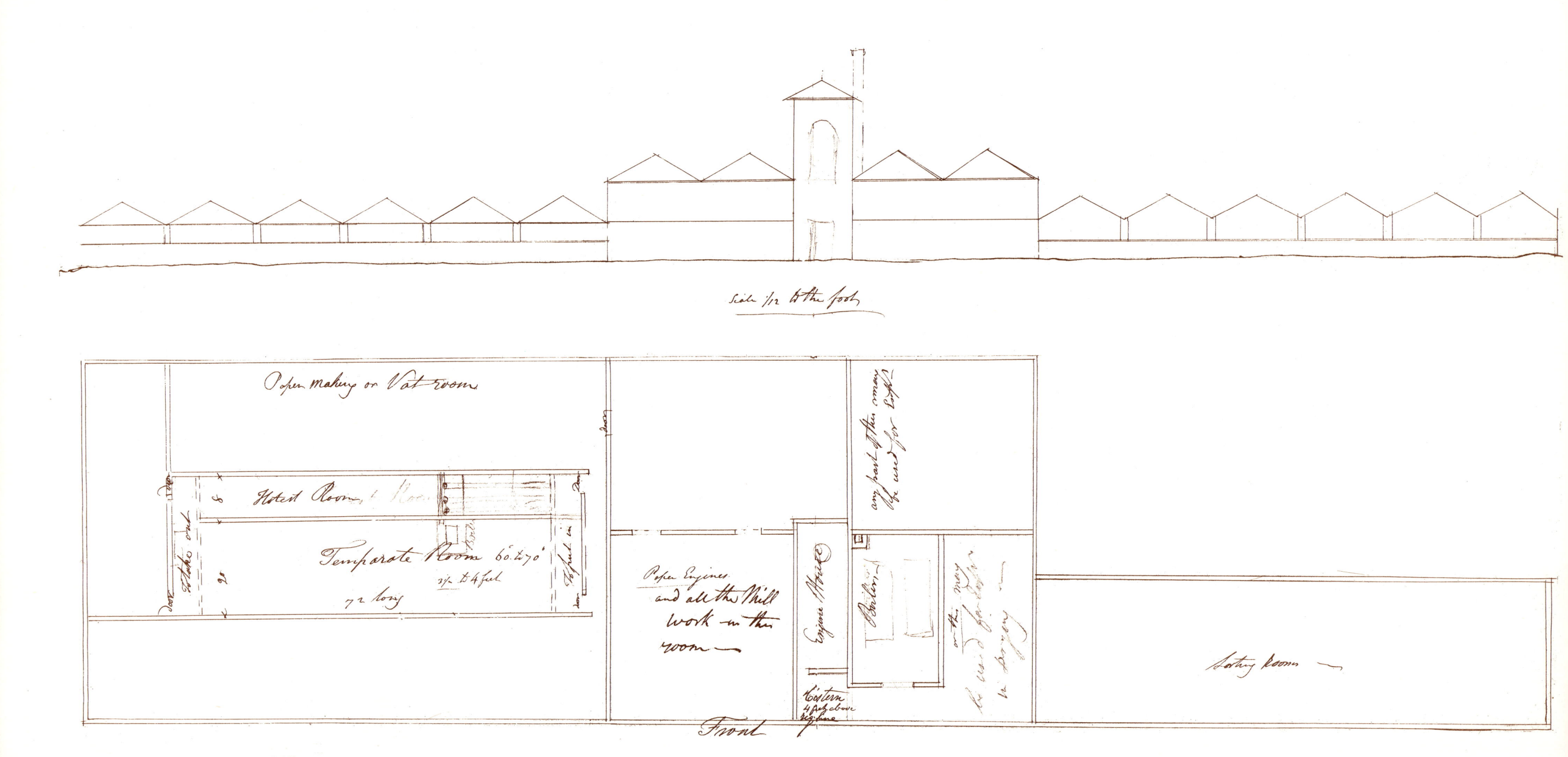

Koops, Tait & Co, Chelsea Paper Mill 1800
This was a short-lived venture established by Matthias Koops to make paper from waste paper, straw and wood instead of rags. The factory was built in about 1800 and probably ceased work about 1802. Rennie was involved either as millwright or as engine erector.
Portf: 294

Ground Plan of Chelsea Paper Mill.

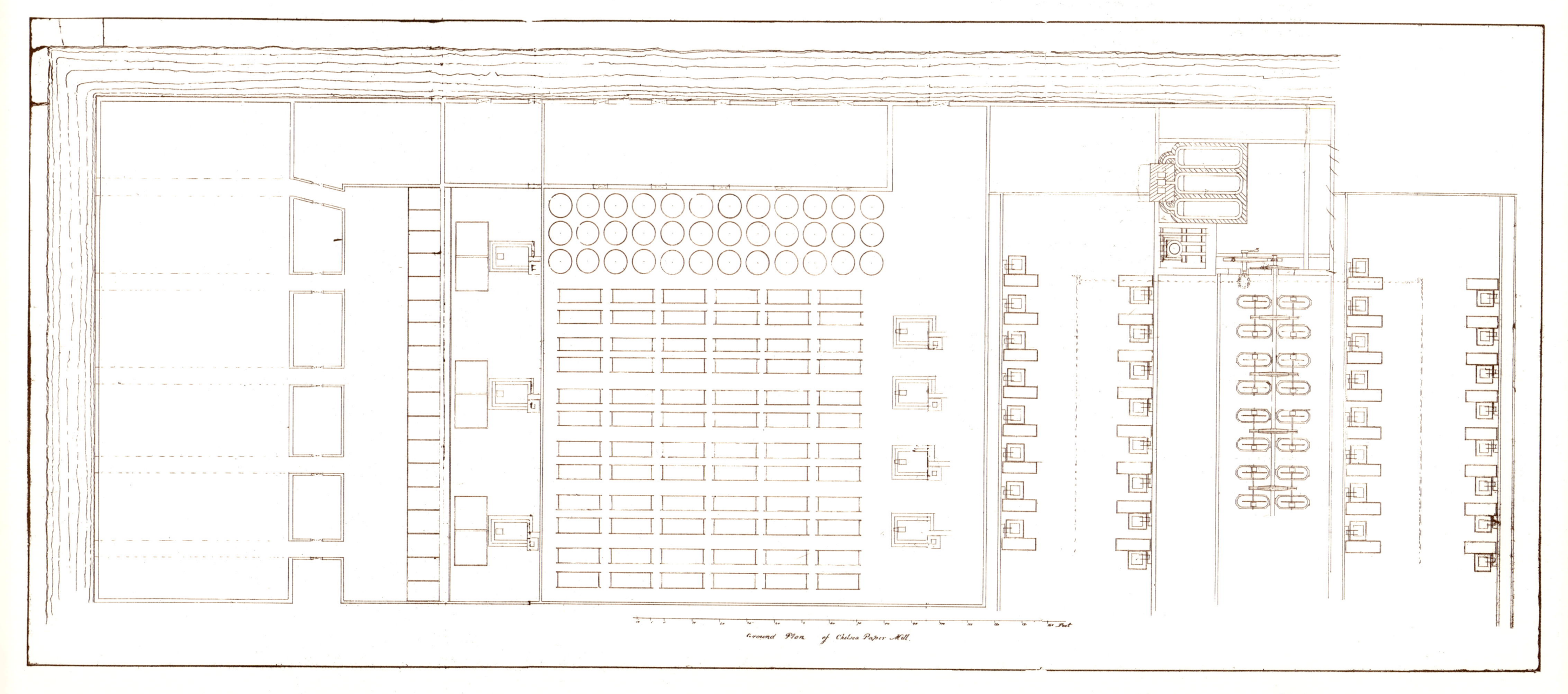

APPENDIX

Some Factory Architects

Name	Factory	Firm	Date
Burton, Decimus	Abersychau Ironworks	–	1826
Byfield, George	Brewery in Knightsbridge	–	1804
Clark, John	Flax Mill, Leeds	Hives & Atkinson	–
	Cotton mill, Leeds	John Wilkinson	–
Edwards, Francis	Imperial Gas Co, Hoxton	–	1823
	Cannon Brewery, Knightsbridge	T Goding	c.1835
	Lion Brewery, Golden Sq.	Goding & Broadwood	–
	Lion Brewery, Lambeth	Goding & Co	1836
	Distillery, Vauxhall	Burnett	1841-57
	factory in Horseferry Rd, Westminster	Broadwood	1856
Foulon, John	Meux's Brewery	Meux	–
Gardner, -	Etruria Pottery (Asst arch)	Wedgwood	1767-73
Johnson, James	Royal Mint	–	1790s-1807
Laxton, W. Robert	Meux's Brewery	Meux	1804
	Mary Brewery, London	J Raymond	1810
	Union Brewery, Wapping	R Bowman	1810
Mullins, J	Imperial Plate Glass Manufactory, London		1826
Pickford, Joseph	Etruria Pottery	Wedgewood	1767-73
Rennie, John	Royal Victualling Yard, Stonehouse	–	1830
Sandby, Thomas	bleachworks in Vale of Clwyd, Denbigh	–	c.1785
Smirke, Sir Robert	Royal Mint	–	1807 –
Stretton, Samuel & William	Goosegate cotton mill, Nottingham	Arkwright	1769
	Dawson's lace factory, Nottingham	Dawson	1791
	brewery, Nottingham	Evans, Storer, Green	1792
	Broadmarsh cotton factory, Nottingham	Green	1792-3
	brewery, Butcher's Close, Nottingham	–	1794-5
	bleach works, Lenton, Notts.	Green & Killingley	1797
Tattersall, Richard	cotton mill, Goulbourne, Lancs.	Samuel Brewis	–
	cotton mill, Carlisle	Peter Dixon & Sons	1836
Tearby, William	brewery, Margate	Cobb	1803-4
Wyatt, Samuel	Albion Mill, London	Albion Mill Co	1784-6
	Soho Manufactory, Birmingham	Boulton & Fothergill	1767

This list refers to factories designed but not necessarily built.
Source: H. M. Colvin.

Bibliography

Manuscript Sources
Birmingham Public Library Boulton & Watt MSS/Wyatt MSS
Birmingham Assay Office Boulton MSS (referred to as A.O. in the notes)
Brotherton Library, Leeds Gott MSS/Marshall MSS
Leeds City Library William Brown, Information Regarding Flax Spinning in Leeds 1821
Shrewsbury Public Library Bage MSS
Derbyshire Record Office Cressbrook Mill Building Accounts/55 watercolour sketches of Derbyshire
Cusworth Hall Museum Gott MSS

Where no location for MSS is given it is to be assumed that the papers are from the Boulton and Watt Collection, Birmingham Reference Library.

Books and Articles
Ackerman, James S, Palladio
Aikin, J.A. Description of the Country from Thirty to Forty Miles Round Manchester, 1795
Ashmore, Owen, Industrial Archaeology of Lancashire, 1969
Aspin, C.& S.D., James Hargreaves and the Spinning Jenny, 1964
Ashton, T.S. Iron and Steel in the Industrial Revolution, 1924
Atkinson, Frank, Some Aspects of the 18th century Woollen and Worsted Trade in Halifax, 1956
Bannister, Turpin, The First Iron Framed Buildings, Architectural Review, 107, 1950
Barker, T.C and Harris, J.R, A Merseyside Town in the Industrial Revolution, St. Helens, 1750-1900, 1959
Barlow, Alfred The History and Principles of Weaving by Hand and by Power, 1878
Boucher, Cyril T. James Brindley, Engineer 1716-1772, 1968
Buchanan, Robertson, An Essay on the Warming of Mills and other Buildings by Steam, 1807
Butt, John, Industrial Archaeology of Scotland, 1967
Butterworth, Edwin, Historical Sketches of Oldham, 1856
Butterworth, James, A Complete History of the Cotton Trade...., 1823
Cardwell, D.S.L, Power Technologies and the Advance of Science 1700-1825, Technology and Culture, 6, 1965
Chaloner, W.H, The Stockdale Family, The Wilkinson Brothers and the Cotton Mills of Cark in Cartmel, Trans. Cumberland & Westmorland Antiq. & Arch. Soc. LXIV, 1964
Chaloner, W.H. & Musson, A.E, Industry and Technology, 1963
Chapman, Stanley D., The Early Factory Masters, 1967
Chapman, Stanley D., The Pioneers of Worsted Spinning by Power, Business Hist. VII, 1965
Chapman, Stanley D., The Peels in the Early English Cotton Industry, Business History, XI, 1969
Collier, F., Samuel Gregg and Styall Mill. Mem. Proc. Manchester Lit. & Phil. 1941-3
Craig, Sir John, The Mint, A History of the London Mint from AD 287 to 1948, 1953
Crump, W.B, The Leeds Woollen Industry, 1931
Crump, W.B. & Ghorbal, Gertrude, History of the Huddersfield Woollen Industry, 1935
Dickinson, H.W, A Short History of the Steam Engine, 2nd ed. 1963
Edwards, Michael M., The Growth of the British Cotton Trade, 1780-1815, 1967
Evans, Oliver, The Young Millwright and Miller's Guide, 1853
Fairbairn, W.F., On the Application of Cast and Wrought Iron to Building Purposes, 1854
Fairbairn, W.F., Mills and Millwork, 1867
Farey, J.A., Treatise on the Steam Engine, 1827
Farey, John, General View of the Agriculture of Derbyshire
Finer, Ann & Savage, George, The Selected Letters of Josiah Wedgwood, 1965
Finlay, James, and Co Ltd, privately printed, 1951
Fitton, R.S. & Wadsworth, A.P., The Strutts and the Arkwrights, 1958
Foulkes, Edward J., The Cotton Spinning Factories of Flintshire 1777-1866, Flints. Hist. Publications, XXI, 1964
French, Gilbert J., The Life and Times of Samuel Crompton, 1859
Gauldie, Enid, Mechanical Aids to Bleaching in Scotland, Textile History 2, 1969
Gloag, John, The English Tradition in Architecture, 1963
Hacker, C. L., William Strutt of Derby, Derbys Arch. Journ 80, 1960
Harold, H. A, Mill Construction, Official Record of Annual Conference of the Textile Institute, 1927

Harris, J. R., The Employment of Steam Power in the 18th Century History, 52, 1967
Heaton, Herbert, The Yorkshire Woollen and Worsted Industries, 1965 (new ed)
Henry, W. C., A Biographical Notice of the late Peter Ewart, Trans Manchester Lit and Phil Soc 7, 1841
Hills, Richard L., Power in the Industrial Revolution, 1970
Hunt, Charles, A History of the Introduction of Gas Lighting, 1907
James, John, History of the Worsted Manufacture, 1857
Jenkins, J. Geraint, The Welsh Woollen Industry, 1969
Jennings, Bernard ed, A History of Nidderdale, 1967
Johnson, B. L. C., The Foley Partnerships: The Iron Industry at the end of the Charcoal Era, Econ. Hist Rev. 4, 1951-2
Johnson, H. R. & Skempton, A. W., William Strutt's Cotton Mills, Trans Newcomen Soc, 30, 1955-7
Kennedy, John, A Brief Memoir of Samuel Crompton, Trans Manchester Lit and Phil Soc 1830
Kennedy, John, Observations on the Rise and Progress of the Cotton Trade in Great Britain, Trans Manchester Lit and Phil Soc, 3, 1819
Mackenzie, M. H., The Bakewell Cotton Mill and the Arkwrights, Derbys. Arch. Journ., LXXIX, 1959
Mackenzie, M. H., Calver Mill and its Owners, Derbys Arch Journ LXXXIII, 1963
Mackenzie, M. H., Calver Mill and its Owners a Supplement, Derbys Arch Journ, LXXXIV, 1964
Mackenzie, M. H., Cressbrook and Litton Mills, Derbys Arch Journ LXXXVIII 1968
Matthews, W., An Historical Sketch of the Origin & Progress of Gas Lighting, 1823
McKendrick, N., Josiah Wedgwood and Thomas Bentley an Inventor-Entrepreneur Partnership in the Industrial Revolution, Trans Roy Hist Soc, 14, 1964
Marshall, J. D., Furness and the Industrial Revolution, 1958
Mathias, Peter, The Brewing Industry in England 1700-1830, 1959
Miller, W. T., The Water Mills of Sheffield, 1949
Moir, Esther, The Discovery of Britain, 1964
Mosse, John, The Albion Mills 1784-92, paper read to Newcomen Soc 3 Jan 1968
Musson, A. E. & Robinson, Eric, Science and Technology in the Industrial Revolution, 1969
Pacey, A. J, Earliest Cast Iron Beams, Architectural Review, 1969, 145
Pelham, R. A., Corn Milling and the Industrial Revolution in England in the 18th Century, University of Birmingham Hist Journ 6, 1957-8
Pierson, William Harvey jun. Notes on Early Industrial Architecture in England, Journ Soc Architectural Historians, 8, 1949
Podmore, Frank, Robert Owen a Biography, 1906
Pole; Wed, The Life of Sir William Fairbairn, 1877
Pollard, Sidney, The Genesis of Modern Management, 1965
Pugin, A. W., The True Principles and Revival of Christian Architecture, 1895
Raistrick, A., The Steam Engine on Tyneside, Trans Newcomen Soc, 27 1936-7
Richards, J. M., The Functional Tradition in Early Industrial Buildings, 1958
Rimmer, W. G., Marshalls of Leeds Flax Spinners 1788-1886, 1960
Rimmer, W. G., Castle Foregate Flax Mill, Shrewsbury, Trans Salop Arch Soc, LVI, 1957-8
Robertson, J., Robert Owen and the Campbell Debt, Business History, XI, 1969
Robinson, Eric & Musson, A. E., James Watt and the Steam Revolution, 1969
Roll, Sir Eric, An Early Experiment in Industrial Organisation, 1930
Rowlands, Marie B., Stonier Parrott and the Newcomen Engine, paper read to Newcomen Soc I Jan 1969
Shepherd, W. D., Early Industrial Buildings, 1700-1850, thesis RIBA, 1950
Sigsworth, Eric M., Black Dyke Mills, 1958
Skempton, A. W. & Johnson H. R., The First Iron Frames, Architectural Review, 131, 1962
Stroud, Dorothy, Humphry Repton, 1962
Summerson, Sir John, The Classical Country House, Journ Roy Soc Arts, 1959, CVII
Summerson, Sir John, Architecture in Britain 1530-1830
Sutcliffe, John, A Treatise on Canals and Reservoirs, 1816
Tann, Jennifer, Gloucestershire Woollen Mills, 1967
Tann, Jennifer, Some Problems of Water Power, A Study of Mill siting in Gloucestershire, Trans Bristol and Glos Arch Soc, 84, 1965
Temin, Peter, Steam and Water Power in the Early 19th Century, Journ Econ Hist. XXVI, 1966
Tredgold, Thomas, Principles of Warming and Ventilating Public Buildings and Dwelling Houses, Manufactories, Hospitals, Hot Houses, Conservatories etc, 1824
Turnbull, G., A History of the Calico Printing Industry... 1951
Unwin, George, Samuel Oldknow and the Arkwrights, 1924

Ure, Andrew, The Philosophy of Manufactures, 1835
Ure, Andrew, The Cotton Manufacture of Great Britain, 1836
Wadsworth, A. P. & Mann, J. de Lacy, The Cotton Trade and Industrial Lancashire, 1931
Ware, Isaac, A Complete Body of Architecture, 1756
Warner, Sir Frank, The Silk Industry of the United Kingdom
Westworth, A. O., The Albion Steam Flour Mill, Econ Hist 2, 1930-3
Williamson, F., George Sorocold of Derby, Derbys Arch Journ 10, 1937
Wilson, Paul N., Water Power and the Industrial Revolution, Water Power, Aug 1954
Wilson, Paul N., The Water Wheels of John Smeaton, Trans Newcomen Soc, 30, 1955-7
Yarranton, Andrew, England's Improvement by Sea and Land to Outdo the Dutch without Fighting, 1677

Index

Abbey Hilton	95
Abbott (Nash &)	85,91
Aberdeen	129,131
Ainsworth, R.	91
Aitcheson	81,83
Albion Fire & Life Insurance Co.	131
Albion Mills	29,37,39,79,81,83,97,103,135,137,141,155,161
Alcock, S.	13
Aldred, E.	31,81
Allingham, T.	39
Ambleside	33
America	73
Ames	77
Anderston Old Mill	113
Andrew, S.	145
Ango, M.	135
Arkwright R.	5,7,9,27,29,31,37,47,59,65,67,69,75,81,83,87,91,99,103,113,133,135,149,153,155,157,161
Arkwright & Co.	27
Armley Mill	129,131
Arnold	75
Asburner, T.	91
Ashbourne	59,135
Ashbourne Mill	95
Ashover	75
Ashton-in-Makerfield	91
Aston	65
Atherstane	91
Atherton, Peter	29,31,79,81,83,99
Attingham Park	59,95
Austin, H.	65
Avening	67
Ayrshire	35
Aysgarth	59
Bacon, J.	75
Bage, C.	109,111,127,131,137,139,141,143,145,147,149
Bage, R.	137
Bakers, Corporation of	79
Bakewell Mill	31,59,103
Ball, T.	91
Banbury	79
Bank of England	135
Banks, Sir J.	155
Barclay	47,83
Barnes	39,87
Barrow	35
Barton, (Dumbell & Co.)	91
Bateman, J.	73,75,77,85,87,91
Bath	59
Baxter	79,155
Baynes & Co.	33
Bean Ing, Leeds	9,75,79,101,161
Beard	101
Beaulieu, Duke of	155
Beighton, H.	71
Bell	33,65
Belper	31,59,61,63,135
Belper Mills	109
Belper North Mill	139,143
Belper South Mill	145
Belper West Mill	63,65,103,123,137,145,147
Benet	77
Bennett	103
Benson & Braithwaite's Mill	33
Benthams	135
Benyon, B.	137,139
Benyon, T.	137,139
Berkeley	49
Berry, G. & N.	91
Bersham Ironworks	11,75
Beverley Cross & Co.	31,91
Birch	59
Birley and Co.	33,103,113,123,131
Birley (Swainson) & Co.	159
Birmingham	7,11,37,65,73,79,103,129,131,151
Birmingham Canal	63
Birmingham Small Arms Factory	7
Bishop's Stortford	7
Bissett (Robert) & Co.	33
Black, Dr. J.	73
Blackburn	91
Blagborough	91
Blakeway, E.	59
Bogle (Monteith & Co.)	127
Bolton	91,99
Bolton le Moors	91
Bonsall Brook	65
Bourne, D.	7
Bowbridge	65,155
Bowen	99
Bowling Ironworks	11,75,77,85,87,101
Boyes	111,113
Boyle	113
Brades	37
Bradford	11,60,103,145,161
Bradley Ironworks	11,59
Braithwaite	33
Brande, J.	91
Brewis, S.	161
Bridge Mills, Trowbridge	33
Brindle, R.	91
Brindley, J.	63,95,101,103
Bristol	73,75,77,79,95,103
Britannia, The, Nail Co.	131
British Plate Glass Co.	99
Broadford Mill, Aberdeen	131
Broadmarsh	161
Brockles, J.	79
Brock Mill, nr. Wigan	7
Brodie, A.	73
Bromwich Forge	37
Brookhouse (Parkes, B & Crompton)	99
Broseley	11
Brown (Aitcheson &)	81,83
Brown, W.	33,145
Bryon	101
Buchanan, P.	49
Buchanan, Robertson	111,113,115
Budd	83
Budgein	113
Buildwas Bridge	139
Burley	131
Burley Mill	131
Burlington, Lord	161
Burnley	91,161
Burslem	11,95,155
Burton, D.	91,131
Burton-on-Trent	81,123
Bury	27
Bury (Houldsworth) & Co.	131
Butterley Co.	11,75
Butterworth	7,47
Byerley	99,113
Byfield, G.	161
Byng	155
Byrne	113
Byron, Lord	67
Cadell, W.	11
Calver Mill	9,113
Calvert, F.	99
Cameron	73,77
Campbell	91
Campbell, Spier & Co.	91
Cannon, W.	91
Capley, W.	91
Capstick	99
Cardigan Foundry	63
Cardwell	113
Cark Mill	29,75
Carlile, J.	91
Carlill, (Boyes &)	111
Carlisle	147,161
Carmarthen	73
Carr, J.	91,101
Carr, W.	29
Carron Ironworks	11,61,75
Carshalton	97
Cartwright, Edmund	49,75
Cartwright, John	31,39,79,87,91
Castle (& Ames)	77
Castle Fields Mill, Shrewsbury	139,143
Castle Foregate Mill, Shrewsbury	143,149
Castle Mill, Papplewick	155
Caton Mill	99
Catrine Mills	61,67,79,99,113
Cave, E.	7,105
Chadwick, J.	91
Chadwick (&Whyte)	49
Chambers, J.	91
Chambers, Sir W.	161
Champion, J.	75
Chance Bros.	103
Chapel of Port, Glasgow	113
Charrington	77
Cheetham, G.	91,103,145
Cheshire	47,79
Chester	91
Chesterfield	5,7,11,75,137
Chesterfield, Lord	161
Chorley	91
Chorlton	91,99
Chorlton-on-Medlock	73
Christian Malford Mill	33
Church Street factory, Birmingham	11
Churwell Mill	91
Clark, J.	161
Clarke,Maze & Co.	103
Clarke, W.	49
Claytons (& Walshman)	75,87,91
Clegg, S.,	125,127,129
Clegg, S., Jun.	125
Cliffe	37
Clowes, C.	77,85,105
Coalbrookdale	135
Coalbrookdale Co.	11,39,59,71,75
Coalport	11
Coates	33
Coates, H. & Co.	83
Colchester	49
Coltham Mill	29
Compagnie, La, D'Ours	113
Coneygre Furnace	11
Constitution Brewery	47
Cook (Hagues, C. & Wormald)	63,67
Cook (Wood &)	91
Cook, B.	129
Cooksen (Markland C & Fawcett)	29,31,75,92,99,101
Cooper (Williams & Boyle)	113
Cooper, James & John	67
Cooper, T.	77,97
Cork	79
Cornbrook	75,91
Corse, W., & Co.	49
Cotchett, T.	7
Cotes & Co.	47
Coupland (& Wilkinson)	75,91
Coupland, T. & Sons	131

171

Covent Garden Theatre	113
Coventry	129
Cradley	59
Cradley Forge	37
Crag Print Works	65
Creighton, W.	2,129,143
Cressbrook Mill	59,155
Cromford	7,31,59,65,69,123,153,155,157
Crompton, S.	5,9,49,59,99
Cross, (Beverley, Cross & Co.)	31,91
Crowder, I.	91
Crowley	3
Cullompton	33
Cuming, T.	33
Currie	91
Curtis	155
Cyfartha Ironworks	11
Daintry, M.	95
Daintry, Ryle & Co.	29,31,75
Daintry, (Wood) & Co.	123
Dale, D.	67,79,111,153,155
Dale Abbey	75
Darby, A.	11,59
Darlington	33
Dalyed, John & Co.	49
Dane, River	95
Darley Abbey	145
Darwin, E.	135
Davidson, J.	49
Davis	113
Davison	75
Dawson	161
Dayus, R.	81
Deansgate Mill	91
Deanston	67
Dearman, John Petty	39
Delafield, J.	47
Denison, R.	27
Deparcieux	60
Derby	7,95,135,139,149,161
Derby Calico Mill	129
Derby Infirmary	109,135
Derbyshire	7,31,47,75,137,153,155,159
Derwent	7,61
Desaguliers	71
De St. Fart	135
Devon	33
Dewsbury	79,91
Dewsbury Mills	63,67
Diderot	157
Dixon, Greenhalgh	31
Dixon, Peter & Sons	161
Doncaster	49,75,91
Dorset	7
Douglas, W.	29,99,123,131,151
Doveridge	65
Dowlais Ironworks	11,35
Downes, D.	27
Driglington	91
Drinkwater, P.	9,27,75,99
Dublin	47
Dudley	145
Duffy (Byrne & Hamill)	113
Dumbell	91
Dundee	65
Dundonald, Lord	35
Dunkerley	91
Dunkin, J.	79
Dunkirk Factory	33
Durham,	99
Dursley	9,65
East India Company	155
Eastwood, J. & Co.	91
Ebley Mills	63
Edale	31
Edgeworth, R.L.	135
Edgeworth, W.	111
Edmondstone Colliery	85
Edwards, F.	161
Edwards, Miles	77
Edwards, T.	77
Eggintons	153
Egremont	33
Elford	137
Ellerby	79
Elliott, J. & W.	99
England, Bank of	135
Etruria	11
Evans, O.	73
Evans, W.	145
Evans Storer & Green	161
Ewart, P.	31,63,95,99,101,103,105,129,131
Ewell Powder Mills	47
Fairbairn, W.	61,67,95,99,103,105,137,139,143,145,147,153,159,161
Falkirk	79
Farey, J.	75,77,85
Farnley	91
Faulkner, S. & Co.	91
Fawcett, (Markland, Cooksen &)	9,29,31,75,83,91,99,101
Felkin	11
Fellows, S.	3,5
Fenton	77,87,89
Fenton Park Colliery	75
Ferrybridge Pottery	39
Fielden	2
Fielding	91
Fife	33
Finlay, Daniel & Co.	49
Firth	91
Fisher	91
Fisherrow	35
Fletcher	35
Flint	91
Foley	3
Forge Mill	65
Foster, J.	37,39
Fothergill	11
Fountaine	99
Fowler, R.	27
Fox, T.	67
France	91
Fromebridge Mill	37
Gainsborough	87
Garbett	11
Gardner, H.	29,97,99,101,103
Gardom (Pares & Co)	9,113
Garnet, J. & R.	91
Garrat	73,75
Gateshead	157
Gibbons, T.	37
Gildersome	91
Gillespie, Richard & Co.	111,131
Gilpin, G.	11
Gimblet, J.	11
Glasgow	49,77,79,91,111,113,129,145
Gloucestershire	5,33,35,49,63,67,95,145,155
Golden Lane Brewery	129
Gomersall	91
Gondsby	83
Goodier, J.	91
Goodrich, S.	61,89,137,143
Goodwyn & Co.	47,77,83
Gorton	81
Gosport	47
Gott, B.	3,9,27,29,31,33,37,39,75,77,79,83,97,99,101,123,129,131,145,157,161
Goulbourne	161
Grange Mills	61
Green	77
Green, Ald.	161
Green, J.	109
Greg, S.	103,105,129,131
Greyhound Inn, Cromford	153
Grieve, J.	35
Griffin Brewery	77
Grime, G.	91
Grimshaw, (Pearson &)	47,87
Grimshaw, J.	91
Guest, J.	11
Gyfford	47,77
Hagues (Cook & Wormald)	63,67
Haigh Ironworks	35,91
Halesowen	59
Halifax	127
Hallam	85,91
Halmer Mill	33
Halsall	29
Halsey, E.	3
Hamill	113
Hamilton, Hon. Charles	153
Hampton Court	71
Hanley	91
Hardman, G.	91
Hare, R.	77
Hargreaves, J.	5,7,9,59
Harper, W.	81
Harris	129
Harris, T.	47
Harris, (Nightingale), & Co.	73
Harrison, D.	65
Harrison, (Topping &)	91
Harrison, T.	145
Harrowgate	37
Harvey, T.	59
Haslington	33
Haugh Mill	33
Haughton, J.	91
Hawkesley (Davison &)	75
Hazledine, W.	59,139
Head, Sir G.	157
Heathfield	91,131
Henery, A.	91
Henry, Dr.	125
Hereford	79
Hertfordshire	7
Heslop	73
Hewes, T.	61,63,67,83,103
Hiblington	29
Hick	109
Hicks, H. & Son	113
Hilberts Mill	97
Hill	49,91
Hilton	91
Hives & Atkinson	161
Hives, Marshall & Co	127,131
Hockley Mill	5
Hodgkinson, R.	91
Hodgson	99
Holbeck	29
Hollings	101
Holroyd	91
Holroyds	75,91
Holt, David & Co.	91,99

172

Holywell	59,123,131	Laxton, W.R.	161	Mann	7
Hooke, R.	71	Lebond	125	Manser	97
Hooper, C.	63	Lee, George A.	27,31,39,111,	Market Place	135
Hope Mill	155		113,115,123,125,127,129,131,139,141,143,145	Markland, (Cooksen & Fawcett)	9,29,31,75,83,91,99,101
Hopetown, Earl of	153	Lee, Mrs. & Miss	125	Marklove, D.	49
Horbury	91	Leeds	27,29,31,33,39,67,73,75,77,79,81,83,85,91,95,99,	Marshall, James	159
Hornblower	73,77		101,105,115,129,131,137,139,141,143,145,157,159,161	Marshall, John	33,35,67,73,75,77,81,91,99,105,
Hornby	113,131	Leek	59,95		111,115,137,139,141,143,145,147
Hornby, Bell & Birley	65	Lees, B.	39	Marshall & Co.	137
Horridge, J.	91	Lees, C.	59,83,99	Marshall, Hives & Co.	131
Horrocks, J.	91,123,131	Lees, D.	39,91	Marshall Hutton & Co.	131
Horrocks, J. & S.	75	Lees, James	91,123,131	Marsland, P.	123,131
Horsehay	11	Lees, John	7,47,91	Marsland, S.	39,67,91
Houghton, T.	35	Lees, Cheetham & Co.	91	Masson Mills	61,153,155
Houldsworth, H.	113,145	Leicester	77	Matlock Bath	153
Houldsworth, J.	115	Leicestershire	59,75	Matthews	129
Houldsworth, T.	113,123,129	Leith	35	Maze (Clarke, M & Co)	103
Houldsworth, W., T. & H.	91	Leominster	7	Meadow Lane	143
Houldsworth, Bury & Co.	131	Level Ironworks	145	Measham	75,77,79,81,85
Houston	111	Levern	33	Meikle, A.	95,97
Hoyle, J.	109	Lewis, E.	131	Mellor Mill	109,113,115,155
Huddart & Co.	131	Ley	79	Meredith, J.	83
Huddersfield	5	Leyland	91	Metcalf	91
Hull	47,79,83,111,155	Lichfield	137	Meux & Co.	161
Hunslet	75,91	Lightholler & Hilton	91	Middle Mills	61
Hunt	37	Lillie	67,103,145	Middx.	29
Hunterian Museum	111	Limehouse	131	Midlothian	85
Huntingford Mill	33	Linby Mill	31	Milcote	149
Hutton	131	Lincolnshire	35	Milford Mills	61,123,131,135,143,145
		Lindsay, Robert & Co.	73	Mill B.	137
Ibberson & Co.	91	Linwood	113	Mill Holme	33
Ireland	103,111	Lion Brewery	161	Milltown Mills	65
Irk	59	Liptrap, D.	47,83,85	Milnthorpe	31
Ironbridge	139	Lister, J.	77	Milton Furnace	35
Irwell	59	Lister Ellis & Co.	131	Mitton, H.	79
Ivy House Works	11	Liverpool	77,79,83,85,91,99,101,103	Mold Twist Co.	103
		Livesey, T.	39	Monach, J.	49
James, J.	99	Lockwood	161	Monteith Bogle & Co.	127
James, T.	5,7	Lodge	27,31,63,127	Moon	29, 31
Jenkinson	33	Lombe, T.	5,7,95,149	Moore, T.	135
Jessop, W.	101,105	London	71,75,77,79,87,95,99,101,129,155,159	Mordaunt, C.	29,157
Jewsbury, T.	77	London Institution	111	Morehouse, C. & J.	87
Johnson, I.C. & Co.	157	London Mint	97	Moreton Corbet Forge	59
Johnson, J.	161	Longfords Mill	67	Morgan, I.	79
Johnston	113	Longsdon, J.	47	Morley	91
Jones	131	Longstone	47	Morris	7
Jordain, T.	73,91	Lonsdale, Lord	113	Mottram	47
Joulet, W. & Son	75,91	Loudon	29	Muirkirk	35
Jubb, J.	99	Lowe, T.	29,87,99,101,103,105	Murdock, W.	79,123,125,127,129
		Lower Furnace	11	Murray, A. & G.	103
Kay, J.	5	Low Moor Ironworks	11,75,77	Murray, M.	39,73,75,87,137,141
Keighley	75,87,91	Lowther, Sir J.	125	Murray, (Fenton), & Wood	77,89
Kelly, W.	9,49,111	Lowther Castle	113	Myers, (Kent & Co.)	85,91
Kendrew	33,137	Lucas, S.	37		
Kennedy, J. & J.	103,113,123,124,131,145	Lum, J.	31	Naish, F.	49
Kennedy, James	75,123,129,131,145	Lyon, T. et. al.	47	Nash	91,161
Kennedy, John	9,27,31,73,115,147	Lythorpe, W.	91	Nash (& Abbott)	85
Kent Myers & Co.	85			Nash & Co.	75
Kenyon, J.	27	Maberley, J.	77,129,131	Navy (Commissioners of the)	81,97
King, W., & Sons	113	McArthur, R.	99	Neath Abbey Ironworks	125
King & Queen Foundry	39	McCleod, Twigg & Co.	91	Neilson, A.	65
Kingswood	33	Macclesfield	7,27,59,65,75,81,95	Newark	105
Ketley	11,139	McConnel, J.	9,27,31,103,123,125,129,131,145	Newbury, Jack of	3
Kier, P.	75	McCracken, J.	113	Newcombe's Mill	33
Kirkman, J. & R.	91	McIntosh	111	Newcomen, T.	71,73,75,77,81,85,87,91,97
Knight	59,127	McKerle	49	New (The) Economical Compressed Grave Cemetry Company	159
Knightsbridge brewery	161	McTaggart, J.	49,77	New England	149
		Madder	47	New Harlem	149
Lambeth	161	Madeley Wood Furnaces	39	New Lanark Mills	67,111,153
Lancashire	33,77,79,91,95,99,103,105,157,161	Malmesbury	3	New Lanark Twist Company	67
Lansdowne, Marquis of	109,155	Manchester	9,27,29,67,68,73,75,	New River Head	71
Lawley, Sir R.	155		77,79,81,83,85,91,99,103,113,115,131,145,157	New Tarbet	65
Lawrence	27,29,63			Newton	153
Lawson	67,75,137,141				

Newtown	49	Preston	75,91,159	Sheppard, E.	33
New Willey	35	Pugin, A.W.	159	Sherborne	7,59
Nielson & Co.	131	Radcliffe, W.	5,111,123	Sherratt, W.	73,75,77,85,87,91
Nightingale, Harris & Co.	73	Radcliffe & Ross	131	Shipscar	75,91
Northampton	7,105	Ramsden	101	Shrewsbury	33,129,131,137,139,141,143,147,149
Northamptonshire	59	Ransom	79	Shropshire	11,35,59,95
Norton	91	Ravenhead	99	Shudehill	27,75
Norwich	59	Redditch	5	Shute	109
Norwich (The) Yarn Co.	103	Redruth	125	Shuttleworth	59
Nottingham	5,7,29,47,59,68,75,77,81,87,99,161	Reeling Mill	145	Silcoats	91
Nottinghamshire	59,61,67,73,75,81,97,155	Reid (Meux)	77	Silk Mill	109
		Rennie, J.	29,47,61,77,81,95,97,99,101,103,111,161	Simpson	29,75,91
Oakenrod	27,63	Repton, H.	155	Simpson, J. & S.	79
Ogden	75	Retford	39, 79	Skipton	33
Oldham	47,91,123,131	Revolution Mill	31,39	Slater, S.	135
Oldknow, S.	39,49,87,105,109,113,115,155	Reynolds J.	139	Smalley Pottery	135
Orr	111	Rhodes	91	Smeaton, J.	11,47,49,60,61,63,65,71,75,81,95,97,99,103,137
Orrell	105,147,159	Ridgway, T.	99	Smirke, Sir R.	159,161
Osborne, W.	83	Rigby, W.	91	Smith (Dyker)	79
Otley	131	Rimmer, J.	47	Smith, E.	73
Owen, R.	2,9,27,39,67,99,155	Ripon	33	Smith, Ebenezer & Co.	137
Owen & Co.	67	Roberts	9,103	Smith & Co.	75
Oxford	79	Roberts, D.	31	Smith (& Currie)	91
Oxford Road Twist & Co.	103	Roberts & Co.	91	Smith (& Townley)	91
		Robertson	77	Smiths	11,75
Paisley	49,77,85,91	Robinsons	7,29,31,61,65,67,73,75,81,97,99,103,155	Snaith	79
Palais Royale	135	Robison, Dr., J.	71,97	Snedshill	35
Paley, R.	27,75,85,91	Rocester Mill	31,59	Snodgrass	111
Papin, D.	71	Roch	59	Soane, Sir J.	135
Papplewick	7,61,65,67,81,97,99	Rochdale	63	Soho	27,75,101,125,127,141,157
Pappworth, J.	67	Roebuck	11	Soho Foundry	11,37,39,83,99,125
Parent, A.	60	Rogerson & Co.	71,91	Soho House	153,161
Pares (Gardom, P & Co)	9,113	Rope Manufactory	2	Soho Manufactory	11,47,97,125,153,157,159
Paris	135	Rose, J.	37	Soho Mint	97
Park Mill	129,131	Ross	111,123	Somerset	33
Parkers	11	Rotherham	35	Somerset, E.	71
Parkes(Brookhouse & Crompton)	99	Rotherhithe	39	Sorocold, G.	95
Parr	91	Rowbottom, J.	27	Southampton	79
Parrott, Stonier	71	Rowe	75, 91	Southern	31,35,37,67,111,125,137,143,161
Patten	75,91	Roxburghe, Duke of	155	Southwark	77
Pattison, J.	77,79,91	Royal Mint	129,161	Sparrow, G.	71
Paul, L.	7	Royal Society	71,125,129	Spearman, C.	87
Paul (& Wyatt)	47	Royal Society of Arts	125	Spedding	125
Pearson	47,87	Royton	47	Speyside	111
Peel, R.	29,31,81,91,103,123,131,157	Rumford, Count	109	Spier	91
Peel, Ainsworth & Co.	91	Runcorn, R.	91	Spitalfields	59
Peel, Williams & Co.	103	Ruthglen	49	Spode, J.	13,75,81,113
Peel, Yates & Co.	91	Ryle	29	Spooner, I.	75
Pemberton, E.	91			Staffordshire	7,9,11,29,31,59,99,137,145,153
Pemberton, J.	129			Stalybridge	91,145
Pendleton	29,91,99,123,125,131			Stancliffe, Dr.	125
Pen-y-Daren Ironworks	35	St. Annes Church	135	Stanley Mill	145,155
Perkins	47	St. Mirran Co.	91	Stanton, C.	29
Philae	159	St. Petersburgh	135	Stanton, G.	29
Philips, (& Lee)	27,31,111,113,115,123,125,127,129,131,139,141	Salford	31,75,91,99,129,131	Stevenson	111
Philips, J. & N.	145	Salford Twist Mill	27,39,111,123,125,127,129,131,137,139,141,143,145	Stewart, Cdr. K.	35
Phipps, W.	131			Stockdale	91
Phoenix, Brewhouse	3	Salt, T.	105,159	Stockport	7,27,49,59,91,105,147,159
Picadilly	9,27	Saltaire	105,159,161	Stonard	155
Pickard, J.	73,79,81	Salvin, G.	63,81,99	Stonehouse	109
Pickford, J.	161	Salvin's Factory	27	Stour	59
Pilkington	7	Sandford, B. & W.	27,73,91	Stratford upon Slainy	111
Playne	67	Sandon	29	Stretton S. & W.	161
Pollock, A. and Co.	49	Saunders, G.	135	Stroud (Valley)	63
Pollokshaws	129	Saunders, M.	129	Strutt, J.	135,139
Poncelet	61	Savery, T.	71,73,75,77,81,85,87,89,91	Strutt, W.	7,61,63,81,83,99,109,111,123,129,
Pooley	85,91,111,123,131	Scotland	29,67,79,99		131,135,137,139,143,145
Pope	161	Scott Stevenson & Co.	77,91	Stumpe, W.	3
Pt. Adelphi Cotton Works	113	Sedgwick	145	Sturgess	77,85
Portland, Duke of	155	Severn	135	Styal	99,103
Portugal	135	Shaddon Mill	147,161	Styal Mill	27,61
Potter	79	Shaw, Stebbing	157	Subscription Flour Co.	79
Prestnall, Oldham et al.	91	Sheffield	7,11,91	Sunderland	79

174

Sun Fire Office	131	Walker, Joshua	35,77	Yarmouth	95
Surrey	47	Walkers	11,79	Yarranton, A.	3,149
Sutcliffe, J.	2,29,47,63,77,99,101,105,145,161	Walshman	87,91	Yates	27,29,63,91
Sutton	155	Wardles	27	Yeoman, T.	7,95,105
Sutton-in-Ashfield	7,73,75,155	Ware, I.	161	York	91
Sutton Scarsdale	153	Warrington	91	Yorkshire	5,7,9,79,91,103
Swainson	159	Warwick	99	Young, J.	73,85
Swann	87	Warwickshire	59	Young, E. & Co.	79
Sword, J.	77	Wasborough, M.	73,79,81		
Sylvester	109	Water Lane Mills	33,137		
Symington	73	Watford	109		
		Watson	91		
Tambouring Mill	113	Watson, J.	91		
Tarbottom	91	Watt, G.	125		
Tattersall, R.	161	Weatherill, J., J., & J.	91		
Tatton	95	Webster	77		
Taylor	91	Wedgwood, J.	11,39,99,111,113,157,161		
Taylor	27,47	Wedgwood, T.	11		
Taylor, J.	11,77	Weevil Brewery	47		
Tean Hall	31,145	Welchman	31		
Telford, T	139	Wells, Heathfield & Co.	91		
Temple Mill	159	Westmorland	31,59		
Tern Forge	59	West of England	33,79		
Tern Mills	95	Weston	91		
Tern Works	11,59,139	Wheelock	95		
Thackery, J.	73,85,91	Whitbread, S.	47,77		
Thackery, Stockdale & Co.	91	White, J.	5		
Thackery, Whitehead & Co.	91	Whitehaven	33,125		
Thomas, J.	131	Whitehead	73,75		
Thomason, E.	11	Whitenbury	75,91		
Thompson & Baxter	155	Whitewell Brook	59		
Thompson (Gorton &)	81	Whither	67		
Thompson F.	73,75,85,91	Whyte	49		
Thompson, J.	99	Wigan	7,91		
Thore	101	Wildboarclough	95		
Thorneley, J.	91	Wilkes, J.	29,75,77,79,81,85		
Thorner	91	Wilkinson (Coupland &)	75		
Thrale	77	Wilkinson, J.	11,35,59,73,85,161		
Tipton	11	Wilkinson & Paley	91		
Todd Fletcher & Co.	35	Wilkinson, W.	35		
Topping	91	Willersley Castle	155		
Townley	91	Willey Furnace	59		
Townsend Factory	9	Williams	103,113		
Tredgold	115	Williams, Cooper & Boyle	113		
Trevithick, R.	73,75,7789	Williams, Jones & Co.	131		
Troughton Bros.	91	Wilson	91		
Trowbridge	33	Wilson, Ald. W.	3		
Truman	47	Wiltshire	33		
Turton, J.	91	Wirksworth	59,87,95		
Turton Mill	157	Withering, Dr.	109		
Tutbury	61,99	Wood, Ald.	129		
Twigg, J.	77	Wood	77		
		Wood (& Cook)	91		
Uley	49	Wood, E.	13,155		
Ulster	59	Wood, J.	103,145		
Underwood Spinning Co.	85	Wood, Daintry & Co.	123		
Union Co.	79	Wood (Murray, Fenton &)	87,89		
Union Plate Glass Co.	103	Woodchester Mill	155		
Unsworth, E.	91	Woolf, A.	73,77		
Unwin, S.	73,77,85,155	Woolwich Smithery	97		
Upcott, W.	33	Worcester, Marquis of	71		
Upper Cam Mill	33	Worcestershire	5,59,95		
Upper Leen Valley	155	Wormald, J.	63,67,91,99		
Upton Forge	59	Wormald, Gott & Wormald	131		
Ure	2,3,159	Worsley, J.	91		
		Wren	103		
Vaughan, P.	73	Wright	87,91		
		Wrigley, Joshua	73,75,77,81,85,91,105		
Wakefield	91	Wrigley, J. & Co.	91		
Wakefield Lunatic Asylum	109	Wyatt, James	153,159,161		
Walk Mill	33	Wyatt, John	7,29,39,47,101		
Walker, A.	75,91	Wyatt, S.	153,159,161		
Walker, J.	135	Wye Mill	155